Algebraic Number Theory

Edwin Weiss

DOVER PUBLICATIONS, INC.
Mineola, New York

Bibliographical Note

This Dover edition, first published in 1998, is an unabridged republication of the work originally published in 1963 by McGraw-Hill Book Company, Inc., New York.

Library of Congress Cataloging-in-Publication Data

Weiss, Edwin.
Algebraic number theory / Edwin Weiss.
p. cm.
Originally published: New York : McGraw-Hill, 1963.
Includes bibliographical references and index.
ISBN 0-486-40189-8
1. Algebraic number theory. I. Title.
QA247.W4 1998
512'.74—dc21 98-7733
CIP

Manufactured in the United States of America
Dover Publications, Inc., 31 East 2nd Street, Mineola, N.Y. 11501

For Janice, Ariel, and Rena

...ולא המדרש עקר אלא המעשה...פרקי אבות

Preface

It is somewhat surprising that, although algebraic number theory has from its inception attracted the efforts and enthusiasm of the most eminent mathematicians (as may be seen by an examination of both the bibliography and preface to Hilbert's "Zahlbericht"—see reference [8]) and although it has throughout its history served as one of the main streams in the development of algebra, there does not exist an adequate, easily accessible introduction to the subject in English. Thus, it is the purpose of this book to provide a leisurely, fairly detailed, and reasonably self-contained exposition of the basic results of classical algebraic number theory from a relatively modern point of view. The contents correspond, more or less, to the material normally covered in a one-semester graduate-level course in algebraic number theory; in fact, this book has its origins in such a course given by the author at Harvard University in the spring of 1958. No attempt is made to treat any advanced topics, but it is to be hoped that the reader will find here most of the number-theoretic prerequisites for a study of either class field theory as formulated by Artin and Tate (see reference [2]) or the contemporary treatment of analytic questions as found, for example, in Tate's thesis (see reference [10]). The reader who has completed a full year van der Waerden type course in algebra should encounter no difficulty with this book.

There are several ways of approaching our subject, and for various reasons we have chosen to emphasize the valuation-theoretic approach. This implies that, of necessity, we have borrowed heavily from the well-known Artin notes (see reference [1]) both as to techniques and organization. Some other sources which have been used without explicit acknowledgment are lectures or unpublished notes of Artin, Brauer, Cohen, Iwasawa, and Tate.

The structure of this book is outlined in the Table of Contents, and the reader should find it rather transparent; there is little profit in providing further details here. However, a few comments would seem to be in order. Our interest is exclusively with algebraic number fields, but the formulation is axiomatic and, therefore, considerably more general in application. In particular, most of the results are valid for algebraic function fields, but the matter is not pursued. In spite of the emphasis on modern abstract techniques, an attempt has been made not to lose sight of the concrete and the classical entirely. The heart of the book is to be found in Chapters 4 and 5; the first two chapters are introductory, Chapter 3 is preparatory and also of independent interest, while Chapters 6 and 7 give some examples and applications of the general theory. The theory of ideals, around which Chapter 4 revolves, may be developed directly within the framework of adèles and idèles of Chapter 5; however, this method has obvious pedagogical disadvantages. Some topics whose omission should be noted are: ζ-function, L-series, cohomology, density theorems, class number formulas, the conductor, Gauss sums, and Kummer theory. The exercises that appear in the text are, as a rule, of greater interest than those at the end of the chapters, but they are almost never essential for what follows.

Thanks and appreciation are due to E. Cohen, D. Schneider, R. Segal, and S. Shatz, who read parts of the manuscript and made suggestions; to my parents and to S. Shufro for less direct forms of assistance; to Judith Crotty for unfailing courtesy and efficiency in typing the manuscript; and to Lincoln Laboratory for the continuous cooperation which made this book possible.

Edwin Weiss

References

1. E. Artin, "Algebraic Numbers and Algebraic Functions," lecture notes, Princeton University and New York University, 1950–1951.
2. E. Artin and J. T. Tate, "Class Field Theory," seminar notes prepared by S. Lang, Institute for Advanced Study, Princeton, N.J., 1960.
3. E. Artin and G. Whaples, Axiomatic Characterization of Fields by the Product Formula for Valuations, *Bulletin of the American Mathematical Society*, vol. 51, no. 7, pp. 469–492, July, 1945.
4. P. G. L. Dirichlet and R. Dedekind, "Vorlesungen über Zahlentheorie," 4th ed., supplement 11, Friedr. Vieweg & Sohn, Brunswick, Germany, 1894.
5. P. Furtwängler, Allgemeine Theorie der algebraischen Zahlen, "Enzyklopädie der Mathematischen Wissenschaften," vol. I_2, no. 8, part 2, B. G. Teubner Verlagsgesellschaft, mbH, Leipzig, 1953.
6. H. Hasse, "Zahlentheorie," Akademie-Verlag GmbH, Berlin, 1949.
7. E. Hecke, "Vorlesungen über die Theorie der algebraischen Zahlen," Akademische Verlagsgesellschaft, Leipzig, 1923.
8. D. Hilbert, Bericht über die Theorie der algebraischen Zahlkörper, *Jahresbericht der Deutschen Mathematiker-Vereinigung*, vol. 4, 1894–1895.
9. K. Iwasawa, On the Rings of Valuation Vectors, *Annals of Mathematics*, vol. 57, no. 2, pp. 331–356, March, 1953.
10. J. T. Tate, "Fourier Analysis in Number Fields and Hecke's Zeta-functions," thesis, Princeton University, 1950.
11. H. S. Vandiver, Fermat's Last Theorem, *American Mathematical Monthly*, vol. 53, pp. 555–578, 1946.
12. H. Weber, "Lehrbuch der Algebra," Friedr. Vieweg & Sohn, Brunswick, Germany, 1896.
13. H. Weyl, Algebraic Theory of Numbers, *Annals of Mathematics Studies*, no. 1, Princeton University Press, Princeton, N.J., 1940.

An attempt has been made to keep this list of references as short as possible. It should be noted, however, that the bibliographies in [5] and [8], taken together, provide a rather comprehensive survey of the literature up to about 1950. [11] contains a detailed but specialized bibliography.

Contents

1

Elementary Valuation Theory

The central feature of the subject commonly known as ***algebraic number theory*** is the problem of factorization in an algebraic number field, where by an ***algebraic number field*** we mean a finite extension of the rational field $\mathbf{Q}$. Of course, it will take some time before the full meaning of this statement will become apparent.

There are several distinct approaches to our subject matter, and we have chosen to emphasize the valuation-theoretic approach. In this chapter, we begin to learn the language.

1-1. Valuations and Prime Divisors

Let us begin with a definition. A ***valuation*** of the field F is a function φ from F into the nonnegative reals such that

(i) $$\varphi(a) = 0 \Leftrightarrow a = 0$$

(ii) $$\varphi(ab) = \varphi(a)\varphi(b)$$

(iii) There exists a real constant C such that

$$\varphi(a) \leq 1 \Rightarrow \varphi(1 + a) \leq C$$

Some obvious examples of valuations are the following: (1) Let F denote the field of real numbers $\mathbf{R}$ or the field of complex numbers $\mathbf{C}$, and put $\varphi(a) = |a|$, $C = 2$. It may be remarked that the original impetus for the study of valuations of an arbitrary field arose from this example—that is, as a generalization of the notion of absolute value. (2) Let

F be any field, and put $\varphi(0) = 0$, $\varphi(a) = 1$ for $a \neq 0 \in F$. This is known as the ***trivial valuation*** of F.

In view of the numerous examples that will be considered later, let us be content for the moment with the simple examples above. The impatient reader may turn to Section 1-4.

A valuation determines a homomorphism of F^*, the multiplicative group of F, into the multiplicative group of positive real numbers ($\varphi: F^* \to \mathbf{R}_{>0}$); therefore, we have $\varphi(1) = 1$, $\varphi(a^{-1}) = 1/\varphi(a)$, and $\varphi(b/a) = \varphi(b)/\varphi(a)$ for $a \neq 0$. Also, since $\varphi(1) = \varphi(-1)\varphi(-1)$, it follows that $\varphi(-1) = 1$ and $\varphi(-a) = \varphi(-1)\varphi(a) = \varphi(a)$. It is also immediate that if ζ is any root of unity belonging to F then $\varphi(\zeta) = 1$ and that a finite field admits only the trivial valuation.

1-1-1. Proposition. In the definition of a valuation, (iii) may be replaced by

(iii′) $$\varphi(a + b) \leq C \max \{\varphi(a), \varphi(b)\}$$

Proof. (iii) ⇒ (iii′). Suppose $\varphi(a) = \max \{\varphi(a), \varphi(b)\}$. If $\varphi(a) = 0$ then $a = b = 0$ and (iii′) holds. If $\varphi(a) \neq 0$, then

$$\varphi(a + b) = \varphi\left[a\left(1 + \frac{b}{a}\right)\right] = \varphi(a)\varphi\left(1 + \frac{b}{a}\right) \leq C\varphi(a)$$

(iii′) ⇒ (iii). If $\varphi(a) \leq 1$, then

$$\varphi(1 + a) \leq C \max \{\varphi(1), \varphi(a)\} = C$$

In the more usual definition of valuation, condition (iii) is replaced by the so-called ***triangle inequality***

$$\varphi(a + b) \leq \varphi(a) + \varphi(b)$$

We shall soon see that, for all practical purposes, our valuations may be taken to satisfy the triangle inequality.

1-1-2. Proposition. A valuation φ of F determines a Hausdorff topology T_φ on F. For each $a \in F$, a fundamental system of neighborhoods of a is given by the set of all

$$U(a, \varepsilon) = \{b \in F | \varphi(a - b) < \varepsilon\}$$

Proof. Let Δ denote the diagonal of $F \times F$, and for subsets S and T of $F \times F$ let

$$S^{-1} = \{(b, a) \in F \times F | (a, b) \in S\}$$
$$T \circ S = \{(a, b) \in F \times F | \exists c \in F \ni : (a, c) \in S \text{ and } (c, b) \in T\}$$

For each $\varepsilon > 0$ we put

$$U(\varepsilon) = \{(a, b) \in F \times F | \varphi(a - b) < \varepsilon\}$$

From the properties of φ, we have

(1) $U(\varepsilon) \supset \Delta$

(2) $U(\varepsilon_1) \cap U(\varepsilon_2) = U(\varepsilon)$ where $\varepsilon = \min \{\varepsilon_1, \varepsilon_2\}$

(3) $U(\varepsilon)^{-1} = U(\varepsilon)$

(4) $U\left(\frac{\varepsilon}{C}\right) \circ U\left(\frac{\varepsilon}{C}\right) \subset U(\varepsilon)$

(5) $\bigcap_{\varepsilon>0} U(\varepsilon) = \Delta$

In fact, (1), (2), and (3) are trivial; (5) is clear since $\varphi(a - b) = 0 \Leftrightarrow a = b$; finally, the assertion that (a, b) belongs to the left side of (4) means that there exists $c \in F$ with $\varphi(a - c) < \varepsilon/C$ and $\varphi(c - b) < \varepsilon/C$, so that $\varphi(a - b) < \varepsilon$ and (4) holds.

Properties 1 to 4 guarantee that $\{U(\varepsilon)\}$ is a filter base on $F \times F$. This filter base defines a uniform structure on F, and according to (5) the uniform structure is separated. Therefore, the topology of F "deduced" from the uniform structure in the canonical way has fundamental neighborhoods as described and is a Hausdorff topology. There is no particular significance attached to the method by which we have arrived at the topology of F. It is based on facts and notation as found in N. Bourbaki's "Topologie Générale" (Hermann & Cie, Paris, 1940).

1-1-3. Corollary. Let φ be a valuation of F, and let $\{x_n\}$ be a sequence in F; then

$$x_n \to 0 \text{ for } T_\varphi \Leftrightarrow \varphi(x_n) \to 0$$

We say that two valuations φ_1 and φ_2 are ***equivalent*** (and denote this by $\varphi_1 \sim \varphi_2$) when they determine the same topology on F—that is, if and only if $T_{\varphi_1} = T_{\varphi_2}$. The equivalence classes with respect to this equivalence relation are called ***prime divisors of*** F and are denoted by P, Q, etc. The prime divisor to which the trivial valuation of F belongs is called the ***trivial prime divisor;*** all the others are ***nontrivial prime divisors.***

Let φ be a valuation of F with constant C, and let $\alpha > 0$ be a real number; then the function φ^α on F given by $\varphi^\alpha(a) = [\varphi(a)]^\alpha$ is a valuation of F with constant C^α.

1-1-4. Theorem. Let φ_1, φ_2 be nontrivial valuations of F, and let a denote an arbitrary element of F; then the following statements are equivalent:

$$\varphi_2 = \varphi_1^{\alpha} \qquad \text{with } \alpha > 0 \tag{1}$$

$$T_{\varphi_2} = T_{\varphi_1} \tag{2}$$

$$T_{\varphi_1} \text{ is a stronger topology than } T_{\varphi_2} \tag{3}$$

$$\varphi_1(a) < 1 \Rightarrow \varphi_2(a) < 1 \tag{4}$$

$$\varphi_1(a) \leq 1 \Rightarrow \varphi_2(a) \leq 1 \tag{5}$$

$$\begin{cases} \varphi_1(a) < 1 \Leftrightarrow \varphi_2(a) < 1 \\ \varphi_1(a) > 1 \Leftrightarrow \varphi_2(a) > 1 \\ \varphi_1(a) = 1 \Leftrightarrow \varphi_2(a) = 1 \end{cases} \tag{6}$$

Proof. (1) $\Rightarrow$ (2). For $i = 1, 2$ we write

$$U_i(\varepsilon) = \{(a, b) \in F \times F \mid \varphi_i(a - b) < \varepsilon\}$$

Then $U_2(\varepsilon^{\alpha}) = U_1(\varepsilon)$, so that

$$\{U_2(\varepsilon) \mid \varepsilon > 0\} = \{U_1(\varepsilon) \mid \varepsilon > 0\}$$

Thus the uniform structures are equivalent, and the topologies are the same—that is, $T_{\varphi_1} = T_{\varphi_2}$.

(2) $\Rightarrow$ (3). This is trivial since T_{φ_1} stronger than T_{φ_2} means that every open set for the topology T_{φ_2} is also open for the topology T_{φ_1}.

(3) $\Rightarrow$ (4). Making use of (1-1-3), we have

$$\begin{aligned} \varphi_1(a) < 1 &\Rightarrow \varphi_1(a^n) \to 0 \Rightarrow a^n \to 0 \text{ for } T_{\varphi_1} \\ &\Rightarrow a^n \to 0 \text{ for } T_{\varphi_2} \Rightarrow \varphi_2(a^n) \to 0 \\ &\Rightarrow \varphi_2(a) < 1 \end{aligned}$$

(4) $\Rightarrow$ (5). Since φ_1 is nontrivial, there exists $b \in F^*$ such that $\varphi_1(b) \neq 1$, and we may then take $\varphi_1(b) < 1$. Consequently,

$$\begin{aligned} \varphi_1(a) = 1 &\Rightarrow \varphi_1(a^n b) < 1 \text{ for all } n > 0 \\ &\Rightarrow \varphi_2(a^n b) < 1 \\ &\Rightarrow \varphi_2(a) < \frac{1}{[\varphi_2(b)]^{1/n}} \\ &\Rightarrow \varphi_2(a) \leq 1 \end{aligned}$$

In the same way, $\varphi_2(a^{-1}) \leq 1$, and we conclude that $\varphi_2(a) = 1$.

(5) $\Rightarrow$ (6). Since φ_2 is nontrivial, there exists $c \in F^*$ such that

$0 < \varphi_2(c) < 1$, and then

$$\begin{aligned}\varphi_1(a) < 1 &\Rightarrow \varphi_1(a)^n \leq \varphi_1(c) \text{ for sufficiently large } n\\ &\Rightarrow \varphi_1\left(\frac{a^n}{c}\right) \leq 1\\ &\Rightarrow \varphi_2\left(\frac{a^n}{c}\right) \leq 1\\ &\Rightarrow \varphi_2(a^n) \leq \varphi_2(c) < 1\\ &\Rightarrow \varphi_2(a) < 1\end{aligned}$$

Thus (4) holds, and as in the proof that (4) $\Rightarrow$ (5) we have

$$\varphi_1(a) = 1 \Rightarrow \varphi_2(a) = 1$$

Of course, $\varphi_1(a) > 1 \Rightarrow \varphi_1(a^{-1}) < 1 \Rightarrow \varphi_2(a^{-1}) < 1 \Rightarrow \varphi_2(a) > 1$. The validity of (6) is now clear.

(6) $\Rightarrow$ (1). Fix $a \in F$ such that $\varphi_1(a) > 1$. Then $\varphi_2(a) > 1$, and we may put

$$\alpha = \frac{\log \varphi_2(a)}{\log \varphi_1(a)} > 0$$

We show that $\varphi_2 = \varphi_1^\alpha$; for this it suffices to show that if for each $b \in F^*$ we write $\gamma_i = [\log \varphi_i(b)]/[\log \varphi_i(a)]$ $(i = 1, 2)$ then $\gamma_1 = \gamma_2$. Let $r = m/n$ with $n > 0$ denote a rational number; then

$$\begin{aligned}r = \frac{m}{n} \geq \gamma_1 &\Leftrightarrow m \log \varphi_1(a) \geq n \log \varphi_1(b) \qquad [\text{since } \log \varphi_1(a) > 0]\\ &\Leftrightarrow \varphi_1(a^m) \geq \varphi_1(b^n)\\ &\Leftrightarrow \varphi_1\left(\frac{b^n}{a^m}\right) \leq 1\\ &\Leftrightarrow \varphi_2\left(\frac{b^n}{a^m}\right) \leq 1\\ &\Leftrightarrow m \log \varphi_2(a) \geq n \log \varphi_2(b)\\ &\Leftrightarrow \frac{m}{n} \geq \gamma_2 \qquad [\text{since } \log \varphi_2(a) > 0]\end{aligned}$$

1-1-5. Corollary. Let P be a prime divisor of F; then for any $\varphi \in P$

$$P = \{\varphi^\alpha | \alpha > 0\}$$

Proof. (1-1-4) takes care of the case where P is nontrivial. Since it is easy to see that the valuation φ of F is trivial if and only if T_φ is the discrete topology, it follows that a trivial prime divisor P consists solely of the trivial valuation; thus, the assertion holds in this case also.

In order to clarify the connection between our valuations and those satisfying the triangle inequality, it is convenient to introduce a simple definition. Given a valuation φ, we define $\|\varphi\|$ (the ***norm*** of φ) by $\|\varphi\| = \inf C$, where C runs over all constants that may be used in (iii) of the definition of valuation. $\|\varphi\|$ is the smallest possible value for C, and (1-1-1) applies with $C = \|\varphi\|$. For any real number $\alpha > 0$, it follows immediately that $\|\varphi^\alpha\| = \|\varphi\|^\alpha$. In view of (1-1-5) we see that, given a prime divisor P of F, there exists $\varphi \in P$ with $\|\varphi\| \leq 2$. We may also note that, for any φ, $\|\varphi\| \geq 1$; in fact, $1 = \varphi(1) = \varphi(1 + 0) \leq \|\varphi\|$.

1-1-6. Exercise. Suppose that φ is a valuation for which condition (iii) is replaced by the triangle inequality. Give an example to show that under such circumstances φ^α need not be a valuation (in the sense that it need not satisfy the triangle inequality).

1-1-7. Lemma. Let φ be a valuation of F; then

$$\|\varphi\| \leq 2 \Rightarrow \varphi(a_1 + \cdots + a_n) \leq 2n \max \{\varphi(a_1), \ldots, \varphi(a_n)\}$$

Proof. By (1-1-1) we have $\varphi(a_1 + a_2) \leq 2 \max \{\varphi(a_1), \varphi(a_2)\}$, and by induction

$$\varphi(a_1 + \cdots + a_{2^r}) \leq 2^r \max \{\varphi(a_1), \ldots, \varphi(a_{2^r})\}$$

For any $n > 1$ there exists a unique $r > 0$ such that $2^r \leq n < 2^{r+1}$, and then

$$\begin{aligned} \varphi(a_1 + \cdots + a_n) &= \varphi(a_1 + \cdots + a_n + 0 + \cdots + 0) \qquad (2^{r+1} \text{ terms}) \\ &\leq 2^{r+1} \max \{\varphi(a_1), \ldots, \varphi(a_n)\} \\ &\leq 2n \max \{\varphi(a_1), \ldots, \varphi(a_n)\} \end{aligned}$$

1-1-8. Proposition. Let φ be a real-valued function on F which satisfies conditions (i) and (ii) of the definition of valuation; then $\varphi(a + b) \leq \varphi(a) + \varphi(b)$ for all $a, b \in F \Leftrightarrow \varphi$ is a valuation and $\|\varphi\| \leq 2$.

Proof. $\Rightarrow$: $\varphi(a + b) \leq \varphi(a) + \varphi(b) \leq 2 \max \{\varphi(a), \varphi(b)\}$; so, by (1-1-1), φ is a valuation, and $\|\varphi\| \leq 2$.

$\Leftarrow$: For each $n > 1$, we have

$$\begin{aligned} [\varphi(a + b)]^n &= \varphi\left[a^n + \binom{n}{1} a^{n-1}b + \cdots + \binom{n}{n-1} ab^{n-1} + b^n\right] \\ &\leq 2(n + 1) \max_i \left\{\varphi\left[\binom{n}{i} a^{n-i}b^i\right]\right\} \\ &\leq 4(n + 1) \sum_i \left[\binom{n}{i} \varphi(a)^{n-i}\varphi(b)^i\right] \\ &= 4(n + 1)[\varphi(a) + \varphi(b)]^n \end{aligned}$$

Taking nth roots and letting $n \rightarrow \infty$, we arrive at the desired conclusion.

1-1-9. Theorem. Let P be a prime divisor of F, and let $\varphi \in P$ have $\|\varphi\| \leq 2$. Then T_φ is a metric topology with respect to which F is a topological field and φ is uniformly continuous.

Proof. For $a, b \in F$ put $d(a, b) = \varphi(a - b)$. By using the fact that $\varphi(x + y) \leq \varphi(x) + \varphi(y)$, it is immediate that d is a metric on F. In the metric topology T_d determined by d, a fundamental system of neighborhoods of $a \in F$ is given by

$$V(a, \varepsilon) = \{b \in F | d(a, b) < \varepsilon\} = U(a, \varepsilon)$$

Hence, $T_\varphi = T_d$. We see from $\varphi(a) \leq \varphi(a - b) + \varphi(b)$ and $\varphi(b) \leq \varphi(b - a) + \varphi(a)$ that

$$|\varphi(a) - \varphi(b)| \leq \varphi(a - b)$$

Therefore, it follows that φ is uniformly continuous on F. It is now easy to verify that the maps $(a, b) \to a - b$, $(a, b) \to ab$, and $b \to b^{-1}$ are continuous; that is, F is a topological field with respect to T_φ.

Henceforth, whenever we consider a valuation $\varphi \in P$, it is to be understood—unless explicit mention is made to the contrary—that $\|\varphi\| \leq 2$. Of course, there will be no harm in writing T_P in place of T_φ.

1-2. The Approximation Theorem

Having determined the relations between equivalent nontrivial valuations in (1-1-4), we proceed now to a description of the connections between inequivalent nontrivial valuations.

1-2-1. Lemma. Let $\varphi_1, \ldots, \varphi_n$ be inequivalent nontrivial valuations of F; then there exists an element $a \in F$ with

$$\varphi_1(a) > 1, \varphi_2(a) < 1, \ldots, \varphi_n(a) < 1$$

Proof. By induction on n. Suppose $n = 2$; then by (1-1-4) there exist $b, c \in F$ such that $\varphi_1(b) > 1$, $\varphi_2(b) \leq 1$ and $\varphi_1(c) \leq 1$, $\varphi_2(c) > 1$—so that $a = b/c$ works. Suppose now that the lemma holds for $n - 1$. Choose $b \in F$ such that

$$\varphi_1(b) > 1, \varphi_2(b) < 1, \ldots, \varphi_{n-1}(b) < 1$$

and $c \in F$ such that

$$\varphi_1(c) > 1, \varphi_n(c) < 1$$

If $\varphi_n(b) \leq 1$, then surely $a = b^m c$, for sufficiently large m, will do. It remains to consider the case where $\varphi_n(b) > 1$.

In general, if φ is any valuation, then

$$\varphi(b) < 1 \Rightarrow b^m \to 0 \text{ for } T_\varphi \Rightarrow \frac{b^m}{1+b^m} \to 0 \text{ for } T_\varphi \Rightarrow \varphi\left(\frac{b^m}{1+b^m}\right) \to 0$$

and also

$$\varphi(b) > 1 \Rightarrow \frac{1}{b^m} \to 0 \text{ for } T_\varphi \Rightarrow \frac{b^m}{1+b^m} = \frac{1}{1+1/b^m} \to 1 \text{ for } T_\varphi$$
$$\Rightarrow \varphi\left(\frac{b^m}{1+b^m}\right) \to 1$$

Therefore, for sufficiently large m, $a = [b^m/(1+b^m)]c$ satisfies the required conditions.

1-2-2. Lemma. Let $\varphi_1, \ldots, \varphi_n$ be inequivalent nontrivial valuations of F; then given any $\varepsilon > 0$, there exists $b \in F$ such that

$$\varphi_1(1-b) < \varepsilon, \ \varphi_2(b) < \varepsilon, \ \ldots, \ \varphi_n(b) < \varepsilon$$

Proof. Choose a as in (1-2-1), and put $b = a^m/(1+a^m)$ —with m to be determined. For $i > 1$,

$$\varphi_i(b) = \varphi_i\left(\frac{a^m}{1+a^m}\right) \to 0$$

while, for $i = 1$,

$$\varphi_1(1-b) = \varphi_1\left(\frac{1}{1+a^m}\right) = \varphi_1\left(\frac{1}{a^m}\right)\varphi_1\left(\frac{1}{(1/a)^m+1}\right) \to 0$$

Hence, we need only take m sufficiently large.

1-2-3. Approximation Theorem. Let $\varphi_1, \ldots, \varphi_n$ be inequivalent nontrivial valuations of F, and let $a_1, \ldots, a_n$ be elements of F. Given any $\varepsilon > 0$, there exists an element $a \in F$ such that

$$\varphi_i(a - a_i) < \varepsilon \qquad i = 1, \ldots, n$$

Proof. Let $M = \max\{\varphi_i(a_j)\}$, where $i, j = 1, \ldots, n$. According to (1-2-2) there exist $b_1, \ldots, b_n \in F$ such that,

$$\varphi_i(1-b_i) < \frac{\varepsilon}{Mn} \qquad i = 1, \ldots, n$$
$$\varphi_i(b_j) < \frac{\varepsilon}{Mn} \qquad j \neq i$$

Clearly [in virtue of (1-1-8)], $a = a_1b_1 + a_2b_2 + \cdots + a_nb_n$ is an element of the desired type.

The approximation theorem may also be expressed in the following way: Suppose $P_1, \ldots, P_n$ are distinct nontrivial prime divisors of F; and for each $i = 1, \ldots, n$ let $F^{(i)}$ denote F with the topology determined by the prime divisor P_i. The set F may then be imbedded along the diagonal in the product space $F^{(1)} \times \cdots \times F^{(n)}$. That F is dense in $F^{(1)} \times \cdots \times F^{(n)}$ is the assertion of the approximation theorem. To put it still another way, the approximation theorem implies that any finite number of inequivalent nontrivial valuations of F are completely "independent"; however, we shall see in Section 1-4 that this need not be valid when the number of valuations is infinite.

It is surprising perhaps that the validity of the approximation theorem in its full generality (i.e., as we have stated it) was not discovered until 1945.

1-2-4. Exercise. For nontrivial inequivalent valuations $\varphi_1, \ldots, \varphi_n$ the topology T_{φ_1} is not comparable with the topology generated by the T_{φ_i} $(i = 2, \ldots, n)$.

1-3. Archimedean and Nonarchimedean Prime Divisors

Let P be a prime divisor of F. We say that P is ***archimedean*** when $\|\varphi\| > 1$ for every $\varphi \in P$ and that P is ***nonarchimedean*** when $\|\varphi\| = 1$ for every $\varphi \in P$. For a valuation φ of F it is only natural to call φ archimedean (nonarchimedean) when the prime divisor P to which it belongs is archimedean (nonarchimedean). Note that an archimedean valuation and a nonarchimedean one cannot be equivalent.

It is clear that the trivial valuation is nonarchimedean and that the absolute value valuation of **R** or **C** is archimedean.

The next result gives equivalent and more standard formulations of our definition.

1-3-1. Proposition. Let φ be a valuation of F; then

$$\begin{aligned}\varphi \text{ is nonarchimedean} &\Leftrightarrow \varphi(a+b) \le \max\{\varphi(a), \varphi(b)\} \text{ for all } a, b \in F \\ &\Leftrightarrow \{\varphi(n \cdot 1) | n \in \mathbf{Z}\} \text{ is bounded above}\end{aligned}$$

Proof. The first equivalence is immediate from (1-1-1) and the properties of $\|\varphi\|$. As for the remaining implications, we have:

$\Rightarrow$: By induction, $\varphi(a_1 + \cdots + a_n) \le \max\{\varphi(a_1), \ldots, \varphi(a_n)\}$, so that $\varphi(na) \le \varphi(a)$ and $\varphi(n \cdot 1) \le \varphi(1)$.

$\Leftarrow$: Suppose that $\varphi(n \cdot 1) \leq M$ for all $n \in \mathbf{Z}$, and suppose $\varphi(a) \geq \varphi(b)$; then

$$\begin{aligned}[\varphi(a+b)]^n &= \varphi[(a+b)^n] \\ &= \varphi\left[a^n + \binom{n}{1} a^{n-1}b + \cdots + \binom{n}{1} ab^{n-1} + b^n\right] \\ &\leq 2(n+1) \max\left\{\varphi\left[\binom{n}{i} a^{n-i}b^i\right]\right\} \qquad \text{[by (1-1-7)]} \\ &\leq 2(n+1)M \max\{\varphi(a)^{n-i}\varphi(b)^i\} \\ &\leq 2(n+1)M[\varphi(a)]^n\end{aligned}$$

Therefore, $\varphi(a+b) \leq \sqrt[n]{2(n+1)M}\ \varphi(a)$; and letting $n \to \infty$ gives $\varphi(a+b) \leq \varphi(a) = \max\{\varphi(a), \varphi(b)\}$.

1-3-2. Corollary. The archimedean or nonarchimedean character of a valuation is completely determined by its action on the prime subfield. In particular, a field with characteristic $p \neq 0$ can have only nonarchimedean prime divisors.

We may now prove an elementary but extremely useful property of nonarchimedean valuations.

1-3-3. Proposition. Let φ be a nonarchimedean valuation of F; then

$$\varphi(a) \neq \varphi(b) \Rightarrow \varphi(a+b) = \max\{\varphi(a), \varphi(b)\}$$

Proof. Suppose $\varphi(a) < \varphi(b)$; then

$$\begin{aligned}\varphi(b) = \varphi(a+b-a) &\leq \max\{\varphi(a+b), \varphi(-a)\} \\ &\leq \max\{\varphi(b), \varphi(a)\} \\ &= \varphi(b)\end{aligned}$$

1-3-4. Corollary. Let φ be a nonarchimedean valuation of F, and consider $a_1, \ldots, a_n \in F$; then

(i) If $\varphi(a_1) > \varphi(a_i)$ for $i = 2, \ldots, n$ then

$$\varphi(a_1) = \varphi(a_1 + a_2 + \cdots + a_n)$$

(ii) $a_1 + \cdots + a_n = 0 \Rightarrow \varphi(a_i)$ is maximal for at least two of the a_i.

1-3-5. Exercise. Let $\varphi_1, \ldots, \varphi_n$ be nontrivial inequivalent nonarchimedean valuations of F, and consider $a_1, \ldots, a_n \in F^*$; then

i. There exists $a \in F^*$ such that

$$\varphi_i(a) = \varphi_i(a_i) \qquad i = 1, \ldots, n$$

ii. If $\varepsilon_i > 0$ is a value taken on by φ_i for $i = 1, \ldots, n$, then there exists $a \in F$ such that

$$\varphi_i(a - a_i) = \varepsilon_i \qquad i = 1, \ldots, n$$

Suppose that P is a nonarchimedean prime divisor of F, and choose any $\varphi \in P$. Put

$$\begin{aligned} O &= \{a \in F | \varphi(a) \leq 1\} \\ \mathcal{P} &= \{a \in F | \varphi(a) < 1\} \\ U &= \{a \in F | \varphi(a) = 1\} \end{aligned}$$

From (1-1-5) it follows that O, $\mathcal{P}$, and U are independent of the choice of $\varphi \in P$.

Using the properties of φ, we see that O is an integral domain with 1; it is called the ***valuation ring at*** P or the ***ring of integers at*** P. We may note that $F = O \cup O^{-1}$, where $O^{-1} = \{a^{-1} | a \in O,\ a \neq 0\}$. Furthermore, $\mathcal{P}$ is a prime ideal in O; it is called the ***prime ideal at*** P. Clearly U is the (multiplicative) group of units of O, and $U = O \cap O^{-1}$; it is called the ***group of units at*** P. Since $O = \mathcal{P} \cup U$ and $\mathcal{P} \cap U = \emptyset$, we see that $\mathcal{P}$ consists of all nonunits of O. Consequently, $\mathcal{P}$ is the unique maximal ideal of O, and the residue class ring $O/\mathcal{P}$ is a field. We shall write $\bar{F} = O/\mathcal{P}$ and call it the ***residue class field at*** P. The canonical map

$$O \twoheadrightarrow \bar{F} = \frac{O}{\mathcal{P}}$$

will be denoted by ψ or simply by a bar, so that it maps $a \to \psi(a) = \bar{a}$. This map is known as the ***residue class map at*** P or as the ***place at*** P. There are alternative versions of the development of tools for algebraic number theory in which the notions of valuation ring and place, rather than that of valuation, play a central role.

For archimedean P, the objects O, $\mathcal{P}$, U are not defined.

In case P is trivial, we have $O = F$, $\mathcal{P} = (0)$, $U = F^*$, $\bar{F} = F$. Before going on to more significant examples, it is convenient to set up another way of looking at a nonarchimedean prime divisor P of F. We recall that for $\varphi \in P$ condition (iii) in the definition of valuation may be replaced by $\varphi(a + b) \leq \max \{\varphi(a), \varphi(b)\}$. Let us put

$$\nu(a) = -\log \varphi(a) \qquad a \in F$$

so that $\varphi(a) = e^{-\nu(a)}$. All the properties of φ carry over to ν, and ν is called an ***exponential valuation*** of F because it is a function from

F into $\mathbf{R} \cup \{\infty\}$ such that

$$\text{(i)} \qquad \nu(a) = \infty \Leftrightarrow a = 0$$

$$\text{(ii)} \qquad \nu(ab) = \nu(a) + \nu(b)$$

$$\text{(iii)} \qquad \nu(a + b) \geq \min\{\nu(a), \nu(b)\}$$

Clearly there is a 1-1 correspondence between the set of all nonarchimedean valuations of F and the set of all exponential valuations of F. To carry the connection further, we say that the exponential valuations ν and ν' are ***equivalent*** (and again denote this by $\nu \sim \nu'$) $\Leftrightarrow$ the corresponding valuations $\varphi = e^{-\nu}$ and $\varphi' = e^{-\nu'}$ are equivalent. Thus, $\nu \sim \nu' \Leftrightarrow \nu' = \alpha\nu$ for some $\alpha > 0$, $\alpha \in \mathbf{R}$. No confusion can arise if we write $\nu \in P$ when $e^{-\nu} = \varphi \in P$. Hence, if $\nu \in P$, then $P = \{\alpha\nu | \alpha > 0\}$. In terms of an exponential valuation $\nu \in P$ we have

$$O = \{a \in F | \nu(a) \geq 0\}$$
$$\mathcal{P} = \{a \in F | \nu(a) > 0\}$$
$$U = \{a \in F | \nu(a) = 0\}$$

Of course, O, $\mathcal{P}$, and U are independent of the choice of $\nu \in P$.

Any $\nu \in P$ determines a homomorphism $\nu : F^* \to \mathbf{R}^+$ (the additive group of reals). The image $\nu(F^*)$ is a subgroup of $\mathbf{R}^+$; it is called the ***value group of*** ν and is often written as $G(\nu)$. If also $\nu' \in P$, then $\nu'(F^*) = \alpha\nu(F^*)$ for some $\alpha > 0$, and $G(\nu') \simeq G(\nu)$—an order isomorphism. It is well known that any subgroup of $\mathbf{R}^+$ is either discrete or dense in $\mathbf{R}^+$ and that a discrete subgroup is either trivial (consisting of $\{0\}$ alone) or infinite cyclic (generated by the smallest positive element that it contains). The prime divisor P is said to be ***discrete*** or ***nondiscrete*** according as $G(\nu) = \nu(F^*)$ (for $\nu \in P$) is discrete or nondiscrete in $\mathbf{R}^+$; naturally, this does not depend on the choice of $\nu \in P$.

Suppose that P is a discrete prime divisor of F. If, for some $\nu \in P$, $G(\nu) = \nu(F^*) = \{0\}$, then $\varphi = e^{-\nu}$ is the trivial valuation, so that P is the trivial prime divisor. On the other hand, if P is discrete and nontrivial, then there exists in P a unique exponential valuation, which we shall invariably denote by ν_P, such that $G(\nu_P) = \nu_P(F^*) = \mathbf{Z}$. ν_P is then known as the ***normalized exponential valuation*** belonging to P. In such a situation we have

$$O = \{a \in F | \nu_P(a) \geq 0\} = \{a \in F | \nu_P(a) > -1\}$$
$$\mathcal{P} = \{a \in F | \nu_P(a) > 0\} = \{a \in F | \nu_P(a) \geq 1\}$$
$$U = \{a \in F | \nu_P(a) = 0\}$$

1-4. The Prime Divisors of Q

In this section, we determine all the nontrivial prime divisors of the rational field $\mathbf{Q}$. In particular, we shall give many concrete examples of valuations and related objects.

First of all, the ordinary absolute value function, which we write as φ_∞ [that is, $\varphi_\infty(a) = |a|$ for $a \in \mathbf{Q}$], is a valuation. The prime divisor which it determines is archimedean and will be denoted by ∞. It will usually be called "the" ***infinite prime*** of $\mathbf{Q}$, and φ_∞ will be known as the ***normed infinite valuation.***

To construct nonarchimedean valuations of $\mathbf{Q}$, we make use of a procedure which applies to any field F which is the quotient field of a unique factorization domain. Let $p \in \mathbf{Z}$ be any prime number. Since $\mathbf{Z}$ is a unique factorization domain, every element of its quotient field $\mathbf{Q}$ can be expressed uniquely as a product of primes from $\mathbf{Z}$—where, of course, negative exponents may appear. We define a function

$$\nu_p : \mathbf{Q}^* \to \mathbf{Z}$$

by putting for $a \in \mathbf{Q}^*$

$$\begin{aligned} \nu_p(a) &= \operatorname{ord}_p(a) \qquad \text{(called the } \textbf{\textit{ordinal of } a \textit{ at } p}\text{)} \\ &= \text{the exponent to which } p \text{ appears in the factorization of } a \end{aligned}$$

Thus

$$a = \varepsilon \prod_p p^{\nu_p(a)} = \varepsilon \prod_p p^{\operatorname{ord}_p(a)}$$

where $\varepsilon = \pm 1$ and p runs over all primes of $\mathbf{Z}$. Since all the primes $p \in \mathbf{Z}$ are taken to be > 0, ε is $+1$ or -1 according as a is > 0 or < 0. If we also put $\nu_p(0) = \infty$, it is immediate that $\nu_p : \mathbf{Q} \to \mathbf{Z} \cup \{\infty\}$ is an exponential valuation. The prime divisor to which ν_p belongs will be denoted simply by p. Clearly p is nonarchimedean—in fact it is a discrete nontrivial prime divisor of $\mathbf{Q}$—and ν_p is the normalized exponential valuation belonging to p. Note that an element of $\mathbf{Q}$ is near 0 (for p) when it is divisible by a high power of p.

From Section 1-3 we know that $\varphi = e^{-\nu_p}$ is a nonarchimedean valuation belonging to p; however, for reasons which will appear soon, we define a function φ_p by

$$\varphi_p(a) = \left(\frac{1}{p}\right)^{\nu_p(a)} \qquad a \in \mathbf{Q}$$

In other words, $\varphi_p = \varphi^\alpha$, where $\alpha = \log_e p > 0$ and $\varphi_p : \mathbf{Q} \to \mathbf{R}$ is a valuation belonging to p—it is called the ***normed,*** or ***normalized, p-adic valuation.*** If p and q are distinct primes of $\mathbf{Z}$, then the prime divisors p and q are distinct; this follows from (1-1-4) and the fact that $\varphi_p(p) = 1/p < 1$, while $\varphi_q(p) = 1$.

For each prime p, we have O, $\mathcal{P}$, and U defined. From their description (as given in Section 1-3) one sees easily that

$$O_p = \left\{\frac{m}{n} \in \mathbf{Q} \middle| m, n \in \mathbf{Z}, (m, n) = 1, (n, p) = 1\right\}$$

$$\mathcal{P}_p = \left\{\frac{m}{n} \in \mathbf{Q} \middle| m, n \in \mathbf{Z}, (m, n) = 1, (n, p) = 1, p|m\right\}$$

$$U_p = \left\{\frac{m}{n} \in \mathbf{Q} \middle| m, n \in \mathbf{Z}, (m, n) = 1, (n, p) = (m, p) = 1\right\}$$

Note that $0 = 0/1 \in \mathcal{P}_p \subset O_p$.

1-4-1. Exercise. Show that the residue class field of $\mathbf{Q}$ at the prime p is a field with p elements which may be written as

$$\{\bar{0}, \bar{1}, \bar{2}, \ldots, \overline{p-1}\} = \{0, 1, \ldots, p-1\}$$

1-4-2. Theorem. Let $\mathfrak{M}(\mathbf{Q})$ denote the set of all nontrivial prime divisors of $\mathbf{Q}$; then

$$\mathfrak{M}(\mathbf{Q}) = \{\infty, 2, 3, 5, 7, \ldots, p, \ldots\}$$

For each prime divisor p in $\mathfrak{M}(\mathbf{Q})$, let φ_p be the normed valuation belonging to p; then for each $a \neq 0 \in \mathbf{Q}$ we have

(i) $$\varphi_p(a) = 1 \text{ for almost all } p \in \mathfrak{M}(\mathbf{Q})$$

(ii) $$\prod_{p \in \mathfrak{M}(\mathbf{Q})} \varphi_p(a) = 1$$

Proof. Before proceeding to the details of the proof, let us remark that the assertion of (i) and (ii) is usually stated in the form: *The product formula holds in* $\mathbf{Q}$. It is the product formula which accounts for the way the normed valuations were selected. The fact that the product formula can be carried over to arbitrary algebraic number fields may be used as the central theme in the proof of some of the most important theorems of algebraic number theory; this will be done in Chapter 5.

Coming back to the proof, let P be any nontrivial prime divisor of $\mathbf{Q}$, and choose $\varphi \in P$ with $\|\varphi\| \leq 2$. Since $\mathbf{Q}$ is the quotient field of $\mathbf{Z}$, it suffices to determine φ on $\mathbf{Z}$; in fact, if $\varphi|\mathbf{Z} = \varphi_p^\alpha|\mathbf{Z}$ for some $\alpha > 0$, then $\varphi = \varphi_p^\alpha$ on $\mathbf{Q}$.

If P is nonarchimedean, then $\|\varphi\| = 1$ and $\varphi(n) \leq 1$ for all $n \in \mathbf{Z}$. Consider

$$I = \{n \in \mathbf{Z} | \varphi(n) < 1\}$$

We have $I \neq (0)$ because φ is nontrivial, and $I \neq \mathbf{Z}$ because $\varphi(1) = 1$. Moreover, I is a prime ideal of $\mathbf{Z}$ and is therefore generated by a prime p. For any $n \neq 0 \in \mathbf{Z}$ we have a unique expression $n = p^{\nu_p(n)}n'$ with $(n', p) = 1$. Since $n' \notin I$, we have $\varphi(n') = 1$ and, therefore, $\varphi(n) = \varphi(p)^{\nu_p(n)}$ where $0 < \varphi(p) < 1$. Therefore, $\varphi(p) = (1/p)^\alpha$ for some $\alpha > 0$, and then $\varphi(n) = \varphi_p(n)^\alpha$ for all $n \in \mathbf{Z}$. It follows that $\varphi = \varphi_p^\alpha$ on $\mathbf{Q}$.

Suppose then that P is archimedean. Fix any integers m and n both > 1. For any integer $t > 0$ we can write

$$m^t = a_0 + a_1 n + \cdots + a_s n^s$$

where $a_i \in \mathbf{Z}$, $0 \leq a_i < n$, and $a_s \neq 0$. Since $a_s \geq 1$, we have $n^s \leq m^t$ and $s \leq t(\log m/\log n)$. Upon using the fact that $\varphi(a_i) \leq a_i < n$, it follows that

$$\begin{aligned} \varphi(m^t) &\leq \varphi(a_0) + \varphi(a_1)\varphi(n) + \cdots + \varphi(a_s)\varphi(n)^s \\ &\leq n\{1 + \varphi(n) + \cdots + \varphi(n)^s\} \\ &\leq n(s+1)M \qquad \text{where } M = \max\{1, \varphi(n)\} \\ &\leq n\left(1 + t\frac{\log m}{\log n}\right)[\max\{1, \varphi(n)\}]^{t(\log m/\log n)} \end{aligned}$$

Taking tth roots and letting $t \to \infty$ yields

$$\varphi(m) \leq [\max\{1, \varphi(n)\}]^{\log m/\log n}$$

We assert that $\varphi(n) > 1$ for every $n > 1$; for if there exists $n_0 > 1$ with $\varphi(n_0) \leq 1$, then from the above relation $\varphi(m) \leq 1$ for all $m \in \mathbf{Z}$, which contradicts the archimedean character of φ. The relation therefore becomes $\varphi(m)^{1/\log m} \leq \varphi(n)^{1/\log n}$, and because the assumptions are symmetric in m and n, it follows that

$$\varphi(m)^{1/\log m} = \varphi(n)^{1/\log n} = e^\alpha$$

for some $\alpha > 0$ and all $m, n > 1$. Consequently, $\varphi(n) = n^\alpha$ for $n > 1$, and then, for any $n \in \mathbf{Z}$, $\varphi(n) = |n|^\alpha = \varphi_\infty^\alpha(n)$. We conclude that $\varphi = \varphi_\infty^\alpha$ on $\mathbf{Q}$, and thus complete the proof of the first part of the theorem.

The remaining part of the theorem presents no difficulties. For $a \neq 0 \in \mathbf{Q}$, $\nu_p(a) = 0$ for almost all $p \in \mathfrak{M}(\mathbf{Q})$; hence $\varphi_p(a) = 1$ for almost all p. Finally, we note that it suffices to verify the product formula for an arbitrary prime number q—and then $\varphi_\infty(q) = q$, $\varphi_q(q) = 1/q$, $\varphi_p(q) = 1$ for $p \neq q$. This completes the proof.

It should be observed that we now know that the arithmetic of $\mathbf{Q}$ may be described completely in terms of its prime divisors.

1-4-3. Exercise. i. The product formula in $\mathbf{Q}$ (or in any field in which there is one) is essentially unique in the sense that, if for each $p \in \mathfrak{M}(\mathbf{Q})$ there exists a real number r_p such that

$$\prod_{p \in \mathfrak{M}(\mathbf{Q})} \varphi_p(a)^{r_p} = 1$$

for all $a \in \mathbf{Q}^*$, then all the r_p are equal.

ii. A finite product formula cannot hold. More precisely, let $\varphi_1, \ldots, \varphi_n$ be inequivalent nontrivial valuations of an arbitrary field F, and suppose $r_1, \ldots, r_n$ are real numbers; then $\Pi \varphi_i(a)^{r_i} = 1$ for all $a \in F^* \Rightarrow r_1 = r_2 = \cdots = r_n = 0$.

The full story about the norm of a valuation may now be told.

1-4-4. Proposition. Let φ be a valuation of an arbitrary field F; then

$$\|\varphi\| = \max \{\varphi(1), \varphi(2)\}$$

Proof. Suppose φ is nonarchimedean; then $\|\varphi\| = 1$, and

$$\varphi(2) = \varphi(1 + 1) \leq \max \{\varphi(1), \varphi(1)\} = 1$$

so that $\|\varphi\| = \max \{\varphi(1), \varphi(2)\} = 1$.

Suppose φ is archimedean. Since for any $\alpha > 0$ we have $\|\varphi^\alpha\| = \|\varphi\|^\alpha$ and $\max \{\varphi^\alpha(1), \varphi^\alpha(2)\} = [\max \{\varphi(1), \varphi(2)\}]^\alpha$, it suffices to prove the assertion when $\|\varphi\| \leq 2$. Now the archimedean character of φ forces F to have characteristic 0, and we may take $\mathbf{Q} \subset F$. By (1-4-2), $\varphi|\mathbf{Q} = \varphi_\infty^\alpha$ for some $\alpha > 0$—that is, $\varphi(a) = |a|^\alpha$ for all $a \in \mathbf{Q}$. Furthermore, we have for $a, b \in F$,

$$\begin{aligned} \varphi[(a + b)^n] &\leq 2(n + 1) \max_{=0, \ldots, n} \left\{\varphi\left[\binom{n}{i} a^{n-i}b^i\right]\right\} \\ &\leq 2(n + 1)2^{n\alpha}[\max \{\varphi(a), \varphi(b)\}]^n \end{aligned}$$

since $2^n = (1 + 1)^n = \sum_{i=0}^{n} \binom{n}{i}$ gives $\varphi\left[\binom{n}{i}\right] = \binom{n}{i}^\alpha \leq 2^{n\alpha}$. If we take nth roots and let $n \to \infty$, then we have $\varphi(a + b) \leq 2^\alpha \max \{\varphi(a), \varphi(b)\}$, so that $\|\varphi\| \leq 2^\alpha$. However,

$$2^\alpha = \varphi(2) = \varphi(1 + 1) \leq \|\varphi\| \max \{\varphi(1), \varphi(1)\}$$

Hence, $\|\varphi\| = 2^\alpha = \varphi(2)$, and the proof is complete.

1-4-5. Exercise. Let x be an indeterminate over the field F. Determine all nontrivial prime divisors of $F(x)$ which are trivial on F.

(*Hint:* There is a prime divisor corresponding to each irreducible polynomial in $F[x]$ and also one corresponding to $1/x$.) Select normalized valuations in such a way that the product formula holds. It is convenient here to work with exponential valuations.

1-5. Fields with a Discrete Prime Divisor

What can be said about the structure of an arbitrary field with a prime divisor P of the type possessed by **Q**? In the case where P is archimedean, the answer will become clear in Section 1-8 when we have proved the theorem of Ostrowski. We shall therefore suppose in this section that P is a discrete nontrivial prime divisor of the field F and that $\nu = \nu_P$ is the normalized exponential valuation belonging to P. Choose an element $\pi \in F$ such that $\nu(\pi) = 1$; for obvious reasons, π is known as a ***prime element*** of F with respect to ν or P and sometimes as a ***local uniformizing parameter.*** Note that, for $\varphi \in P$, $\varphi(\pi)$ is the maximal value < 1 taken on by φ.

1-5-1. Ideal Theory. Clearly $\mathcal{P} = \pi O$, and for each integer $r > 0$

$$\mathcal{P}^r = (\pi O)^r = \pi^r O = \{a \in F \mid \nu(a) \geq r\}$$

Each $\mathcal{P}^r$ is an ideal of O. If $a \in F$ has $\nu(a) = r$, then we have $\mathcal{P}^r = aO$. It now follows easily that $\{\mathcal{P}^r \mid r > 0\}$ is the set of all proper ideals of O. Thus, O has a unique prime ideal—namely, the ideal $\mathcal{P}$ of nonunits of O. $\mathcal{P}$ is the unique maximal ideal of O, and it contains every proper ideal of O; in fact, every proper ideal is a power of it. O is a principal ideal domain; in particular, O is a unique factorization domain with π as its only prime element (upto units).

Every $a \in F^*$ can be written in the form $a = \pi^r \varepsilon$, where r is an integer and ε is a unit, and such an expression is unique since $r = \nu(a)$ and $\varepsilon = a\pi^{-r}$. The elements $a \neq 0 \in O$ are precisely those for which $r \geq 0$.

It is also convenient to put

$$\mathcal{P}^r = \pi^r O = \{a \in F \mid \nu(a) \geq r\} \text{ for } r \leq 0$$

These are additive subgroups of F^+; in fact, they are O submodules of F^+. Now suppose, in general, that O is an integral domain with quotient field $F \neq O$. If $\mathfrak{A} \neq (0)$ is an O submodule of F^+ and there exists an element $c \neq 0 \in F$ such that $c\mathfrak{A} \subset O$, then we say that $\mathfrak{A}$ is an O ***ideal*** of F, or a ***fractional ideal*** of F, or simply an ***ideal*** of F. Note that, because F is the quotient field of O, the element c may be taken from O. The fractional ideals contained in O are called ***integral ideals;*** thus the integral ideals of F are precisely the $\neq(0)$ ideals of the ring O. We shall have a great deal to say about ideals. In our case here one merely points out that

$$\{\mathcal{P}^r | r = 0, \pm 1, \pm 2, \ldots\}$$

is the set of all O ideals of F and that these form a group under multiplication. (By the product of two ideals $\mathfrak{A}$ and $\mathfrak{B}$ one means all finite sums of form $\Sigma a_i b_i$ with $a_i \in \mathfrak{A}$, $b_i \in \mathfrak{B}$.) Thus, $\mathcal{P}^r\mathcal{P}^s = \mathcal{P}^{r+s}$, the identity element is $\mathcal{P}^0 = O$, and $(\mathcal{P}^r)^{-1} = \mathcal{P}^{-r}$.

1-5-2. Additive Structure. In view of the above, we have the chain of additive groups

$$F^+ \supset \cdots \supset \mathcal{P}^{-2} \supset \mathcal{P}^{-1} \supset O \supset \mathcal{P} \supset \mathcal{P}^2 \supset \cdots \supset \{0\}$$

Since, for each r, $\mathcal{P}^r = \{a \in F | \nu(a) > r - 1\}$, it follows from the "continuity" of ν that each $\mathcal{P}^r$ is open, and hence closed too. According to (1-1-9), $\{\mathcal{P}^r | r = 0, \pm 1, \pm 2, \ldots\}$ is a fundamental system of neighborhoods of 0 in F. In particular, F is totally disconnected.

For any r, $s \in \mathbf{Z}$, we have a topological isomorphism of additive groups

$$\mathcal{P}^r \approx \mathcal{P}^s$$

In fact, such a map is given explicitly by $a \to \pi^{s-r}a$ for $a \in \mathcal{P}^r$. It follows that for all $n > 0$ and all r there is an isomorphism of additive (discrete) groups

$$\frac{\mathcal{P}^r}{\mathcal{P}^{r+n}} \approx \frac{O}{\mathcal{P}^n}$$

In particular, we have, for every r,

$$\frac{\mathcal{P}^r}{\mathcal{P}^{r+1}} \approx \frac{O}{\mathcal{P}} = \bar{F}^+$$

where $\bar{F}^+$ is the additive group of the residue class field and the isomorphism is additive.

If, in addition, $\bar{F}$ is finite with q elements, then, for all r and all $n > 0$,

$$(\mathcal{P}^r : \mathcal{P}^{r+n}) = (\mathcal{P}^r : \mathcal{P}^{r+1}) \cdots (\mathcal{P}^{r+n-1} : \mathcal{P}^{r+n}) = q^n$$

For $a, b \in F$ and any r, we define $a \equiv b \pmod{\mathcal{P}^r}$ to mean that $a - b \in \mathcal{P}^r$, or, what is the same, $\nu(a - b) \geq r$. Of course, if $a, b \in O$ and $r > 0$, this definition coincides with the usual notion of congruence modulo an ideal of O.

1-5-3. Multiplicative Structure. For each $r > 0$, $1 + \mathcal{P}^r$ is a multiplicative subgroup of F^*. In fact, if $a, b \in 1 + \mathcal{P}^r$, we have $a = 1 + \pi^r\mu$, $b = 1 + \pi^r\mu'$, with $\mu, \mu' \in O$, so that

$$ab = 1 + \pi^r(\mu + \mu' + \pi^r\mu\mu') \in 1 + \pi^r O = 1 + \mathcal{P}^r$$

Also, $\nu(a - 1) = \nu(\pi^r\mu) \geq r > 0$ so that $\nu(a) = 0$ and $1 + \mathcal{P}^r \subset U$. Since $a^{-1} \in U$, we have $1 = a^{-1} + \pi^r\mu a^{-1}$ and $a^{-1} = 1 + \pi^r(-\mu a^{-1}) \in 1 + \mathcal{P}^r$. Thus $1 + \mathcal{P}^r$ is indeed a multiplicative group.

Consider the chain of multiplicative groups

$$F^* \supset U \supset 1 + \mathcal{P} \supset 1 + \mathcal{P}^2 \supset \cdots \supset \{1\}$$

A fundamental system of open and closed neighborhoods of 1 in the topological group F^* is given by $\{1 + \mathcal{P}^r | r > 0\}$. In particular, F^* is totally disconnected. It is standard to write $U_0 = U$ and $U_r = 1 + \mathcal{P}^r$ for $r > 0$.

The factor groups formed from this chain are discrete; and the individual steps in the chain may be analyzed algebraically as follows:

(i) $$\frac{F^*}{U_0} \approx \mathbf{Z}^+$$

(ii) $$\frac{U_0}{U_1} \approx \bar{F}^*$$

(iii) $$\frac{U_r}{U_{r+1}} \approx \bar{F}^+ \qquad \text{for all } r > 0$$

The proof of (i) is trivial, since $\nu: F^* \to \mathbf{Z}^+$ is a homomorphism onto with kernel U_0. To verify (ii), consider the exact sequence of rings and ring homomorphisms

$$\{0\} \to \mathcal{P} \xrightarrow{i} O \xrightarrow{\psi} \bar{F} \to \{0\}$$

where i is the inclusion map and $\psi: O \to \bar{F}$ is the residue class map. By extracting the multiplicative part of this sequence, we get the exact sequence of multiplicative groups

$$\{1\} \to U_1 \xrightarrow{i} U_0 \xrightarrow{\psi} \bar{F}^* \to \{1\}$$

Finally, (iii) is proved by setting up an "exponential-logarithmic" correspondence. For each $r > 0$, one observes that there is a 1-1 correspondence $a \leftrightarrow 1 + \pi^r a$ of

$$O \leftrightarrow U_r = 1 + \pi^r O$$

Moreover, under this correspondence

$$\mathcal{P} \leftrightarrow U_{r+1} = 1 + \pi^{r+1}O$$

Let $f:U_r \to \bar{F}^+$ be the composite map $U_r \twoheadrightarrow O \twoheadrightarrow \bar{F}^+$; in other words, $f(1 + \pi^r a) = \bar{a}$. Since

$$\begin{aligned}\{f(1 + \pi^r a)(1 + \pi^r b)\} &= f\{1 + \pi^r(a + b + \pi^r ab)\}\\ &= \overline{a + b + \pi^r ab} = \bar{a} + \bar{b}\end{aligned}$$

we see that f is a homomorphism; and f is clearly onto and has kernel U_{r+1}. Thus $U_r/U_{r+1} \approx O/\mathcal{P}$.

In the case where $\bar{F}$ is finite with q elements, we have then

$$(U_0:U_r) = (q - 1)q^{r-1} \qquad \text{for all } r > 0$$

One may also introduce multiplicative congruences with respect to the subgroups U_r for $r > 0$. For $a, b \in F^*$, we take $a \underset{\times}{\equiv} b \pmod{\mathcal{P}^r}$ to mean that $a/b \in U_r$. Note that $\underset{\times}{\equiv}$ is an equivalence relation; after all, $a \underset{\times}{\equiv} b \pmod{\mathcal{P}^r} \Leftrightarrow a$ and b belong to the same coset of F^* modulo U_r. Another way of phrasing the definition is

$$a \underset{\times}{\equiv} b \pmod{\mathcal{P}^r} \Leftrightarrow \nu\left(\frac{a}{b} - 1\right) \geq r$$

If we write $a = \pi^{\nu(a)}\varepsilon$ and $b = \pi^{\nu(b)}\eta$, where ε and η are units, then it is easy to check that

$$a \underset{\times}{\equiv} b \pmod{\mathcal{P}^r} \Leftrightarrow \nu(a) = \nu(b) \text{ and } \varepsilon \equiv \eta \pmod{\mathcal{P}^r}$$

A bad notation (which is often used) is the result of extending the above to the case $r = 0$; namely,

$$\begin{aligned}a \underset{\times}{\equiv} b \pmod{\mathcal{P}^0} &\Leftrightarrow \frac{a}{b} \in U_0 \Leftrightarrow \nu(a) = \nu(b)\\ &\Leftrightarrow a \text{ and } b \text{ are in the same coset of } F^* \text{ modulo } U\end{aligned}$$

1-5-4. Exercise. Let P be a discrete nontrivial prime divisor of F, and suppose the residue class field has characteristic $p \neq 0$:

i. If $r \geq 1$ and $a \equiv b \pmod{\mathcal{P}^r}$, then $a^{p^s} \equiv b^{p^s} \pmod{\mathcal{P}^{r+s}}$ for all $s \geq 0$.

ii. If $r \geq 1$ and $a \underset{\times}{\equiv} b \pmod{\mathcal{P}^r}$, then $a^{p^s} \underset{\times}{\equiv} b^{p^s} \pmod{\mathcal{P}^{r+s}}$ for all $s \geq 0$.

1-6. *e* and *f*

We begin this section by setting up, rather carefully, the notation and terminology to be used in studying the connection between prime divisors of a field and prime divisors of an extension field.

Suppose that F and E are fields and that $F \subset E$. If φ is a valuation of E, then $\varphi' = \varphi|F$ is a valuation of F; naturally, φ' is called the ***projection,*** or ***restriction,*** of φ on F, and φ is called an ***extension*** of φ' to E. Let Q denote the prime divisor of E to which φ belongs. Since $\varphi^\alpha|F = (\varphi|F)^\alpha$ for $\alpha > 0$, it follows that the restrictions to F of all the valuations belonging to Q run over a full prime divisor P of F—that is, there is a 1-1 correspondence between the valuations belonging to Q and those belonging to P. We say that P is the ***projection*** of Q on F and that Q is an ***extension*** of P to E. For reasons which will appear later, it is convenient to generalize the preceding remarks slightly.

Let F and E be fields, and let $\mu: F \to E$ be a monomorphism. In the preceding case $\mu = i$, the inclusion map. Let $\mathfrak{M}(E)$ denote the set of all prime divisors of E. Consider $Q \in \mathfrak{M}(E)$, and choose any $\varphi \in Q$. We may define a valuation φ^μ of F by

$$\varphi^\mu = \varphi \circ \mu$$

We shall often say that φ^μ is a ***restriction*** of φ or that φ is an ***extension*** of φ^μ. Any other element of Q is of form φ^α with $\alpha > 0$; hence, from

$$(\varphi^\alpha)^\mu(a) = (\varphi^\alpha)(\mu a) = [\varphi(\mu a)]^\alpha = [\varphi^\mu(a)]^\alpha \qquad a \in F$$

it follows that $(\varphi^\alpha)^\mu = (\varphi^\mu)^\alpha$. In other words, the prime divisor Q leads to a full prime divisor P of F; we write $P = \mu^*(Q)$ and also $\varphi^\mu = \mu^*(\varphi) = \varphi \circ \mu$. Thus, a monomorphism $\mu: F \to E$ has associated with it a mapping $\mu^*: \mathfrak{M}(E) \to \mathfrak{M}(F)$.

Clearly, if $\mu: F \to E$ and $\lambda: E \to K$ are monomorphisms of fields, then $(\lambda \circ \mu)^* = \mu^* \circ \lambda^*$. Also, if $F = E$ and $\mu = 1$, the identity map, then $\mu^* = 1^*$ is the identity map. Consequently, if μ is an isomorphism onto, then $\mu^*: \mathfrak{M}(E) \to \mathfrak{M}(F)$ is 1-1 onto and $(\mu^*)^{-1} = (\mu^{-1})^*$. It should be pointed out that, in general, μ^* is not 1-1.

We shall write $\langle E, Q, \mu \rangle \supset \langle F, P \rangle$ and say that $\langle E, Q, \mu \rangle$ is an ***extension*** of $\langle F, P \rangle$ to describe the situation in which $\mu: F \to E$ is a monomorphism and $\mu^*(Q) = P$. The obvious restriction and projection terminology will also be used. In the case where $\mu = i$, this definition coincides with the old one.

Suppose that $\langle E, Q, \mu\rangle \supset \langle F, P\rangle$; then E has a topology T_Q determined by Q, and so does $\mu(F)$. On the other hand, F has a topology T_P determined by $P = \mu^*(Q)$. Directly from the definitions, it is clear that $\mu:F \to E$ is continuous with respect to the topologies T_P and T_Q; moreover, $\mu:F \to \mu(F)$ is a homeomorphism. In particular, $\{a_n\}$ is a Cauchy sequence in F (for T_P) $\Leftrightarrow$ $\{\mu(a_n)\}$ is a Cauchy sequence in E (for T_Q).

Suppose again that $\langle E, Q, \mu\rangle \supset \langle F, P\rangle$. Since μ maps the prime subfield of F isomorphically onto the prime subfield of E, it is immediate that Q is archimedean (nonarchimedean) $\Leftrightarrow P = \mu^*(Q)$ is archimedean (nonarchimedean). In the nonarchimedean situation, let us write O_Q for the valuation ring at Q and O_P for the valuation ring at P. The corresponding prime ideals will be denoted by $\mathcal{Q}$ and $\mathcal{P}$, respectively, or else by $\mathcal{P}_Q$ and $\mathcal{P}_P$, respectively. More explicitly, if $\varphi \in Q$ and $\varphi' = \varphi^\mu \in P$, then

$$\begin{aligned} O_Q &= \{\alpha \in E \mid \varphi(\alpha) \leq 1\} \\ \mathcal{P}_Q &= \mathcal{Q} = \{\alpha \in E \mid \varphi(\alpha) < 1\} \\ O_P &= \{a \in F \mid \varphi'(a) \leq 1\} \\ \mathcal{P}_P &= \mathcal{P} = \{a \in F \mid \varphi'(a) < 1\} \end{aligned}$$

We have clearly $a \in O_P \Leftrightarrow \mu(a) \in O_Q$, so that $\mu(O_P) \subset O_Q$, and also $a \in \mathcal{P} \Leftrightarrow \mu(a) \in \mathcal{Q}$, so that $\mu(\mathcal{P}) \subset \mathcal{Q}$. Consequently, there arises in the canonical way a mapping $\bar{\mu}$ of the residue class fields $\bar{F} = O_P/\mathcal{P}$ and $\bar{E} = O_Q/\mathcal{Q}$. To be precise, the mapping $\bar{\mu}:\bar{F} \to \bar{E}$ is given by

$$\bar{\mu}(\bar{a}) = \overline{\mu(a)} \qquad \text{for all } a \in O_P$$

To put it still another way, we have the commutative diagram

$$\begin{array}{ccc} O_P & \xrightarrow{\mu} & O_Q \\ {\scriptstyle \psi_P}\downarrow & & \downarrow{\scriptstyle \psi_Q} \\ \bar{F} & \xrightarrow{\bar{\mu}} & \bar{E} \end{array}$$

where ψ_P and ψ_Q are the respective places. Since $\bar{\mu}(\bar{1}) = \bar{1} \neq 0 \in \bar{E}$, $\bar{\mu}$ is nontrivial, so that $\bar{\mu}:\bar{F} \to \bar{E}$ is a monomorphism.

We may also note that $\mu(O_P) = O_Q \cap \mu(F)$ and $\mu(\mathcal{P}) = \mathcal{Q} \cap \mu(F)$. In the case where $\mu = i$, these become $O_P = O_Q \cap F$ and $\mathcal{P} = \mathcal{Q} \cap F$. It is customary to identify and consider $\bar{F}$ as a subfield of $\bar{E}$; that is, $\bar{\imath}:\bar{F} \to \bar{E}$ is the inclusion map.

In order to prevent the notation from becoming too burdensome, we shall as a rule limit the discussion to the case $\mu = i$. Any statements made can be transferred easily to the case of a general μ.

1-6-1. Proposition. Let P be the trivial prime divisor of F, and let E be an algebraic extension of F; then P has a unique extension to E—namely, the trivial prime divisor.

Proof. The trivial prime divisor of E is clearly an extension of P. On the other hand, suppose that Q is an extension of P to E, and let $\varphi_Q \in Q$ be an extension of the trivial valuation φ_P. For any $\alpha \in E$ we have an equation

$$\alpha^n + a_1\alpha^{n-1} + \cdots + a_{n-1}\alpha + a_n = 0 \qquad a_i \in F$$

If Q is not trivial, there exists an α with $\varphi_Q(\alpha) > 1$. Since $\varphi_Q(\alpha^n) > \varphi_Q(a_i\alpha^{n-i})$ for $i = 1, \ldots, n$, (1-3-4) provides a contradiction.

Suppose now that $\langle E, Q, i\rangle$ with Q nonarchimedean is an extension of $\langle F, P\rangle$. For $\nu \in Q$, put $\nu' = \nu|F \in P$; then

$$f = f\left(\frac{Q}{P}\right) = [\bar{E}:\bar{F}]$$

is called the ***degree of*** Q ***over*** P, or the ***residue class degree of*** E ***over*** F ***at*** Q, and

$$e = e\left(\frac{Q}{P}\right) = (\nu(E^*):\nu'(F^*)) = (\nu(E^*):\nu(F^*))$$

is called the ***ramification index of*** Q ***over*** P. Of course, $e(Q/P)$ is independent of the choice of $\nu \in Q$—or for that matter of the choice of $\varphi \in Q$. As a simple example, we may note that, in the situation of (1-6-1), $e = 1$ and $f = [E:F]$.

1-6-2. Proposition. If $\langle E, Q, i\rangle \supset \langle F, P\rangle$ and $\langle K, R, i\rangle \supset \langle E, Q\rangle$, then $\langle K, R, i\rangle \supset \langle F, P\rangle$. Moreover, if R is nonarchimedean, then

$$e\left(\frac{R}{P}\right) = e\left(\frac{R}{Q}\right) e\left(\frac{Q}{P}\right)$$
$$f\left(\frac{R}{P}\right) = f\left(\frac{R}{Q}\right) f\left(\frac{Q}{P}\right)$$

Proof. Since dimensions of field extensions and group indices are both multiplicative, e and f are multiplicative.

An important question will be to study the connections between an extension E/F and the e's and f's that arise from it. We give some elementary steps in this direction.

1-6-3. Lemma. Let $\langle E, Q, i\rangle$ be an extension of $\langle F, P\rangle$ with P non-archimedean.

i. Let $\omega_1, \ldots, \omega_r$ be elements of O_Q such that $\bar{\omega}_1, \ldots, \bar{\omega}_r$ are linearly independent over $\bar{F}$; then $\omega_1, \ldots, \omega_r$ are linearly independent over F. In fact, for $a_1, \ldots, a_r \in F$ and $\nu \in Q$, we have

$$\nu(a_1\omega_1 + \cdots + a_r\omega_r) = \min_{i=1,\ldots,r} \{\nu(a_i)\}$$

ii. Let $\pi_0, \ldots, \pi_s \in E^*$ determine representatives of distinct cosets of $\nu(F^*)$ in $\nu(E^*)$; then $\pi_0, \ldots, \pi_s$ are linearly independent over F. In fact, for $b_0, \ldots, b_s \in F$, we have

$$\nu(b_0\pi_0 + \cdots + b_s\pi_s) = \min_{j=0,\ldots,s} \{\nu(b_j\pi_j)\}$$

Proof. i. It suffices to prove the property of ν; for then

$$a_1\omega_1 + \cdots + a_r\omega_r = 0$$

implies $\nu(a_i) = \infty$ for $i = 1, \ldots, r$, so that $a_1 = \cdots = a_r = 0$. To do this, note first that, if $a_1 = \cdots = a_r = 0$, both sides are ∞. Suppose then that not all a_i are 0 and that, say, $\nu(a_1) < \infty$ is a minimal value. Thus $a_i/a_1 \in O_P$ for $i = 1, \ldots, r$, and

$$\alpha = \omega_1 + (a_2/a_1)\omega_2 + \cdots + (a_r/a_1)\omega_r \in O_Q$$

Since $\bar{\omega}_1, \ldots, \bar{\omega}_r$ are linearly independent over $\bar{F}$, it follows that

$$\bar{\alpha} = \bar{\omega}_1 + \overline{\left(\frac{a_2}{a_1}\right)}\bar{\omega}_2 + \cdots + \overline{\left(\frac{a_r}{a_1}\right)}\bar{\omega}_r \neq 0$$

Therefore, $\alpha \notin \mathfrak{Q}$. This means that $\nu(\alpha) = 0$, so that

$$\nu(a_1\omega_1 + \cdots + a_r\omega_r) = \nu(a_1) = \min\{\nu(a_1), \ldots, \nu(a_r)\}$$

ii. Again it suffices to prove the property of ν; for then

$$b_0\pi_0 + \cdots + b_s\pi_s = 0$$

implies $\nu(b_j\pi_j) = \infty$ for $j = 0, \ldots, s$, so that $b_0 = \cdots = b_s = 0$. Consider $\beta = b_0\pi_0 + \cdots + b_s\pi_s$; for the moment, let us assume that each $b_j \in F^*$. Thus, $\nu(b_j\pi_j) < \infty$ for $j = 0, \ldots, s$. We assert that the values $\nu(b_j\pi_j)$ $(j = 0, \ldots, s)$ are all distinct—in fact, $\nu(b_i\pi_i) = \nu(b_j\pi_j)$ for $i \neq j$ implies $\nu(\pi_j/\pi_i) = \nu(b_i/b_j) \in \nu(F^*)$, contradicting the hypothesis on $\pi_0, \ldots, \pi_s$. From (1-3-4) we know that $\nu(\beta) = \min_j \{\nu(b_j\pi_j)\}$. It is now also clear that

$$\nu(b_0\pi_0 + \cdots + b_s\pi_s) = \min_j \{\nu(b_j\pi_j)\}$$

even if some or all of the $b_j = 0$.

From (1-6-3) we know that $e(Q/P) \leq [E:F]$ and $f(Q/P) \leq [E:F]$. With a slight additional effort, it is possible to go one step further.

1-6-4. Proposition. Let $\langle E, Q, i\rangle$ be an extension of $\langle F, P\rangle$ with P nonarchimedean; then

$$e\left(\frac{Q}{P}\right)f\left(\frac{Q}{P}\right) \leq [E:F]$$

Proof. Let us take $\omega_1, \ldots, \omega_f$ and $\pi_0, \ldots, \pi_{e-1}$ as in (1-6-3); we show that the elements $\omega_i\pi_j$ $(i = 1, \ldots, f; j = 0, \ldots, e - 1)$ are linearly independent over F. To accomplish this, it suffices (by an argument of the type used above) to show that, for $a_{ij} \in F$, we have

$$\nu\left\{\sum_{i,j} a_{ij}\omega_i\pi_j\right\} = \min_{i,j} \{\nu(a_{ij}\pi_j)\}$$

To see this, note first that, for each fixed j having some $a_{ij} \neq 0$, we have

$$\nu\left\{\sum_i a_{ij}\omega_i\pi_j\right\} = \nu(\pi_j) + \nu\left(\sum_i a_{ij}\omega_i\right)$$
$$= \nu(\pi_j) + \min_i \{\nu(a_{ij})\}$$

For the various admissible j's, these terms are all distinct by the choice of the π_j's; therefore, the assertion is true, and the proof is complete.

Our final result will appear as an easy consequence of the work in Chapter 2; the proof here is an extension of the proof of (1-6-4).

1-6-5. Theorem. Let F be a field with nonarchimedean nontrivial prime divisor P, and let E be an extension field of F. If $Q_1, \ldots, Q_g$ are distinct extensions of P to E, then

$$\sum_1^g e\left(\frac{Q_i}{P}\right)f\left(\frac{Q_i}{P}\right) \leq [E:F]$$

Proof. Obviously, this result is of interest only when $[E:F] = n < \infty$, and it then implies that P can have at most n extensions to E.

The assumption of nontriviality for P guarantees that the Q_i's are nontrivial, so that the approximation theorem may be used. Of course, the theorem is also true when P is trivial.

Choose $\varphi \in P$, and let $\psi: O_P \twoheadrightarrow \bar{F} = O_P/\mathcal{P}$ be the residue class map. For each $i = 1, \ldots, g$ there exists a $\varphi^{(i)} \in Q_i$ which is an extension of φ. These valuations $\varphi^{(i)}$ of E are nontrivial and inequivalent. Furthermore, for each $i = 1, \ldots, g$ we have a residue class map $\psi^{(i)}: O_{Q_i} \twoheadrightarrow \bar{E}^{(i)} = O_{Q_i}/\mathcal{Q}_i$. As pointed out earlier, each $\psi^{(i)}$ is an extension of ψ. Let us write $e_i = e(Q_i/P)$ and $f_i = f(Q_i/P)$.

For $i = 1, \ldots, g$ we have, as before, $\{\omega_\nu^{(i)} \in O_{Q_i} | \nu = 1, \ldots, f_i\}$ such that the elements $\psi^{(i)}(\omega_\nu^{(i)}) \in \bar{E}^{(i)}$ are linearly independent over $\bar{F}$ and also $\{\pi_\mu^{(i)} \in E^* | \mu = 0, 1, \ldots, e_i - 1\}$ such that the real numbers $\varphi^{(i)}(\pi_\mu^{(i)})$ are representatives of distinct cosets of $\varphi(F^*)$ in $\varphi^{(i)}(E^*)$. We shall show that the elements $\{\omega_\nu^{(i)}\pi_\mu^{(i)} | i, \nu, \mu\}$ of E are linearly independent over F.

Consider any element of E of form

$$\sum_{i,\nu,\mu} a_{\nu\mu}^{(i)} \omega_\nu^{(i)} \pi_\mu^{(i)}$$

where $a_{\nu\mu}^{(i)} \in F$ and at least one $a_{\nu\mu}^{(i)} \neq 0$. Adjust the notation so that $\varphi(a_{11}^{(j)}) \neq 0$ is maximal. From the proof of (1-6-4), we know that

$$\varphi^{(j)}\left(\sum_{\nu,\mu} a_{\nu\mu}^{(j)} \omega_\nu^{(j)} \pi_\mu^{(j)}\right) = \max_{\nu,\mu} \{\varphi^{(j)}(a_{\nu\mu}^{(j)} \pi_\mu^{(j)})\} = M$$

Suppose that at the start the $\omega_\nu^{(i)}$ for all i and ν could be chosen so as to guarantee that in our situation

$$\varphi^{(j)}(a_{\nu\mu}^{(i)} \omega_\nu^{(i)} \pi_\mu^{(i)}) < M \qquad \text{for } i \neq j \text{ and all } \nu, \mu$$

This would mean that

$$\varphi^{(j)}\left\{\sum_{i,\nu,\mu} a_{\nu\mu}^{(i)} \omega_\nu^{(i)} \pi_\mu^{(i)}\right\} = \max_{\nu,\mu} \{\varphi^{(j)}(a_{\nu\mu}^{(j)} \pi_\mu^{(j)})\} = M$$

and then

$$\sum_{i,\nu,\mu} a_{\nu\mu}^{(i)} \omega_\nu^{(i)} \pi_\mu^{(i)} = 0$$

with $a_{11}^{(j)} \neq 0$ would give $\varphi^{(j)}(a_{11}^{(j)} \pi_1^{(j)}) = 0$, a contradiction.

It remains then to show that the $\omega_\nu^{(i)}$ may be properly chosen; more precisely, if we let

$$N = \min_{\substack{i,j,\mu,\lambda \\ i \neq j}} \frac{\varphi^{(j)}(\pi_\lambda^{(j)})}{\varphi^{(j)}(\pi_\mu^{(i)})}$$

then the $\{\omega_\nu^{(i)}|i, \nu\}$ may be chosen so that

$$\varphi^{(j)}(\omega_\nu^{(i)}) < N \qquad \text{for } j \neq i$$

Of course, it is automatic that $\varphi^{(j)}(\omega_\nu^{(j)}) = 1$.

Let any choice of $\{\omega_\nu^{(i)}|i, \nu\}$ be given, and fix i and ν. By the approximation theorem, there exists $\varpi_\nu^{(i)} \in E$ such that

$$\begin{aligned} \varphi^{(i)}(\varpi_\nu^{(i)} - \omega_\nu^{(i)}) &< 1 \\ \varphi^{(j)}(\varpi_\nu^{(i)}) &< N \qquad \text{for } j \neq i \end{aligned}$$

From (1-3-3) we see that $\varphi^{(i)}(\varpi_\nu^{(i)}) = 1$; so $\varpi_\nu^{(i)} \in O_{Q_i}$. The relation $\psi^{(i)}(\varpi_\nu^{(i)} - \omega_\nu^{(i)}) = 0$ implies $\psi^{(i)}(\varpi_\nu^{(i)}) = \psi^{(i)}(\omega_\nu^{(i)})$. Thus $\{\varpi_\nu^{(i)}|i, \nu\}$ has the desired properties. This completes the proof.

1-7. Completions

Let P be a prime divisor of the field F. A sequence $\{a_n\}$ of elements of F is a ***P-Cauchy sequence*** if $a_m - a_n \to 0$ (in the topology T_P) when $m, n \to \infty$. We say that F is ***P-complete*** (or that $\langle F, P\rangle$ is complete) if every P-Cauchy sequence converges and that $\langle \tilde{F}, \tilde{P}, \mu\rangle$ is a P ***completion*** of F (or a completion of $\langle F, P\rangle$) when

(i) $\qquad \langle \tilde{F}, \tilde{P}, \mu\rangle \supset \langle F, P\rangle$

(ii) $\qquad \langle \tilde{F}, \tilde{P}\rangle$ is complete

(iii) $\qquad \mu(F)$ is dense in $\tilde{F}$

When there is no danger of confusion, we shall often drop the references to P in the above and say that $\tilde{F}$ is a completion of F.

The obvious example of a completion is to take F as the rational field $\mathbf{Q}$, $\tilde{F}$ as the real field $\mathbf{R}$, P and $\tilde{P}$ as the primes determined by the absolute value (both denoted by P_∞), and μ as the inclusion map $i:\mathbf{Q} \to \mathbf{R}$. Then $\langle \mathbf{R}, P_\infty, i\rangle$ is a completion of $\langle \mathbf{Q}, P_\infty\rangle$. It may also be noted that a field is always complete with respect to the trivial prime divisor. Because of both this fact and (1-6-1), trivial prime divisors play no role in algebraic number theory—nor are they of interest, in general. Therefore, for the remainder of this book, it is to be understood that ***all prime divisors are nontrivial.***

1-7-1. Theorem. Let F be a field with prime divisor P. A P completion of F exists, and it is essentially unique. In fact, if both $\langle \tilde{F}, \tilde{P}, \mu\rangle$ and $\langle \tilde{\tilde{F}}, \tilde{\tilde{P}}, \lambda\rangle$ are P completions of F, then there exists a unique isomorphism $\sigma:\tilde{F} \rightarrowtail\!\!\!\twoheadrightarrow \tilde{\tilde{F}}$ which is an extension of the identity map $1:F \to F$ and such that $\sigma^*(\tilde{\tilde{P}}) = \tilde{P}$.

Proof. The proof of the existence of a completion is essentially the same as the construction of the reals from the rationals (except that here we assume that the reals are known); therefore, we shall content ourselves with a sketch of the proof. We remark in passing that the statement that σ is to be an extension of the identity map means that the following diagram must commute:

$$\begin{array}{ccc} \tilde{F} & \xrightarrow{\ \sigma\ } & \tilde{\tilde{F}} \\ \mu\uparrow & & \uparrow\lambda \\ F & \xrightarrow{\ 1\ } & F \end{array}$$

Of course, if σ exists, then its inverse map σ^{-1} is an isomorphism of $\tilde{\tilde{F}}$ onto $\tilde{F}$ which extends the identity $1:F \to F$ and such that

$$(\sigma^{-1})^*(\tilde{P}) = (\sigma^*)^{-1}(\tilde{P}) = \tilde{\tilde{P}}$$

Choose $\varphi \in P$ with $\|\varphi\| \leq 2$. Let $\mathfrak{R}$ denote the set of all P-Cauchy sequences of elements from F. With addition and multiplication defined componentwise, $\mathfrak{R}$ is a ring. Since $|\varphi(a) - \varphi(b)| \leq \varphi(a - b)$ for $a, b \in F$ and since $\mathbf{R}$ is complete, we may define a function $\tilde{\varphi}:\mathfrak{R} \to \mathbf{R}$ as follows:

$$\text{For } A = \{a_n\} \in \mathfrak{R}, \text{ put } \tilde{\varphi}(A) = \lim_{n\to\infty} \varphi(a_n)$$

It is clear that for $A, B \in \mathfrak{R}$ we have

(i) $$\tilde{\varphi}(A) \geq 0$$

(ii) $$\tilde{\varphi}(AB) = \tilde{\varphi}(A)\tilde{\varphi}(B)$$

(iii) $$\tilde{\varphi}(A + B) \leq \tilde{\varphi}(A) + \tilde{\varphi}(B)$$

Let $\mathfrak{N} \subset \mathfrak{R}$ denote the set of all null sequences of elements from F; that is,

$$\mathfrak{N} = \{\{a_n\}|a_n \to 0\} = \{A \in \mathfrak{R}|\tilde{\varphi}(A) = 0\}$$

From (i), (ii), and (iii) it is immediate that $\mathfrak{N}$ is an ideal of $\mathfrak{R}$. Moreover, $\mathfrak{R}/\mathfrak{N} = \tilde{F}$ is a field; in fact, if $A = \{a_n\} \notin \mathfrak{N}$, then the element $B = \{b_n\} \in \mathfrak{R}$ given by

$$b_n = 0 \text{ if } a_n = 0 \qquad \text{and} \qquad b_n = \frac{1}{a_n} \text{ if } a_n \neq 0$$

is such that $AB \equiv 1 \pmod{\mathfrak{N}}$. It should be remarked that, in proving that $B \in \mathfrak{R}$, full use is made of the fact that $\varphi(ab) = \varphi(a)\varphi(b)$—for the rest $\varphi(ab) \leq \varphi(a)\varphi(b)$ suffices.

For $A \in \mathfrak{R}$ and $N \in \mathfrak{N}$ we have

$$\bar{\varphi}(A) = \bar{\varphi}(A + N - N) \leq \bar{\varphi}(A + N) + \bar{\varphi}(N) = \bar{\varphi}(A + N) \leq \bar{\varphi}(A)$$

Thus $\bar{\varphi}$ is constant on the cosets of $\mathfrak{N}$ in $\mathfrak{R}$, and we may consider $\bar{\varphi}$ to be defined on $\mathfrak{R}/\mathfrak{N} = \tilde{F}$. Denoting the residue class of $A \in \mathfrak{R}$ by $[A]$, we have $\bar{\varphi}([A]) = 0 \Leftrightarrow [A] = 0$. Since properties (i), (ii), and (iii) carry over to $\tilde{F}$, it follows that $\bar{\varphi}$ is a valuation of $\tilde{F}$. Denote the prime divisor of F to which $\bar{\varphi}$ belongs by $\tilde{P}$.

A map $\mu:F \to \tilde{F}$ may be defined as follows: For $a \in F$, let $\mu(a)$ denote the residue class of the "constant" sequence $\{a_n = a\}$. It is clear that μ is a monomorphism; furthermore, $\mu^*(\tilde{P}) = P$—since

$$\bar{\varphi}(\mu(a)) = \lim_{n \to \infty} \varphi(a_n = a) = \varphi(a)$$

Suppose next that $[A] \in \tilde{F}$, where $A = \{a_n\} \in \mathfrak{R}$. For each n, $[A] - \mu(a_n)$ is represented by the sequence $(a_1 - a_n, a_2 - a_n, \ldots)$; consequently,

$$\lim_{n \to \infty} \bar{\varphi}([A] - \mu(a_n)) = \lim_{n \to \infty} \{\lim_{m \to \infty} \varphi(a_m - a_n)\} = 0$$

and $\mu(F)$ is dense in $\tilde{F}$.

Suppose finally that $\{\alpha_n\}$ is a Cauchy sequence of elements from $\tilde{F}$. By density, there exist $a_n \in F$ such that $\bar{\varphi}(\alpha_n - \mu(a_n)) < 1/n$ for each n. Thus, $\{\alpha_n - \mu(a_n)\}$ is a null sequence in $\tilde{F}$, and $\{\mu(a_n)\}$ is then a $\tilde{P}$-Cauchy sequence in $\tilde{F}$. We conclude, therefore, that $\{a_n\}$ is a P-Cauchy sequence in F (since $\bar{\varphi}^\mu = \varphi$). Let $\alpha \in \tilde{F}$ be the residue class of $\{a_n\}$. As before, we see that $\lim \bar{\varphi}(\alpha - \mu(a_n)) = 0$. Hence, $\lim \bar{\varphi}(\alpha - \alpha_n) = 0$, and $\tilde{F}$ is $\tilde{P}$-complete. This takes care of the existence of completions. Uniqueness follows easily from our next result.

1-7-2. Proposition. Let F and E be fields with prime divisors P and Q, respectively, and suppose that there exists a monomorphism $\sigma:F \to E$ such that $\sigma^*(Q) = P$. If $\langle \tilde{F}, \tilde{P}, \mu \rangle$ and $\langle \tilde{E}, \tilde{Q}, \lambda \rangle$ are completions of $\langle F, P \rangle$ and $\langle E, Q \rangle$, respectively, then there exists a unique monomorphism $\bar{\sigma}:\tilde{F} \to \tilde{E}$ such that

(i) $$(\bar{\sigma})^*(\tilde{Q}) = \tilde{P}$$

(ii) $$\bar{\sigma}\mu = \lambda\sigma$$

Proof. Condition ii expresses the commutativity of the diagram

$$\begin{array}{ccc} \tilde{F} & \xrightarrow{\tilde{\sigma}} & \tilde{E} \\ {\scriptstyle \mu}\uparrow & & \uparrow{\scriptstyle \lambda} \\ F & \xrightarrow{\sigma} & E \end{array}$$

Uniqueness. Suppose that a $\tilde{\sigma}$ exists. From $(\tilde{\sigma})^*(\tilde{Q}) = \tilde{P}$, it follows that $\tilde{\sigma}:\tilde{F} \to \tilde{E}$ is continuous. Since $\mu(F)$ is dense in $\tilde{F}$, the relation $\tilde{\sigma}\mu = \lambda\sigma$ shows that $\tilde{\sigma}$ is completely determined by its behavior on $\mu(F)$.

Existence. Given $\alpha \in \tilde{F}$, there exists a sequence $\{a_n\}$ from F such that $\mu(a_n) \to \alpha$ in $\tilde{F}$. Since $\mu^*(\tilde{P}) = P$, $\sigma^*(Q) = P$, $\lambda^*(\tilde{Q}) = Q$, it follows from the discussion in Section 1-6 that $\{a_n\}$ is Cauchy in F, $\{\sigma(a_n)\}$ is Cauchy in E, and $\{\lambda(\sigma(a_n))\}$ is Cauchy in the complete field $\tilde{E}$. Put then

$$\tilde{\sigma}(\alpha) = \lim_{n\to\infty} [\lambda(\sigma(a_n))]$$

Now, $\tilde{\sigma}$ is well defined—in fact, if $\{b_n\}$ is another sequence from F such that $\mu(b_n) \to \alpha$, then $\mu(a_n - b_n) \to 0$ and we have

$$\lim_{n\to\infty} [\lambda\sigma(a_n)] - \lim_{n\to\infty} [\lambda\sigma(b_n)] = \lim_{n\to\infty} [\lambda\sigma(a_n - b_n)] = 0$$

Clearly $\tilde{\sigma}$ is a homomorphism, and $\tilde{\sigma}\mu = \lambda\sigma$. Since σ is a monomorphism, so is its extension $\tilde{\sigma}$.

Because the relation $\mu^*\tilde{\sigma}^* = \sigma^*\lambda^*$ in itself does not permit the conclusion $(\tilde{\sigma})^*(\tilde{Q}) = \tilde{P}$, we must be more explicit. Consider any $\varphi \in \tilde{Q}$; then $\varphi^{\tilde{\sigma}} \in (\tilde{\sigma})^*(\tilde{Q})$, and $\varphi^{\lambda\sigma} \in P$. Furthermore, there exists $\theta \in \tilde{P}$ such that $\theta^{\mu} = \varphi^{\lambda\sigma}$. Given $\alpha \in \tilde{F}$, we write $\alpha = \lim_{n\to\infty} \mu(a_n)$ (with respect to $\tilde{P}$) with $a_n \in F$, and compute—

$$\begin{aligned} \varphi^{\tilde{\sigma}}(\alpha) &= \varphi(\tilde{\sigma}\alpha) = \varphi\{\lim_{n\to\infty} (\lambda\sigma a_n)\} \\ &= \lim_{n\to\infty} \varphi(\lambda\sigma a_n) = \lim_{n\to\infty} \varphi^{\lambda\sigma}(a_n) \\ &= \lim_{n\to\infty} \theta^{\mu}(a_n) = \lim_{n\to\infty} \theta[\mu(a_n)] \\ &= \theta\{\lim_{n\to\infty} \mu(a_n)\} = \theta(\alpha) \end{aligned}$$

Hence $(\tilde{\sigma})^*(\tilde{Q}) = \tilde{P}$, and the proof is complete.

In view of (1-7-1) we shall often refer to *the completion* of a field.

1-7-3. Exercise. Let F be a field with prime divisor P; prove the existence of a P completion of F by completing the uniform structure.

1-7-4. Proposition. Suppose that $\langle E, Q, i_{F\to E}\rangle$ is an extension of $\langle F, P\rangle$ and that $\langle E, Q\rangle$ is complete. Then the closure $\bar{F}$ of F in E is a field, and if we put $\bar{P} = i^*_{\bar{F}\to E}(Q)$, then $\langle \bar{F}, \bar{P}, i_{F\to\bar{F}}\rangle$ is a completion of F.

Proof. From the continuity of the field operations of E in the topology T_Q, it follows easily that $\bar{F}$ is a field. Of course, $\bar{F}$ is complete in the topology T_Q. Hence, $\bar{F}$ is complete in the topology $T_{\bar{P}}$. Since $i^*_{F\to E}(Q) = i^*_{F\to\bar{F}} i^*_{\bar{F}\to E}(Q)$, we have $P = i^*_{F\to\bar{F}}(\bar{P})$.

1-7-5. Proposition. Let $\langle E, Q, i\rangle$ be a completion of $\langle F, P\rangle$, where P is nonarchimedean; then

$$e\left(\frac{Q}{P}\right) = f\left(\frac{Q}{P}\right) = 1$$

In particular, if P is discrete, then so is Q. Moreover, in general, O_Q is the closure of O_P in E, and $\mathcal{P}_Q$ is the closure of $\mathcal{P}_P$ in E.

Proof. Let ν be an exponential valuation belonging to Q, so that $\nu|F$ is an exponential valuation belonging to P. Consider $\alpha \neq 0 \in E$. Since F is dense in E, there exists $a \in F$ with $\nu(a - \alpha) > \nu(\alpha)$. Therefore, $\nu(a) = \nu(\alpha)$. Hence, $\nu(E^*) = \nu(F^*)$, and $e(Q/P) = 1$.

Furthermore, any nonzero element of $\bar{E}$ is of form $\bar{\alpha}$, where $\alpha \in O_Q$. Now, there exists $a \in F$ with $\nu(a - \alpha) > 0$. Since $\nu(\alpha) = 0$, we have $\nu(a) = 0$, so that $a \in O_P \subset O_Q$ and $a - \alpha \in \mathcal{P}_Q$. Hence $\bar{\alpha} = \bar{a}$, and $f(Q/P) = 1$.

The remaining statements are easy to prove, particularly if one is aware of the first assertion of (1-7-6).

1-7-6. Exercise. Let φ be a nonarchimedean valuation of F.

i. If $x_n \to a \neq 0$, then $\varphi(x_n) = \varphi(a)$ for all sufficiently large n.

ii. $\{s_n\}$ is a Cauchy sequence if and only if $(s_{n+1} - s_n) \to 0$.

iii. If the infinite series $\sum_1^\infty a_n$ converges to s, then so does every reordering.

iv. If $\Sigma a_n = s$ and $\Sigma b_n = t$, then $\Sigma(a_n \pm b_n) = s \pm t$ and

$$\sum_{k=1}^{\infty} \sum_{n+m=k} a_n b_m = st$$

v. If F is φ-complete, then Σa_n converges $\Leftrightarrow a_n \to 0$.

vi. Call Σa_n ***absolutely convergent*** when $\Sigma\varphi(a_n)$ converges. If F is φ-complete, absolute convergence implies convergence.

Consider the rational field $\mathbf{Q}$. For any prime divisor $p \in \mathfrak{M}(\mathbf{Q})$, we have a p completion which is denoted simply by $\mathbf{Q}_p$. When $p = \infty$, we know that $\mathbf{Q}_p$ is the real field with prime divisor determined by the absolute value valuation. When $p \neq \infty$, $\mathbf{Q}_p$ is called the field of ***p-adic numbers,*** or the ***p-adic completion of Q;*** in virtue of (1-7-5) and (1-4-1) the residue class field, which we denote by $\bar{\mathbf{Q}}_p$, is the p-element field.

1-8. The Theorem of Ostrowski

The two obvious examples of fields complete with respect to an archimedean prime divisor are $\mathbf{R}$ and $\mathbf{C}$, where the prime divisor is determined by the absolute value. It is perhaps surprising that there are no others. Even more surprising is the fact that we already have all the tools needed for the proof.

1-8-1. Lemma. Suppose that φ is a valuation of $\mathbf{C}$ such that $\varphi(a) = |a|$ for all $a \in \mathbf{R}$; then $\varphi(\alpha) = |\alpha|$ for all $\alpha \in \mathbf{C}$.

Proof. This lemma asserts that the absolute value valuation of $\mathbf{R}$ has a unique extension to $\mathbf{C}$.

Since $\|\varphi\| = \max\{\varphi(1), \varphi(2)\}$, the triangle inequality holds for φ. Note also that $\varphi(i) = 1$ because $i^4 = 1$. Therefore, for $\alpha = a + bi$ with $a, b \in \mathbf{R}$, we have

$$\varphi(\alpha) \leq \varphi(a) + \varphi(b) = |a| + |b| \leq \sqrt{2}\sqrt{|a|^2 + |b|^2} = \sqrt{2}|\alpha|$$

For $\alpha \neq 0$, let $f(\alpha) = \varphi(\alpha)/|\alpha|$. Thus $0 < f(\alpha) \leq \sqrt{2}$, and $f(\alpha)^n = f(\alpha^n) \leq \sqrt{2}$ for $n = 1, 2, 3, \ldots$. Hence, $f(\alpha) \leq 1$, and in the same way $f(1/\alpha) \leq 1$. This means that $f(\alpha) = 1$ for all $\alpha \neq 0$, so that $\varphi(\alpha) = |\alpha|$.

1-8-2. Lemma. Let F be an extension field of $\mathbf{R}$, and suppose that φ is a valuation of F such that $\varphi(a) = |a|$ for all $a \in \mathbf{R}$; then either $F = \mathbf{R}$ or $F = \mathbf{C}$, and $\varphi(\alpha) = |\alpha|$ for all $\alpha \in F$.

Proof. It suffices to show that F is algebraic over $\mathbf{R}$; for then F may be viewed as a subfield of $\mathbf{C}$ so that $[\mathbf{C}:\mathbf{R}] = 2$ implies $F = \mathbf{R}$ or $F = \mathbf{C}$. (1-8-1) then describes φ.

Consider any $\xi \in F$. We show that ξ is algebraic over $\mathbf{R}$—in fact, that ξ is a root of a monic polynomial of degree 2. For any $\alpha \in \mathbf{C}$, $\alpha + \bar{\alpha}$ and $\alpha\bar{\alpha}$ are real, so that the function $f:\mathbf{C} \to \mathbf{R}$ defined by

$$f(\alpha) = \varphi[\xi^2 - (\alpha + \bar{\alpha})\xi + \alpha\bar{\alpha}]$$

has meaning. (φ is not defined for all elements of $\mathbf{C}$.) Since $|f(\alpha_1) - f(\alpha_2)| \leq \varphi[(\alpha_1\bar{\alpha}_1 - \alpha_2\bar{\alpha}_2) + (\alpha_2 + \bar{\alpha}_2)\xi - (\alpha_1 + \bar{\alpha}_1)\xi]$, it follows that f is continuous. Furthermore, $f(\alpha) \to \infty$ as $\alpha \to \infty$; this follows from

$$f(\alpha) \geq \varphi(\alpha\bar{\alpha}) - \varphi(\xi^2) - \varphi(\alpha + \bar{\alpha})\varphi(\xi)$$
$$\geq |\alpha|^2 - \varphi(\xi^2) - 2|\alpha|\varphi(\xi)$$

Let $m = \operatorname*{g.l.b.}_{\alpha \in \mathbf{C}} f(\alpha)$; so $m \geq 0$. Using the fact that $f(\alpha) \to \infty$ as $\alpha \to \infty$, we see that there exists $\alpha \in \mathbf{C}$ such that $f(\alpha) = m$. Put $\mathfrak{S} = \{\alpha \in \mathbf{C} | f(\alpha) = m\}$. Clearly, $\mathfrak{S}$ is a nonempty, closed, bounded set. Hence $\mathfrak{S}$ is compact, and there exists $\alpha_0 \in \mathfrak{S}$ such that $|\alpha_0| \geqq |\alpha|$ for all $\alpha \in \mathfrak{S}$.

If $m = 0$, then $f(\alpha_0) = 0$ and ξ is algebraic over $\mathbf{R}$. Suppose then that $m > 0$. Choose ε with $0 < \varepsilon < m$. Consider the real polynomial $g(x) = x^2 - (\alpha_0 + \bar{\alpha}_0)x + \alpha_0\bar{\alpha}_0 + \varepsilon$, and let $\alpha_1, \bar{\alpha}_1 \in \mathbf{C}$ be its roots. We have $\alpha_1\bar{\alpha}_1 = \alpha_0\bar{\alpha}_0 + \varepsilon$, so that $|\alpha_1| > |\alpha_0|$ and $\alpha_1 \notin \mathfrak{S}$.

Now, let us fix $n \geq 1$ for the moment, and put

$$G(x) = [x^2 - (\alpha_0 + \bar{\alpha}_0)x + \alpha_0\bar{\alpha}_0]^n - (-\varepsilon)^n$$

This is a real polynomial of degree $2n$; let us denote its roots by $\beta_1, \ldots, \beta_{2n} \in \mathbf{C}$. Thus

$$G(x) = \prod_1^{2n} (x - \beta_i) = \prod_1^{2n} (x - \bar{\beta}_i)$$

Since $G(\alpha_1) = 0$, α_1 is one of the β_i—say, $\alpha_1 = \beta_1$. Because F need not contain $\mathbf{C}$, we may not substitute ξ in $\prod_1^{2n} (x - \beta_i)$. Instead we consider

$$[G(x)]^2 = \prod_1^{2n} (x^2 - (\beta_i + \bar{\beta}_i)x + \beta_i\bar{\beta}_i)$$

We have then

$$\varphi[G(\xi)^2] = \prod_1^{2n} \varphi[\xi^2 - (\beta_i + \bar{\beta}_i)\xi + \beta_i\bar{\beta}_i] = \prod_1^{2n} f(\beta_i)$$

so that

$$\varphi[G(\xi)^2] \geq f(\alpha_1)m^{2n-1}$$

On the other hand

$$\varphi[G(\xi)] \leq \varphi[\xi^2 - (\alpha_0 + \bar{\alpha}_0)\xi + \alpha_0\bar{\alpha}_0]^n + \varphi(-\varepsilon)^n$$
$$= f(\alpha_0)^n + \varepsilon^n$$
$$= m^n + \varepsilon^n$$

Combining these inequalities gives

$$f(\alpha_1)m^{2n-1} \leq \varphi[G(\xi)]^2 \leq (m^n + \varepsilon^n)^2$$

which implies that

$$\frac{f(\alpha_1)}{m} \leq \left[1 + \left(\frac{\varepsilon}{m}\right)^n\right]^2$$

Letting $n \to \infty$, we see that $f(\alpha_1)/m \leq 1$; that is, $f(\alpha_1) = m$, and $\alpha_1 \in \mathfrak{S}$, a contradiction. This completes the proof of the lemma.

1-8-3. Theorem of Ostrowski. Let F be complete with respect to the archimedean prime divisor P; then there exists an isomorphism σ of F onto either $\mathbf{R}$ or $\mathbf{C}$ such that $\sigma^*(P_\infty) = P$, where P_∞ is the prime divisor determined by the absolute value.

Proof. In virtue of (1-8-2) the proof requires little more than "fussing" with notation.

Since P is archimedean, F has characteristic 0, and its prime field is therefore $\mathbf{Q}$. The projection of P on $\mathbf{Q}$ is ∞ because $\mathbf{Q}$ has only one archimedean prime divisor. Let $\bar{\mathbf{Q}}$ denote the closure of $\mathbf{Q}$ in F, and let $\bar{P}$ denote the projection of P on $\bar{\mathbf{Q}}$; then $\langle \bar{\mathbf{Q}}, \bar{P}, i_{\mathbf{Q}\to\bar{\mathbf{Q}}}\rangle$ is a completion of $\langle \mathbf{Q}, \infty\rangle$. Since $\langle \mathbf{R}, P_\infty, i_{\mathbf{Q}\to\mathbf{R}}\rangle$ is also a completion of $\langle \mathbf{Q}, \infty\rangle$ (there is nothing lost in assuming that the same $\mathbf{Q}$ is contained in both $\mathbf{R}$ and F), there exists a $\mathbf{Q}$ isomorphism $\tau: \bar{\mathbf{Q}} \rightarrowtail\!\!\twoheadrightarrow \mathbf{R}$ such that $\tau^*(P_\infty) = \bar{P}$. Now there exists a field F' containing $\mathbf{R}$ such that τ can be extended to an isomorphism $\sigma: F \rightarrowtail\!\!\twoheadrightarrow F'$. Put $P' = (\sigma^{-1})^*P$; then

$$i^*_{\mathbf{R}\to F'}(P') = (i^*_{\mathbf{R}\to F'}(\sigma^{-1})^*)P = ((\tau^{-1})^* i^*_{\bar{\mathbf{Q}}\to F})P = (\tau^{-1})^*\bar{P} = P_\infty$$

and P' is an extension of P_∞. In particular, there exists $\varphi \in P'$ with $\varphi(a) = |a|$ for all $a \in \mathbf{R}$. An application of (1-8-2) shows that $P' = P_\infty$.

1-9. Complete Fields with Discrete Prime Divisor

Suppose that F is a field complete with respect to the discrete prime divisor P; then somewhat more can be said about the structure of F than was possible in Section 1-5.

Let ν_P denote the normalized exponential valuation belonging to P, and fix a prime element $\pi \in F$—that is, $\nu(\pi) = 1$. More generally, for each $n \in \mathbf{Z}$ fix an element $\pi_n \in F$ such that $\nu(\pi_n) = n$. The most natural choice of the π_n is to take $\pi_n = \pi^n$. Of course, completeness does not enter here in any way.

Choose also a mapping $\eta:\bar{F} = O/\mathcal{P} \rightarrow O$ such that

(i) $$\overline{\eta\alpha} = \alpha \text{ for every } \alpha \in \bar{F}$$

(ii) $$\eta 0 = 0$$

Such an η surely exists, since it involves nothing more than a selection of a representative for each of the cosets of $\mathcal{P}$ in O, with the additional requirement that the representative of $\mathcal{P}$ be 0; in fact, such η exist for any field F with a nonarchimedean prime divisor. The set $\mathcal{R} = \eta(\bar{F}) \subset O$ will be called a ***complete system of representatives*** for $\bar{F}$; the mapping η will be called a ***representative map*** of $\bar{F}$.

1-9-1. Theorem. Let F be complete with respect to the discrete prime divisor P, and let $\mathcal{R} = \eta(\bar{F})$ be a complete system of representatives of $\bar{F}$. Then every element b of F can be expressed uniquely in the form

(*) $$b = \sum_{n=r}^{\infty} a_n\pi_n$$

where $a_n \in \mathcal{R}$, $a_r \neq 0$, $\nu(\pi_n) = n$. Moreover, every such series represents an element of F for which $\nu(b) = r$.

Proof. Consider first any series of form $\sum_{n=r}^{\infty} a_n\pi_n$ where it is required only that all $a_n \in O$. If all $a_n = 0$, the series converges to 0. We may suppose then that $a_r \neq 0$ (r may be negative). Since $\nu(a_n\pi_n) = \nu(a_n) + \nu(\pi_n) \geq n$ we see that $a_n\pi_n \rightarrow 0$ as $n \rightarrow \infty$. Therefore, (1-7-6v) guarantees that $\sum_{n=r}^{\infty} a_n\pi_n$ converges to some element $b \in F$. In other words, the sequence of partial sums

$$\left\{s_m = \sum_{r}^{m} a_n\pi_n\right\}$$

where $m \geq r$, converges to

$$b = \sum_{n=r}^{\infty} a_n\pi_n$$

Since $\nu(a_n\pi_n) \geq r$ for all n and $\mathcal{P}^r$ is closed, it follows that $\nu(b) \geq r$. If further $a_r \notin \mathcal{P}$, then $\nu(b) = r$ since $\nu(a_r\pi_r) = r$ and

$$\nu\left(\sum_{n=r+1}^{\infty} a_n\pi_n\right) \geq r + 1$$

In particular, the element 0 of F has a unique expansion in the form (*); in fact, an expression

$$0 = \sum_{n=r}^{\infty} a_n \pi_n$$

with $a_n \in O$, $a_r \notin \mathcal{P}$ is impossible.

Consider now any $b \neq 0 \in F$; we find an expansion of form (*) for it. Let $\nu(b) = r \neq \infty$. Thus $b\pi_r^{-1} \in U$, and there exists $a_r \neq 0 \in \mathcal{R}$ such that $\overline{b\pi_r^{-1}} = \overline{a_r}$. This means that $\nu(b/\pi_r - a_r) > 0$ or $\nu(b - a_r\pi_r) > r$. Suppose inductively that we have found $a_r, \ldots, a_m \in \mathcal{R}$ such that $s_m = a_r\pi_r + \cdots + a_m\pi_m$ satisfies $\nu(b - s_m) > m$. It follows that $(b - s_m)/\pi_{m+1} \in O$, and there exists $a_{m+1} \in \mathcal{R}$ with $\nu[(b - s_m)/\pi_{m+1} - a_{m+1}] > 0$. Hence $\nu(b - s_{m+1}) > m + 1$, where $s_{m+1} = a_r\pi_r + \cdots + a_{m+1}\pi_{m+1}$. Since $\nu(b - s_m) > m$ for each $m \geq r$, the series $\sum_{n=r}^{\infty} a_n\pi_n$ converges to b.

It remains to show that the expansion for $b \neq 0$ is unique. Suppose we have two expansions of form (*):

$$b = \sum_{n=r_1}^{\infty} a_n\pi_n = \sum_{n=r_2}^{\infty} a_n'\pi_n$$

Then $r_1 = \nu(b) = r_2$, and

$$0 = \sum_{n=r_1}^{\infty} (a_n - a_n')\pi_n$$

with $a_n - a_n' \in O$. If t is the first subscript for which $a_t - a_t' \neq 0$, then

$$0 = \sum_{t}^{\infty} (a_n - a_n')\pi_n$$

with $a_t - a_t' \notin \mathcal{P}$; this is impossible, and so $a_n = a_n'$ for all n. This completes the proof.

1-9-2. Corollary. If $\pi \in F$ is a prime element of a field complete with respect to a discrete prime divisor, then every element of F has a unique expansion as a *formal Laurent series*

$$\sum_{n=r}^{\infty} a_n\pi^n \qquad a_r \neq 0$$

where the a_n come from some complete system of representatives $\mathcal{R}$.

1-9-3. Example. Let F be any field with a discrete nontrivial prime divisor P, and let $\mathcal{R} \subset O_P$ be a complete system of representatives for

$\bar{F}$. Suppose that $\langle E, Q, i\rangle$ is the P completion of F. Since $\bar{E} = \bar{F}$, the same set $\mathfrak{R} \subset O_P \subset O_Q$ may serve as a complete system of representatives for $\bar{E}$. Also, Q is discrete, since $e(Q/P) = 1$. Thus if $\nu_P \in P$ is the normalized exponential valuation, then its unique extension to E is precisely ν_Q, the normalized exponential valuation belonging to Q. Hence, if $\pi \in F$ has $\nu_P(\pi) = 1$, then $\nu_Q(\pi) = 1$, so that a prime element for E may be taken from F.

Now consider the situation where $F = \mathbf{Q}$, the field of rational numbers, and $P = p \neq \infty$. The most natural choice for $\mathfrak{R}$ is the set $\{0, 1, \ldots, p-1\}$, while the element $p \in \mathbf{Q}$ is the most natural choice for a prime element π. We write $\langle \mathbf{Q}_p, \tilde{p}, i\rangle$ for $\langle E, Q, i\rangle$, and under ordinary circumstances the expansion of elements from $\langle \mathbf{Q}_p, \tilde{p}\rangle$ will be taken with respect to these natural choices of $\mathfrak{R}$ and π. According to (1-9-2) an element $b \in \mathbf{Q}_p$ for which $\nu_{\tilde{p}}(b) = r$ has a unique expansion

$$b = \sum_{n=r}^{\infty} a_n p^n$$

with $a_r \neq 0$, $a_n \in \{0, 1, \ldots, p-1\}$. Of course,

$$b \in O_{\tilde{p}} \Leftrightarrow \nu_{\tilde{p}}(b) \geq 0 \Leftrightarrow b \text{ has form } b = a_0 + a_1 p + a_2 p^2 + \cdots$$

It should be pointed out that the fields of p-adic numbers $\mathbf{Q}_p$ were originally introduced directly as fields of formal power series $\sum_{n=r}^{\infty} a_n p^n$ by Hensel. His motivation was the analogy with function fields.

1-9-4. Exercise. i. Find the expansions of ± 5, ± 7, ± 12, $\pm\frac{1}{5}$, $\pm\frac{1}{7}$ in $\mathbf{Q}_2$.

ii. Find the expansions of -1, ± 2, ± 5, $\pm\frac{1}{2}$, $\pm\frac{1}{5}$ in $\mathbf{Q}_3$.

iii. The expansion of $a \in \mathbf{Q}_p$ becomes periodic $\Leftrightarrow a \in \mathbf{Q}$. Moreover, the proof may be arranged so as to determine the expansion of $a \in \mathbf{Q}$.

1-9-5. Exercise. Let F be complete with respect to the discrete prime divisor P, and suppose that the residue class field $\bar{F}$ is a perfect field of characteristic $p \neq 0$.

i. There exists a unique representative map η which is multiplicative. [*Hint:* Use (1-5-4).] In particular, $\mathbf{Q}_p$ contains the $(p-1)$st roots of unity.

ii. The multiplicative group of units U has a direct product decomposition $U = U_1 \times \eta(\bar{F}^*)$, where $U_1 = 1 + \mathfrak{P}$.

iii. Suppose that, in addition, F has characteristic p; then the multiplicative representative map is also additive—so that $\eta:\bar{F} \to F$ is an isomorphism, and $F_0 = \eta(\bar{F})$ is a field on which P is trivial. Let x be an indeterminate over F_0, and consider the field $F_0(x)$ with the prime divisor determined [as in (1-4-5)] by the irreducible polynomial x in $F_0[x]$. Its completion is denoted by $F_0\langle x\rangle$ and is known as the field of all ***formal Laurent*** (or ***power***) ***series*** over F_0. Show that F is isomorphic to $F_0\langle x\rangle$.

1-9-6. Proposition. Let F be a field complete with respect to a discrete prime divisor P, and suppose that the residue class field $\bar{F}$ is finite. Then $\mathcal{P}^r$ is compact and open for every $r \in \mathbf{Z}$.

Proof. Since $\mathcal{P}^r$ and $\mathcal{P}^s$ are homeomorphic for all $r, s \in \mathbf{Z}$, it suffices to prove the result for $O = \mathcal{P}^0$. That O is open has already been observed in (1-5-2). Since $\bar{F}$ is finite, so is any complete system of representatives $\mathcal{R}$. If $\pi \in F$ is a prime element, then

$$O = \{a_0 + a_1\pi + a_2\pi^2 + \cdots \mid a_n \in \mathcal{R},\ n = 0, 1, 2, \ldots\}$$

Put $\mathcal{R}_n = \mathcal{R}$ for each $n = 0, 1, 2, \ldots$; so the product space $\prod_{n=0}^{\infty} \mathcal{R}_n$ is compact. Define a map

$$\lambda\colon \prod_{n=0}^{\infty} \mathcal{R}_n \to O$$

by $$\lambda\{\alpha = (a_0, a_1, a_2, \ldots)\} = a_0 + a_1\pi + a_2\pi^2 + \cdots$$

Clearly λ is 1-1 and onto. We need only show that λ is continuous. Take any neighborhood V of $\lambda(\alpha)$ in O; it suffices to consider V of form $\{b \in O \mid \nu[b - \lambda(\alpha)] > N\}$, where $N > 0$. Then

$$V' = \{a_0\} \times \{a_1\} \times \cdots \times \{a_N\} \times \prod_{N+1}^{\infty} \mathcal{R}_n$$

is a neighborhood of α in $\prod_0^{\infty} \mathcal{R}_n$ with $\lambda(V') \subset V$. Hence λ is a homeomorphism, and O is compact.

It is also clear that $U = O - \mathcal{P}$ is compact and open.

1-9-7. Remark. Suppose that F has a discrete prime divisor P and that $\bar{F}$ is finite. We define $\mathfrak{N}P = \mathfrak{N}\mathcal{P}$ to be the number of elements in the residue class field—that is,

$$\mathfrak{N}P = \mathfrak{N}\mathcal{P} = \#(\bar{F}) = \#\left(\frac{O}{\mathcal{P}}\right)$$

More generally, for any integer $r \geq 1$, we put

$$\mathfrak{N}(P^r) = \mathfrak{N}(\mathcal{P}^r) = \#\left(\frac{O}{\mathcal{P}^r}\right)$$

From (1-5-2) we know that $\mathfrak{N}(\mathcal{P}^r) = (\mathfrak{N}\mathcal{P})^r$.

We may note in passing that, because the residue class field is finite, it is possible to select a ***normalized valuation*** $\varphi \in P$ by putting:

$$\varphi(a) = \left(\frac{1}{\mathfrak{N}P}\right)^{\nu_P(a)} \qquad a \in F$$

This is in keeping with the way the normalized p-adic valuations were defined in Section 1-4.

Let $\langle \tilde{F}, \tilde{P} \rangle$ be the completion of $\langle F, P \rangle$; so, with an obvious choice of notation, $\tilde{\mathcal{P}}^r$ is the closure of $\mathcal{P}^r$ in $\tilde{F}$, $\tilde{\mathcal{P}}^r \cap F = \tilde{\mathcal{P}}^r \cap O = \mathcal{P}^r$ and $\mathcal{P}^r\tilde{O} = \tilde{\mathcal{P}}^r$. It is easy to see that there is an isomorphism of rings

$$(*) \qquad \frac{O}{\mathcal{P}^r} \approx \frac{\tilde{O}}{\tilde{\mathcal{P}}^r}$$

In fact, if $\pi \in F$ is a prime element and $\mathcal{R} \subset F$ is a complete system of representatives for $\bar{F}$, then the set

$$(**) \qquad \{a_0 + a_1\pi + \cdots + a_{r-1}\pi^{r-1} | a_i \in \mathcal{R}, i = 0, \ldots, r-1\}$$

serves as a full set of representatives for the residue classes on either side of $(*)$. In particular,

$$\mathfrak{N}(\tilde{\mathcal{P}}^r) = \mathfrak{N}(\mathcal{P}^r)$$

The groups of units of the rings in $(*)$ are isomorphic. The units may be called the **prime residue classes** (mod $\mathcal{P}^r$) or (mod $\tilde{\mathcal{P}}^r$), respectively, and their number is denoted by

$$\Phi(\mathcal{P}^r) = \Phi(P^r) = \Phi(\tilde{\mathcal{P}}^r) = \Phi(\tilde{P}^r)$$

in other words, Φ is an analogue of the Euler ϕ function. It is clear that a residue class with representative $a_0 + a_1\pi + \cdots + a_{r-1}\pi^{r-1}$ has an inverse if and only if $a_0 \neq 0$; consequently,

$$\Phi(P^r) = \Phi(\tilde{P}^r) = (\mathfrak{N}P - 1)(\mathfrak{N}P)^{r-1}$$

EXERCISES

1-1. How is the approximation theorem in **Q** related to the Chinese remainder theorem?

1-2. How are results on valuations affected if F is a division ring?

In Exercises 1-3 through 1-8 suppose that F is a field with a nonarchimedean prime divisor P.

1-3*a*. Fix $\nu \in P$, and consider $E = F(x)$, where x is an indeterminate. For $f(x) = a_0 + a_1x + \cdots + a_nx^n \in F[x]$ we put

$$\bar{\nu}(f) = \min_i \{\nu(a_i)\}$$

Thus, $\bar{\nu}(f) = \infty \Leftrightarrow f = 0$, and one checks that, for $f,\ g \in F[x]$, $\bar{\nu}(f + g) \geq \min \{\bar{\nu}(f), \bar{\nu}(g)\}$ and $\bar{\nu}(fg) = \bar{\nu}(f) + \bar{\nu}(g)$. Since E is the quotient field of $F[x]$, $\bar{\nu}$ may be made into a valuation of E which extends ν. Of course, $\bar{\nu}$ is discrete if and only if ν is discrete.

b. Let $f(x) \in O[x]$ be monic, and suppose we have a factorization $f(x) = g(x)h(x)$, where $g(x),\ h(x) \in F[x]$ are both monic; then $g(x),\ h(x) \in O[x]$.

c. The polynomial $f(x) \in F[x]$ is said to be ***primitive*** when $\bar{\nu}(f) = 0$; so the product of primitive polynomials is primitive. Any element of E may be put in the form f/g, where $f,\ g \in F[x]$ and g is primitive, and then $f/g \in O_E \Leftrightarrow \bar{\nu}(f) \geq 0$, $f/g \in \mathcal{P}_E \Leftrightarrow \bar{\nu}(f) > 0$. Show that the residue class field $O_E/\mathcal{P}_E$ is isomorphic to $\bar{F}(x)$.

d. In the above $\bar{\nu}(x) = 0$; if, however, one starts with an arbitrary value for $\bar{\nu}(x)$ and puts $\bar{\nu}(f) = \min_i \{\nu(a_i) + i\bar{\nu}(x)\}$, then the same procedure leads to a valuation extending ν. Show that ν has an infinite number of inequivalent extensions to E.

1-4. What are the possible relations between the characteristic of F and the characteristic of $\bar{F}$? Give examples.

1-5. Let $f(x) \in O[x]$ be a monic polynomial; then $f(x)$ is irreducible over O if and only if it is irreducible over F.

1-6. Let $f(x) = x^n + a_{n-1}x^{n-1} + \cdots + a_1x + a_0 \in O[x]$ be a polynomial with $a_0, a_1, \ldots, a_{n-1} \in \mathcal{P}$ and such that a_0 is not the product of two elements of $\mathcal{P}$; then $f(x)$ is irreducible over F. How is this related to the Eisenstein irreducibility criterion?

1-7. Show that O is a maximal subring of F.

1-8*a*. In O every nonzero ideal is of form $\{a \in F | \nu(a) > \delta\}$ or $\{a \in F | \nu(a) \geq \delta\}$, where δ is some nonnegative real number.

b. If P is not discrete, δ is uniquely determined by the ideal. If P is discrete, δ may be chosen as a value taken on by ν, and only $\geq$ is needed.

c. In the discrete case all proper ideals are positive powers of $\mathcal{P}$; in the nondiscrete case, all positive powers of $\mathcal{P}$ equal $\mathcal{P}$.

d. What can be said for the fractional ideals?

1-9. Let O be a subring of an arbitrary field F. If $O \cup O^{-1} = F$, one says that O is a ***valuation ring*** of F. In particular, the ring of integers at a nonarchimedean prime divisor is a valuation ring. Put $U = O \cap O^{-1}$ and $\mathcal{P} = O - U$ (set theoretic).

a. $1 \in O$, and U is the group of units of O.

b. $\mathcal{P}$ is an ideal of O; so it is the unique maximal ideal of O.

c. $F = \mathcal{P} \cup U \cup \mathcal{P}^{-1}$ (disjoint union).

d. Let O_1 and O_2 be valuation rings of F; then

$$O_1 \subset O_2 \Leftrightarrow \mathcal{P}_1 \supset \mathcal{P}_2 \Leftrightarrow U_1 \subset U_2$$

Consequently, $$O_1 = O_2 \Leftrightarrow \mathcal{P}_1 = \mathcal{P}_2 \Leftrightarrow U_1 = U_2$$

1-10. Let F be a field with discrete prime divisor P; then F is locally compact if and only if F is complete and $\bar{F}$ is finite. Moreover, if $\varphi \in P$ is then the normalized valuation and μ is Haar measure on F normalized so that $\mu(O_P) = 1$, then for every $a \neq 0 \in F$ and every measurable set S of F we have

$$\mu(aS) = \varphi(a)\mu(S)$$

1-11. The polynomial $f(x) \in \mathbf{Z}[x]$ has a solution in $\mathbf{Q}_p \Leftrightarrow f(x) \equiv 0 \pmod{p^n}$ has a rational solution for every positive integer n.

2

Extension of Valuations

It has already been observed that the arithmetic of the rational field $\mathbf{Q}$ may be described in terms of its prime divisors. It is natural, therefore, to investigate the extent to which this result may be carried over to an arbitrary algebraic number field, and this requires, first of all, information about the set of all prime divisors of an algebraic number field.

Thus we are led to consider the general "extension problem." More precisely, suppose that F is a field with prime divisor P—P may be archimedean or nonarchimedean—and that E is a finite extension of F; do there exist extensions of P (or of $\varphi \in P$) to E, and if so how many? In this chapter we shall be concerned primarily with the answers to these questions. Of course, if P is trivial, so are the answers [see (1-6-1)] and therefore, in keeping with our convention, we shall always assume that P is nontrivial. It turns out that, when F is P-complete, the answers are especially simple and explicit, and this case will be treated first. The treatment of the general case will then make heavy use of the results for the complete case.

2-1. Uniqueness of Extensions (Complete Case)

Let φ be a valuation of the field F. We shall say that a vector space X over F is ***normed*** over $\langle F, \varphi\rangle$ if there exists a "norm" function $\| \quad \|: X \to \mathbf{R}$ such that

(i) $$\begin{cases} \|\xi\| \geq 0 \text{ for all } \xi \in X \\ \|\xi\| = 0 \Leftrightarrow \xi = 0 \end{cases}$$

(ii) $$\|a\xi\| = \varphi(a)\|\xi\| \qquad a \in F,\ \xi \in X$$

(iii) $$\|\eta + \xi\| \leq \|\eta\| + \|\xi\| \qquad \eta, \xi \in X$$

If $\langle E, \varphi' \rangle$ with $\|\varphi'\| \leq 2$ is an extension of $\langle F, \varphi \rangle$ (this terminology for extensions has not been defined explicitly, but its meaning is clear), then $X = E^+$ becomes a normed vector space over $\langle F, \varphi \rangle$ as soon as we put $\|\ \ \| = \varphi'$. The main reason for introducing the notion of normed vector space is to emphasize the fact that the proof of uniqueness of extensions of valuations (in the complete case) does not make use of all the multiplicative properties of valuations.

In connection with the important result which follows, it is useful to note that any finite-dimensional vector space X over a field F with valuation φ such that $\|\varphi\| \leq 2$ can be made into a normed vector space over $\langle F, \varphi \rangle$. In fact, suppose that we fix a basis $\omega_1, \ldots, \omega_n$ of X over F and express any $\xi \in X$ uniquely in the form

$$\xi = a_1\omega_1 + \cdots + a_n\omega_n$$

with $a_i \in F$. Clearly the function $\|\ \ \|$ given by

$$\|\xi\| = \max_i \{\varphi(a_i)\}$$

is a norm; we shall refer to it as the ***canonical norm***, even though it depends on the choice of $\omega_1, \ldots, \omega_n$ and φ. Of course, the finite dimensionality of X is not essential.

Note also that, if X is any normed vector space, the function

$$d(\eta, \xi) = \|\eta - \xi\| \qquad \eta, \xi \in X$$

is a metric on X, and X is then a topological group in the metric topology.

2-1-1. Theorem. Let F be P-complete, and let $\varphi \in P$ have $\|\varphi\| \leq 2$. Suppose that X is a finite-dimensional vector space over F and that $\|\ \ \|$ denotes the canonical norm of X over $\langle F, \varphi \rangle$ defined via some basis $\omega_1, \ldots, \omega_n$ of X over F. If $|\ \ |$ is any norm on X over $\langle F, \varphi \rangle$, then X is a complete topological group in the metric topology determined by $|\ \ |$ and there exist constants $D_1, D_2 > 0$ such that

(*) $$D_1\|\xi\| \leq |\xi| \leq D_2\|\xi\|$$

for all $\xi \in X$. In particular, the topologies induced on X by any two norms are equivalent.

Proof. Of course, X is complete in the metric defined by $\|\ \ \|$, for then the topology of X is that of the product of n copies of F. Furthermore, since $\xi \in X$ has form $\xi = a_1\omega_1 + \cdots + a_n\omega_n$ with $a_i \in F$, it is clear that we may take

$$D_2 = |\omega_1| + \cdots + |\omega_n|$$

The rest (including completeness) is proved by induction on n. For $n = 0$ everything is trivial; if one is uneasy about this, note that the case $n = 1$ is immediate. Suppose then that the conclusion holds for $(n-1)$-dimensional normed spaces over F. Let

$$Y_i = F\omega_1 + \cdots + \cancel{F\omega_i} + \cdots + F\omega_n$$

for $i = 1, \ldots, n$ so that

$$X = Y_i + F\omega_i$$

By the induction hypothesis, Y_i is complete in the metric induced by $|\ \ |$; hence Y_i is closed in X. Since X is a topological group, $Y_i + \omega_i$ is closed. In addition, $0 \notin Y_i + \omega_i$ and $0 \notin \bigcup_1^n (Y_i + \omega_i)$. Thus there exists a neighborhood of 0 disjoint from the closed set $\bigcup_1^n (Y_i + \omega_i)$; that is, there exists $D_1 > 0$ such that

$$|\eta_i + \omega_i| \geq D_1$$

for all $\eta_i \in Y_i$ and all $i = 1, \ldots, n$. This is the desired D_1. In fact, if $\xi = a_1\omega_1 + \cdots + a_n\omega_n \neq 0$ and $\varphi(a_r) = \max_i \{\varphi(a_i)\}$, then $a_r \neq 0$ and

$$|a_r^{-1}\xi| = \left| \frac{a_1}{a_r}\omega_1 + \cdots + \omega_r + \cdots + \frac{a_n}{a_r}\omega_n \right| \geq D_1$$

so that
$$|\xi| \geq D_1\varphi(a_r) = D_1\|\xi\|$$

Finally, it follows from the inequalities (*) that the topologies determined by $|\ \ |$ and $\|\ \ \|$ are equivalent, because the metrics bound each other. Since X is known to be complete under $\|\ \ \|$, the proof is now complete.

To put it another way, the theorem says that, in both the archimedean and nonarchimedean cases, the map $\xi \to (a_1, \ldots, a_n)$, where $\xi = \Sigma a_i\omega_i$, is a topological and algebraic isomorphism of X with $F^{(n)}$ (the product of n copies of F with the obvious algebraic and topological structure).

2-1-2. Corollary. Let F be P-complete, and let E be an extension field of F with $[E:F] < \infty$. If an extension of P to E exists, then it is unique; more explicitly, if $\varphi \in P$ has an extension to E, then it is unique. Moreover, E is then complete.

Proof. Suppose that $\|\varphi\| \leq 2$ and that φ_1, φ_2 are extensions of φ to E. From (2-1-1) we see that φ_1 and φ_2 determine the same topology on E. Since φ is nontrivial, so are φ_1 and φ_2; therefore, $\varphi_2 = \varphi_1^\alpha$ for some $\alpha > 0$. Hence, on F we have $\varphi = \varphi^\alpha$, which implies that $\alpha = 1$.

The case of $\varphi \in P$ with $\|\varphi\| > 2$ is trivial, since φ is a power of some element of P whose norm is ≤ 2.

2-1-3. Corollary. Let F be P-complete, and let Ω/F be an algebraic extension. If P has an extension to Ω, then it is unique; more explicitly, if $\varphi \in P$ has an extension to Ω, then it is unique.

Proof. Suppose φ_1 and φ_2 are distinct extensions of φ; then, for some $\alpha \in \Omega$, $\varphi_1(\alpha) \neq \varphi_2(\alpha)$. It follows that φ has two distinct extensions to the finite extension $F(\alpha)$ of F, a contradiction.

2-1-4. Corollary. Suppose that $\langle F, P\rangle$ is complete and that $\langle E_1, Q_1, i\rangle$ and $\langle E_2, Q_2, i\rangle$ are algebraic extensions of $\langle F, P\rangle$. If σ is an F isomorphism of E_1 onto E_2 then $Q_1 = \sigma^*(Q_2)$.

2-2. Existence of Extensions (Complete Case)

Suppose that F is P-complete and that E is a finite extension of F. We shall show that an extension of P to E always exists. If P is archimedean, the theorem of Ostrowski takes care of both the existence and uniqueness of an extension; so we restrict ourselves to the situation where P is nonarchimedean.

First, let us fix some notation. Choose an exponential valuation $\nu \in P$ and an indeterminate x. The canonical map $a \to \bar{a}$ of O onto the residue class field $\bar{F}$ then determines a ring homomorphism of $O[x] \twoheadrightarrow \bar{F}[x]$; namely, if $f(x) = a_0 + a_1x + \cdots + a_nx^n \in O[x]$, then

$$f(x) \to \bar{f}(x) = \bar{a}_0 + \bar{a}_1x + \cdots + \bar{a}_nx^n \in \bar{F}[x]$$

When $\bar{f}(x) \neq 0$, we shall say that $f(x)$ is ***primitive.*** Furthermore, for $f(x) = a_0 + a_1x + \cdots + a_nx^n \in F[x]$ let us put

$$\bar{\nu}(f) = \min \{\nu(a_0), \ldots, \nu(a_n)\}$$

It may be noted (although we do not need this fact here) that if we also put $\bar{\nu}(f/g) = \bar{\nu}(f) - \bar{\nu}(g)$ for $f, g \in F[x]$ with $g \neq 0$, then $\bar{\nu}$ is a valuation of $F(x)$ which extends ν. The remarks above have nothing to do with completeness, and they have already appeared in Exercise 1-3.

2-2-1. *Hensel's Lemma.* Let F be complete with respect to the nonarchimedean prime divisor P and let $f(x) \in O[x]$ be primitive. Suppose that $\mathcal{G}(x)$ and $\mathcal{H}(x)$ are polynomials in $\bar{F}[x]$ which are relatively prime and such that

$$\bar{f}(x) = \mathcal{G}(x)\mathcal{H}(x)$$

Then there exist $g(x)$, $h(x) \in O[x]$ such that

$$\text{(i)} \qquad f(x) = g(x)h(x)$$

$$\text{(ii)} \qquad \bar{g}(x) = \mathcal{G}(x) \qquad \bar{h}(x) = \mathcal{H}(x)$$

$$\text{(iii)} \qquad \deg g(x) = \deg \mathcal{G}(x)$$

Proof. We write $\deg f(x) = s$ and $\deg \mathcal{G}(x) = r$, so that therefore $\deg \bar{f}(x) \leq s$ and $\deg \mathcal{H}(x) \leq s - r$.

First of all, there exist $g_1(x)$, $h_1(x) \in O[x]$ such that $\bar{g}_1(x) = \mathcal{G}(x)$, $\bar{h}_1(x) = \mathcal{H}(x)$, $\deg g_1(x) = \deg \mathcal{G}(x)$, $\deg h_1(x) = \deg \mathcal{H}(x)$. Since $(\mathcal{G}(x), \mathcal{H}(x)) = 1$, there exist $a(x)$, $b(x) \in O[x]$ such that

$$\bar{a}(x)\mathcal{G}(x) + \bar{b}(x)\mathcal{H}(x) = 1$$

Thus $f - g_1h_1 \in \mathcal{P}[x]$ and $ag_1 + bh_1 - 1 \in \mathcal{P}[x]$. We put

$$\varepsilon = \min \{\bar{\nu}(f - g_1h_1), \bar{\nu}(ag_1 + bh_1 - 1)\}$$

If $\varepsilon = \infty$, then $f(x) = g_1(x)h_1(x)$ and we are finished. Suppose then that $0 < \varepsilon < \infty$, and choose $\pi \in F$ such that $\nu(\pi) = \varepsilon$. (If P is discrete and $\nu = \nu_P$, one simply takes a prime element π.) We have then

$$\begin{aligned} f &\equiv g_1h_1 \pmod{\pi} \\ \bar{g}_1 &= \mathcal{G} \qquad \bar{h}_1 = \mathcal{H} \\ \deg g_1 &= \deg \mathcal{G} = r \\ \deg h_1 &\leq s - r \end{aligned}$$

(where $f \equiv g_1h_1 \pmod{\pi}$ means $\bar{\nu}(f - g_1h_1) \geq \nu(\pi) = \varepsilon$, or what is the same $f - g_1h_1 \in \pi O[x]$).

Now let us construct, for each $i = 1, 2, 3, \ldots$, polynomials $g_i(x)$, $h_i(x) \in O[x]$ such that

$$\text{(1)} \qquad f \equiv g_ih_i \pmod{\pi^i}$$

$$\text{(2)} \qquad g_i \equiv g_{i-1} \pmod{\pi^{i-1}} \qquad h_i \equiv h_{i-1} \pmod{\pi^{i-1}} \qquad (i > 1)$$

$$\text{(3)} \qquad \bar{g}_i = \mathcal{G} \qquad \bar{h}_i = \mathcal{H}$$

$$\text{(4)} \qquad \deg g_i = \deg \mathcal{G} = r \qquad \deg h_i \leq s - r$$

The case $i = 1$ has already been done; so suppose inductively that the construction has been accomplished for $i = n - 1$, $n \geq 2$. Put

$$g_n(x) = g_{n-1}(x) + \pi^{n-1}u(x)$$
$$h_n(x) = h_{n-1}(x) + \pi^{n-1}v(x)$$

with the polynomials $u(x)$, $v(x) \in O[x]$ still to be determined. It is clear that (2) and (3) are satisfied for $i = n$.

Because $g_nh_n = g_{n-1}h_{n-1} + \pi^{n-1}(g_{n-1}v + h_{n-1}u) + \pi^{2n-2}uv$ and $2n - 2 \geq n$, we have

$$f \equiv g_nh_n \pmod{\pi^n} \Leftrightarrow f \equiv g_{n-1}h_{n-1} + \pi^{n-1}(g_{n-1}v + h_{n-1}u) \pmod{\pi^n}$$

By the induction hypothesis,

$$w = \frac{f - g_{n-1}h_{n-1}}{\pi^{n-1}} \in O[x]$$

so that $f \equiv g_nh_n \pmod{\pi^n} \Leftrightarrow w \equiv g_{n-1}v + h_{n-1}u \pmod{\pi}$

Now, let us choose u and v. From $ag_1 + bh_1 \equiv 1 \pmod{\pi}$ we have $wag_1 + wbh_1 \equiv w \pmod{\pi}$. According to the euclidean algorithm, we may write $w(x)b(x) = q(x)g_1(x) + u(x)$ with $\deg u(x) < \deg g_1(x) = r$. Because there is no loss of generality in assuming at the start that $\mathcal{G}(x)$ is monic and then taking $g_1(x) \in O[x]$ to be monic, it follows that $u(x) \in O[x]$. Thus

$$(wa + qh_1)g_1 + uh_1 \equiv w \pmod{\pi}$$

Let v be the polynomial gotten by replacing every coefficient of $wa + qh_1$ which is divisible by π by 0; then $v \in O[x]$, and

$$vg_1 + uh_1 \equiv w \pmod{\pi} \tag{\#}$$

Since $g_{n-1} \equiv g_1 \pmod{\pi}$ and $h_{n-1} \equiv h_1 \pmod{\pi}$, (1) is satisfied for $i = n$. Only (4) remains. We have $\deg g_n = \deg g_{n-1} = r$ because $\deg u < r$. Also, $\deg h_n > s - r$ implies $\deg v > s - r$. Since $\deg w \leq s$, $\deg uh_1 \leq s$, and g_1 is monic with $\deg g_1 = r$, the relation (#) gives a contradiction. Hence $\deg h_n \leq s - r$, and the construction is complete.

For $i = 1, 2, 3, \ldots$, write

$$g_i(x) = a_0^{(i)} + a_1^{(i)}x + \cdots + a_r^{(i)}x^r$$

with $a_j^{(i)} \in O$. Then for $j = 0, 1, \ldots, r$, $g_{i+1} \equiv g_i \pmod{\pi^i} \Rightarrow a_j^{(i+1)} \equiv a_j^{(i)} \pmod{\pi^i} \Rightarrow \{a_j^{(i)}\}$ is a Cauchy sequence. By completeness $a_j^{(i)} \to a_j \in O$. Put $g(x) = a_0 + a_1x + \cdots + a_rx^r$. Since $a_j \equiv a_j^{(i)} \pmod{\pi^i}$, we have $g(x) \equiv g_i(x) \pmod{\pi^i}$, $(i = 1, 2, 3, \ldots)$. Similarly one gets $h(x) = b_0 + b_1x + \cdots + b_{s-r}x^{s-r} \in O[x]$ with $h(x) \equiv h_i(x) \pmod{\pi^i}$. Hence

$$f(x) \equiv g_i(x)h_i(x) \equiv g(x)h(x) \pmod{\pi^i}$$

for all i, so that $f(x) = g(x)h(x)$. Finally $\bar{g} = \bar{g}_i = \mathcal{G}$, $\bar{h} = \bar{h}_i = \mathcal{H}$, and since $\deg g_1 \leq r$, $\deg h \leq s - r$, $\deg f = \deg g + \deg h$, it follows that $\deg g = r = \deg \mathcal{G}$. This completes the proof.

The details of the above proof indicate that, if $\mathcal{G}(x)$ or $\mathcal{H}(x)$ is monic, then the respective polynomials $g(x)$ or $h(x)$ may be taken to be monic.

There exist more complicated versions of Hensel's lemma in which the requirement that $\mathcal{G}(x)$ and $\mathcal{H}(x)$ be relatively prime is weakened, but we have no need to discuss them. Instead, we describe some of the easy consequences of Hensel's lemma and in this way give some indication of its significance. It should be emphasized that any procedure for arriving at the results of extension theory rests on some result more or less equivalent to Hensel's lemma.

2-2-2. Corollary. If $f(x) = a_0 + a_1x + \cdots + a_nx^n \in F[x]$ is irreducible, then

$$\bar{\nu}(f) = \min \{\nu(a_0), \ldots, \nu(a_n)\} = \min \{\nu(a_0), \nu(a_n)\}$$

Proof. By multiplying $f(x)$ by a suitable element of F, we may assume that $f(x)$ is primitive—that is, $f(x) \in O[x]$ and $\bar{\nu}(f) = 0$. Suppose that $\min \{\nu(a_0), \nu(a_n)\} > 0$. Then there exists an r with $0 < r < n$ such that $\nu(a_r) = 0$ and $\nu(a_i) > 0$ for $i = r + 1, \ldots, n$. Now write $\bar{f}(x) = \bar{f}(x) \cdot 1 = \mathcal{G}(x)\mathcal{H}(x)$. Because $(\mathcal{G}(x), \mathcal{H}(x)) = 1$ and $\deg \mathcal{G}(x) = r$, Hensel's lemma gives a factorization $f(x) = g(x)h(x)$ with $\deg g(x) = r > 0$ and $\deg h(x) = n - r > 0$. This contradicts irreducibility.

2-2-3. Corollary. If $f(x) = x^n + b_1x^{n-1} + \cdots + b_n \in F[x]$ is a monic irreducible polynomial, then $f(x) \in O[x] \Leftrightarrow b_n \in O$.

2-2-4. Exercise. i. Show by example that, if F is not P-complete, then Hensel's lemma need not be true.

ii. Suppose that α belongs to the finite extension E of F, then (2-2-3) says that the norm $N_{E\to F}(\alpha) \in O$ if and only if the minimum polynomial of α over F belongs to $O[x]$. Give an example to show that this need not be true if F is not P-complete.

2-2-5. Corollary. If $f(x) \in O[x]$ is monic and irreducible over F, then $\bar{f}(x)$ is a power of an irreducible polynomial in $\bar{F}[x]$.

Proof. Write

$$\bar{f}(x) = \prod_1^r \mathcal{G}_i(x)^{n_i}$$

where each $\mathcal{G}_i(x)$ is irreducible in $\bar{F}[x]$ and $\mathcal{G}_i(x) \neq \mathcal{G}_j(x)$ for $i \neq j$. Since $\deg \bar{f}(x) = \deg f(x) \geq 1$, the indexing may be taken so that $\mathcal{G}(x) = \mathcal{G}_1(x)^{n_1} \neq 1$. If $\mathcal{G}(x) = \bar{f}(x)$, we are finished. Otherwise, we may take

$$\mathcal{H}(x) = \prod_2^r \mathcal{G}_i(x)^{n_i} \neq 1$$

so that Hensel's lemma gives a nontrivial factorization of $f(x)$.

2-2-6. Corollary. Suppose that $f(x) \in O[x]$ and that $\alpha \in \bar{F}$ is a simple root of $\bar{f}(x)$. Then there exists an element $a \in O$ such that $\bar{a} = \alpha$ and $f(a) = 0$.

Proof. By hypothesis, we have $\bar{f}(x) = (x - \alpha)\mathcal{H}(x)$ with $x - \alpha$ and $\mathcal{H}$ relatively prime. Applying Hensel's lemma gives a factorization $f(x) = g(x)h(x)$ where $g(x)$, $h(x) \in O[x]$ and $g(x)$ is monic of degree 1. It is immediate that $g(x) = x - a$ for some $a \in O$. This is the desired a. The problem of finding such an a explicitly will be considered in Section 3-1.

2-2-7. Corollary. $\mathbf{Q}_p$ contains the $(p - 1)$st roots of unity.

Proof. Apply (2-2-6) to the polynomial $x^{p-1} - 1$.

2-2-8. Corollary. If p is an odd prime with $p \equiv 1 \pmod 4$, then -1 is a square in $\mathbf{Q}_p$.

2-2-9. Digression. In order to be able to round out the statement of (2-2-10) by the inclusion of part iii, and also for later reference, it is convenient to make some remarks about the notion of integral dependence. Proofs of the various assertions are not given, but they are short, and the reader should have little difficulty in supplying them. Otherwise one may consult O. Zariski and P. Samuel's "Commutative Algebra" (vol. I, D. Van Nostrand Co., Inc., Princeton, N. J.).

Let R and S be integral domains (with the same 1) such that $R \subset S$. An element $\alpha \in S$ is said to be ***integrally dependent on*** R or ***integral over*** R if there exists a monic polynomial $f(x) \in R[x]$ such that $f(\alpha) = 0$. In other words, there exist $a_1, \ldots, a_n \in R$ such that

$$\alpha^n + a_1\alpha^{n-1} + \cdots + a_n = 0$$

If every element of S is integral over R, we say that S is ***integral over*** R or ***integrally dependent on*** R.

1. The element $\alpha \in S$ is integral over R if and only if $R[\alpha]$ is finitely generated as an R module.

2. If S is a finitely generated R module, then S is integral over R.

3. If $\alpha_1, \ldots, \alpha_n \in S$ are integral over R, then $R[\alpha_1, \ldots, \alpha_n]$ is a finitely generated R module.

The set of all elements of S that are integral over R is called the ***integral closure of** R **in** S*; we denote it by R^S. According to (2) and (3), R^S is an integral domain. We say that R is ***integrally closed in** S* when $R^S = R$ and that R is ***integrally closed*** when it is integrally closed in its quotient field.

4. Suppose that T is a domain containing S. If $\alpha \in T$ is integral over S and S is integral over R, then α is integral over R.

5. R^S is integrally closed in S.

6. Suppose that S is integral over R; then R is a field if and only if S is a field.

7. Any unique factorization domain is integrally closed.

Suppose now that R is an integral domain with quotient field F. Let E be an extension field of F, and let S be the integral closure of R in E.

8. If $\alpha \in F \cap S$, there exists $c \neq 0 \in R$ such that $c\alpha^n \in R$ for all $n = 1, 2, 3, \ldots$.

9. If R is integrally closed, then $F \cap S = R$.

10. If R is integrally closed and $\alpha \in S$, then its minimum polynomial over F has coefficients in R.

11. Suppose that R is integrally closed and that $f(x) \in R[x]$ is monic; then $f(x)$ is irreducible over R if and only if it is irreducible over F.

12. If E is algebraic over F, then E is the quotient field of S. In fact, any $\alpha \in E$ may be expressed in the form $\alpha = \beta/c$ with $\beta \in S$, $c \in R$.

13. Suppose that E is algebraic over F and that $\omega_1, \ldots, \omega_n \in S$ are such that $S = R\omega_1 \oplus \cdots \oplus R\omega_n$ (direct sum of R modules); then $\omega_1, \ldots, \omega_n$ is a basis for E/F. In this situation, one says that $\omega_1, \ldots, \omega_n$ is an ***integral basis*** for S over R (or for E over F).

2-2-10. Theorem. Let F be complete with respect to the non-archimedean prime divisor P, and let E/F be a finite extension of degree n. Then:

i. P has a unique extension Q to E. In fact, for any $\varphi \in P$, its unique extension φ' to E is given in terms of $N_{E \to F}$, the norm function from E to F, by

$$\varphi'(\alpha) = \{\varphi[N_{E \to F}(\alpha)]\}^{1/n} \qquad \alpha \in E$$

ii. P is discrete $\Leftrightarrow Q$ is discrete.

iii. O_Q is the integral closure of $O_P = F \cap O_Q$ in E.

Proof. In order to prove (i), one must verify that φ' is a valuation of E extending φ. It is equally trivial to check that $\varphi'|F = \varphi$ and that $\varphi'(\alpha) \geq 0$, $\varphi'(\alpha) = 0 \Leftrightarrow \alpha = 0$, $\varphi'(\alpha\beta) = \varphi'(\alpha)\varphi'(\beta)$ for all α, $\beta \in E$. It remains to show that $\varphi'(\alpha) \leq 1 \Rightarrow \varphi'(1 + \alpha) \leq 1$. To start with, we have $\varphi'(\alpha) \leq 1 \Rightarrow \varphi[N_{E\to F}(\alpha)] \leq 1$. Now let

$$f(x) = f(\alpha, F) = x^r + a_1x^{r-1} + \cdots + a_r \in F[x]$$

be the minimum polynomial of α over F [henceforth, the notation $f(\alpha, F)$ will always be used for the minimum polynomial]. It follows that $N_{E\to F}(\alpha) = \pm a_r^m$, where $m = n/r = [E:F(\alpha)]$, so that $\varphi(a_r)^m \leq 1$ and $a_r \in O_P$. According to (2-2-3) we have $a_1, \ldots, a_r \in O_P$. If we put $g(x) = f(x - 1)$, then $g(1 + \alpha) = 0$ and $g(x)$ is the minimum polynomial of $1 + \alpha$ over F. In particular,

$$\begin{aligned} N_{E\to F}(1 + \alpha) &= \{\pm g(0)\}^m = \{\pm f(-1)\}^m \\ &= \pm\{(-1)^r + a_1(-1)^{r-1} + \cdots + a_r\} \in O_P \end{aligned}$$

which implies that $\varphi'(1 + \alpha) \leq 1$.

From the definition of discreteness and the relation

$$\log \varphi'(\alpha) = \frac{1}{n} \log \varphi\, [N_{E\to F}(\alpha)]$$

it follows immediately that P is discrete if and only if Q is discrete.

As for (iii), it is clear that $O_P = O_Q \cap F$. To complete the proof, we show that for $\alpha \in E$

$$\alpha \in O_Q \Leftrightarrow \alpha \text{ is integral over } O_P \Leftrightarrow f(\alpha, F) \in O_P[x]$$

From item 10 of (2-2-9) and from Exercise 2-1, we know that the second equivalence holds. Furthermore, the proof of (i) showed that $\alpha \in O_Q \Rightarrow f(\alpha, F) \in O_P[x]$. Suppose then that α is integral over O_P. This means that there exists a relation of form

$$\alpha^r + a_1\alpha^{r-1} + \cdots + a_r = 0$$

with all $a_i \in O_P$. Consequently,

$$[\varphi'(\alpha)]^r = \varphi'(a_1\alpha^{r-1} + \cdots + a_r) \leq \max_{i=0,\ldots,r-1} \{\varphi'(\alpha)^i\}$$

and $\varphi'(\alpha) \leq 1$.

By making use of (2-2-3) we may note that

$$\alpha \in O_Q \Leftrightarrow N_{E\to F}(\alpha) \in O_P$$

It is often convenient, especially when it is necessary to deal with several extension fields of a given field simultaneously, to index the objects belonging to a field in terms of the field itself. Thus, in the situation of (2-2-10), one may write φ_F for φ, φ_E for the unique extension of φ_F to E, O_F for O_P, and O_E for the integral closure of O_F in E.

2-2-11. Corollary. Let F be complete with respect to a prime divisor P, and let Ω be any algebraic extension of F. Then P has a unique extension to Ω (as does any $\varphi \in P$). Moreover, if P is nonarchimedean, then O_Ω is the integral closure of $O_F = O_\Omega \cap F$ in Ω and

$$O_\Omega = \bigcup_E O_E$$

where E runs over all finite extensions of F contained in Ω.

Proof. The archimedean case is trivial, as has been observed earlier; note also that $[\Omega:F] < \infty$ and that, if φ_F is the absolute value on F, then its unique extension φ_Ω to Ω is also given by the formula in (2-2-10i).

Suppose that P is nonarchimedean; then only the case $[\Omega:F] = \infty$ remains to be discussed. Uniqueness has been proved in (2-1-3). As for the rest, we have $\Omega = \bigcup_E E$, and for each such finite extension there is a unique extension φ_E of $\varphi = \varphi_F$. Given $\alpha \in \Omega$, it belongs to some E, and we put $\varphi_\Omega(\alpha) = \varphi_E(\alpha)$. φ_Ω is well defined; in fact, we have $F \subset F(\alpha) \subset E$ and

$$\begin{aligned}\{\varphi[N_{F(\alpha)\to F}(\alpha)]\}^{1/[F(\alpha):F]} &= \{\varphi[N_{F(\alpha)\to F}(\alpha)]^{[E:F(\alpha)]}\}^{1/[E:F]} \\ &= \{\varphi[N_{E\to F}(\alpha)]\}^{1/[E:F]}\end{aligned}$$

The verification that φ_Ω is a valuation of Ω is immediate, and the remaining statements follow easily.

2-2-12. Corollary. In the situation of (2-2-11), choose $\nu \in P$, and denote its unique extension to Ω by ν also. If σ is any automorphism of Ω which leaves F pointwise fixed, then $\nu \circ \sigma = \nu$.

2-3. Extensions of Discrete Prime Divisors

2-3-1. Proposition. Let F be complete with respect to the discrete prime divisor P. Let E be a finite extension of F, and denote the extension of P to E by Q. If ν_P, ν_Q are the normalized exponential valuations belonging to P and Q, respectively, then for all $a \in F$

$$\nu_Q(a) = e\left(\frac{Q}{P}\right) \nu_P(a)$$

Proof. According to (2-2-10), Q is discrete, and ν_Q is defined. Let $\nu' = \nu_Q|F$, so that $\nu' = \alpha\nu_P$ for some real $\alpha > 0$. If $\pi \in F$ is such that $\nu_P(\pi) = 1$, then $\nu'(\pi) = \alpha \in \nu'(F^*)$. Thus $e(Q/P) = (\nu_Q(E^*):\nu'(F^*)) = (\mathbf{Z}:\alpha\mathbf{Z}) = \alpha$, and on F we have $\nu_Q = e(Q/P)\nu_P$.

2-3-2. Theorem. Let F be complete with respect to the discrete prime divisor P. Let E be a finite extension of F, and denote the extension of P to E by Q. Then

$$e\left(\frac{Q}{P}\right)f\left(\frac{Q}{P}\right) = [E:F]$$

In fact, there exists an integral basis for E over F.

Proof. We exhibit an integral basis for E over F [see item 13 of (2-2-9)]. Let $\omega_1, \ldots, \omega_f \in O_E = O_Q$ be such that $\bar{\omega}_1, \ldots, \bar{\omega}_f$ form a basis for $\bar{E}$ over $\bar{F}$, and let $\pi_0, \pi_1, \ldots, \pi_{e-1}$ be representatives of all the distinct cosets of $\nu_Q(F^*)$ in $\nu_Q(E^*)$. From (1-6-4) we know that $ef \leq [E:F]$; in fact, the ef elements $\omega_i\pi_j$ ($i = 1, \ldots, f$; $j = 0, 1, \ldots, e-1$) are linearly independent over F. We show that these elements are indeed an integral basis for E over F.

Fix a prime element $\pi_E \in E$ and a prime element $\pi_F \in F$—that is, $\nu_Q(\pi_E) = 1$ and $\nu_P(\pi_F) = 1$. By (2-3-1) π_E^e and π_F are associates in O_E; in other words, $\pi_E^e O_E = \pi_F O_E$. We may take $\pi_j = \pi_E^j$ for $j = 0, 1, \ldots, e-1$ and also write $\pi_E = \pi$. Let us put

$$L = \sum_1^f O_F\omega_i$$

We have then

$$O_E = L + \pi O_E$$

In fact, it is clear that $L + \pi O_E \subset O_E$; and if $\alpha \in O_E$, then $\bar{\alpha} = \bar{a}_1\bar{\omega}_1 + \cdots + \bar{a}_f\bar{\omega}_f$ for certain $a_1, \ldots, a_f \in O_F$ so that

$$\alpha = a_1\omega_1 + \cdots + a_f\omega_f + \pi\alpha_0$$

with $\alpha_0 \in O_E$. It follows that

$$O_E = L + \pi(L + \pi O_E) = \cdots = L + \pi L + \cdots + \pi^{e-1}L + \pi^e O_E$$

Now let us put $M = L + \pi L + \cdots + \pi^{e-1}L$, and note immediately that

$$M = \sum_{i,j} \oplus\, O_F\omega_i\pi^j$$

Thus it suffices to show that $M = O_E$.

We have, first of all,

$$\begin{aligned} O_E &= M + \pi^e O_E = M + \pi_F O_E = \cdots \\ &= M + \pi_F M + \cdots + \pi_F^{r-1} M + \pi_F^r O_E \end{aligned}$$

for all $r \geq 1$. Since $\pi_F^s L \subset L$ for all $s \geq 1$, it follows that $\pi_F^s M \subset M$ and consequently $M = M + \pi_F M + \cdots + \pi_F^{r-1} M$ for all $r \geq 1$. This means that

$$O_E = M + \pi_F^r O_E \text{ for all } r \geq 1$$

Since $\{\pi_F^r O_E | r = 1, 2, \ldots\}$ is a fundamental system of neighborhoods of 0 in E, we conclude that M is dense in O_E. On the other hand, $\{\omega_i \pi^j | i = 1, \ldots, f, \; j = 0, \ldots, e-1\}$ is part of some basis of E over F, and O_F is open and closed in F. In view of the remark following the proof of (2-1-1), it is easy to see that $M = \sum_{i,j} \oplus O_F \omega_i \pi^j$ is closed in E and in O_E. Therefore, $M = O_E$, and the proof is complete.

2-3-3. Corollary. Under the hypotheses of (2-3-2) we have

$$\nu_P[N_{E\to F}(\alpha)] = f\left(\frac{Q}{P}\right) \nu_Q(\alpha)$$

for all $\alpha \in E$.

Proof. From (2-2-10) we know that for any $\varphi \in P$ its unique extension $\varphi' \in Q$ is given by $\varphi'(\alpha) = \{\varphi[N_{E\to F}(\alpha)]\}^{1/[E:F]}$. If the notation is arranged so that $\nu_Q = -\log \circ \varphi'$, then, by (2-3-1), $-\log \circ \varphi = e(Q/P)\nu_P$. Combining these facts gives

$$\nu_Q(\alpha) = \frac{e(Q/P)}{e(Q/P)f(Q/P)} \nu_P[N_{E\to F}(\alpha)]$$

2-3-4. Exercise. Classify all fields complete with respect to a discrete prime divisor and having finite residue class field. More precisely, either such a field is a finite extension of some $\mathbf{Q}_p$ (in which case the field is often called a ***p-adic field***), or it is a finite extension of a completion of a field of rational functions $k(x)$ over a finite field k.

2-4. Extensions in the General Case

Having solved the extension problem when the base field is complete, we proceed to consider the general case. Thus, throughout this section, the basic situation is as follows: F is a field with prime divisor P—P may be archimedean or nonarchimedean—and E denotes a finite extension of F.

By applying results for the complete case, it is easy to see that an extension of P to E always exists. In fact, consider a completion $\langle \tilde{F}, \tilde{P}, i_{F\to\tilde{F}} \rangle$ of $\langle F, P \rangle$. Although it is of no particular importance at this point, we may fix it so that $E \cap \tilde{F} = F$. Let Ω denote an algebraic closure of $\tilde{F}$; then $\tilde{P}$ has a unique extension R to Ω. Thus $i^*_{\tilde{F}\to\Omega}(R) = \tilde{P}$ and $\langle \Omega, R, i_{F\to\Omega} \rangle$ is an extension of $\langle F, P \rangle$. Since Ω is an algebraically closed field containing F, and since E is algebraic over F, there exists an F isomorphism $\mu : E \to \Omega$. If we put

$$Q_\mu = \mu^*(R)$$

then Q_μ is a prime divisor of E which is an extension of P. In fact, it follows from $\mu \circ i_{F\to E} = i_{F\to\Omega}$ that

$$i^*_{F\to E}(Q_\mu) = i^*_{F\to E}\mu^*(R) = i^*_{F\to\Omega}(R) = P$$

Of course, the finiteness of $[E:F]$ is not essential, and we have proved the following proposition.

2-4-1. Proposition. Let E be an algebraic extension of the field F. If P is a prime divisor of F, then P has an extension to E.

Naturally, this result is valid for valuations; let us be explicit. Denote the unique extension of $\varphi \in P$ to $\tilde{F}$ by $\tilde{\varphi}$. The unique extension of $\tilde{\varphi}$ to Ω may be denoted by $\tilde{\varphi}$ also. Then for any F isomorphism $\mu : E \to \Omega$ the function $\tilde{\varphi} \circ \mu$ is a valuation of E extending φ. We shall see that all extensions of φ to E arise in this way.

In order to get information about the number of extensions of P to E, it is useful to introduce a general concept—that of a composite extension. The notion of composite extension in its full generality is as follows: Suppose that E and $\tilde{F}$ are arbitrary extensions of a field F and that $E \cap \tilde{F} = F$. A composite extension of E and $\tilde{F}$ over F is a triple $\{\bar{E}, \mu, \lambda\}$, where

(i) $\bar{E}$ is an extension field of F

(ii) μ is an F isomorphism of E into $\bar{E}$

(iii) λ is an F isomorphism of $\tilde{F}$ into $\bar{E}$

(iv) $\bar{E} = (\mu E)(\lambda \tilde{F})$

Furthermore, two such composite extensions $\{\bar{E}_1, \mu_1, \lambda_1\}$ and $\{\bar{E}_2, \mu_2, \lambda_2\}$ are said to be equivalent when there exists an F isomorphism ρ of $\bar{E}_1$ onto $\bar{E}_2$ such that $\rho \circ \mu_1 = \mu_2$ and $\rho \circ \lambda_1 = \lambda_2$.

Since we want to consider composite extensions in the situation that occurs in the proof of (2-4-1), namely, where all fields that arise are imbedded in one large field, some slight simplifications (especially in notation) are possible. Thus, for the discussion here, let E/F denote a finite extension, and let $\tilde{F}/F$ be an arbitrary extension such that $E \cap \tilde{F} = F$. Fix an algebraic closure Ω of $\tilde{F}$. It will be standard for us to denote the set of all F isomorphisms of E into Ω by $\Gamma(F, E \to \Omega)$. Any element $\mu \in \Gamma(F, E \to \Omega)$ will be called a ***composition map*** for $\{E, F, \tilde{F}, \Omega\}$, and the field $\tilde{F}(\mu E)$ will be called the ***composite extension*** belonging to μ. Since F, $\tilde{F}$, Ω will not vary, we shall often say that μ is a composition map for E. Two such composition maps μ_1 and μ_2 will be said to be ***equivalent*** (denoted by $\mu_1 \sim \mu_2$) when there exists $\sigma \in \Gamma(\tilde{F}, \Omega \to \Omega) = \mathcal{G}(\Omega/\tilde{F})$ [where $\mathcal{G}(\Omega/\tilde{F})$ denotes the group of all automorphisms of Ω which leave $\tilde{F}$ pointwise fixed] such that $\sigma \circ \mu_1 = \mu_2$. Naturally, the composite extensions $\tilde{F}(\mu_1 E)$ and $\tilde{F}(\mu_2 E)$ will then be said to be ***equivalent.*** This is slightly more than the statement that these fields are conjugate over $\tilde{F}$. Since any $\tilde{F}$ isomorphism of $\tilde{F}(\mu_1 E)$ onto $\tilde{F}(\mu_2 E)$ can be extended to an element of $\mathcal{G}(\Omega/\tilde{F})$, it follows that $\mu_1 \sim \mu_2$ if and only if there exists an $\tilde{F}$ isomorphism $\rho: \tilde{F}(\mu_1 E) \rightarrowtail\!\!\!\twoheadrightarrow \tilde{F}(\mu_2 E)$ such that $\rho \circ \mu_1 = \mu_2$.

In algebraic number theory, the extension E/F is always separable; however, our method is quite general, and it will not be necessary to make this restriction. The separable part of the extension E/F will be denoted by E^S; thus E^S/F is separable and E/E^S is purely inseparable. As is customary, we write

$$[E:E^S] = [E:F]_{\text{ins}} \qquad \text{and} \qquad [E^S:F] = [E:F]_{\text{sep}}$$

2-4-2. Lemma. Let $\tilde{F}$ be an arbitrary extension of F, and let Ω be an algebraic closure of $\tilde{F}$. Suppose that E is a finite extension of F such that $E \cap \tilde{F} = F$; then:

i. If λ is a composition map for $\{E^S, F, \tilde{F}, \Omega\}$, then λ can be extended uniquely to a composition map μ for $\{E, F, \tilde{F}, \Omega\}$.

ii. $\tilde{F}(\lambda E^S)$ is the separable closure of $\tilde{F}$ in $\tilde{F}(\mu E)$.

iii. If $\mu_1, \mu_2 \in \Gamma(F, E \to \Omega)$ are the unique extensions of $\lambda_1, \lambda_2 \in \Gamma(F, E^S \to \Omega)$, respectively, then

$$\lambda_1 \sim \lambda_2 \Leftrightarrow \mu_1 \sim \mu_2$$

iv. The number

$$g_\mu = \frac{[E:F]_{\text{ins}}}{[\tilde{F}(\mu E):\tilde{F}]_{\text{ins}}}$$

is an integer; in fact, it is 1 or a power of a prime.

Proof. i. This is the statement of the well-known 1-1 correspondence between the finite sets $\Gamma(F, E^S \to \Omega)$ and $\Gamma(F, E \to \Omega)$.

ii. E^S is separable over $F \Rightarrow \lambda E^S$ is separable over $F \Rightarrow \tilde{F}(\lambda E^S)$ is finite separable over $\tilde{F}$. If $E^S \neq E$, then F has characteristic $p \neq 0$. According to a standard criterion for pure inseparability, if an element of E is given, then some p^eth power of it belongs to E^S. From the reverse direction of the same criterion, it now follows easily that $\tilde{F}(\mu E)$ is purely inseparable over $\tilde{F}(\lambda E^S)$.

iii. We have immediately

$$\begin{aligned} \mu_1 \sim \mu_2 &\Leftrightarrow \text{there exists } \sigma \in \mathcal{G}\left(\frac{\Omega}{\tilde{F}}\right) \text{ such that } \sigma \circ \mu_1 = \mu_2 \\ &\Rightarrow \text{there exists } \sigma \in \mathcal{G}\left(\frac{\Omega}{\tilde{F}}\right) \text{ such that } \sigma \circ \lambda_1 = \lambda_2 \\ &\Leftrightarrow \lambda_1 \sim \lambda_2 \end{aligned}$$

Thus just one implication remains, and it, too, is trivial because both $\sigma \circ \mu_1$ and μ_2 in $\Gamma(F, E \to \Omega)$ are extensions of $\lambda_2 \in \Gamma(F, E^S \to \Omega)$.

iv. If $E^S = E$, then $g_\mu = 1$. If $E^S \neq E$, let $p \neq 0$ be the characteristic of F. It is clear that

$$[\tilde{F}(\mu E):\tilde{F}]_{\text{ins}} = [\tilde{F}(\mu E):\tilde{F}(\lambda E^S)] \leq [E:E^S] = [E:F]_{\text{ins}}$$

Since the degree of a purely inseparable finite extension must be a power of the characteristic, g_μ is indeed an integer.

2-4-3. Lemma. Let the situation be as in (2-4-2), and let $\mu_1, \ldots, \mu_r$ be a full set of representatives for the equivalence classes of composition maps for $\{E, F, \tilde{F}, \Omega\}$. If we write $G = \mathcal{G}(\Omega/\tilde{F})$ and $G_i = \mathcal{G}(\Omega/\tilde{F}(\mu_i E))$ for $i = 1, \ldots, r$, then $m_i = (G:G_i)$ is the number of elements in $\Gamma(\tilde{F}, \tilde{F}(\mu_i E) \to \Omega)$. Moreover, if we take a left coset decomposition

$$G = \bigcup_{j=1}^{m_i} \sigma_{i,j} G_i$$

for each $i = 1, \ldots, r$, then

$$\Gamma(F, E \to \Omega) = \{\sigma_{i,j} \circ \mu_i | i = 1, \ldots, r, \quad j = 1, \ldots, m_i\}$$

Proof. Put $\tau_{i,j} = \sigma_{i,j}|\tilde{F}(\mu_i E)$, so that $\tau_{i,j} \in \Gamma(\tilde{F}, \tilde{F}(\mu_i E) \to \Omega)$. Since

$$\tau_{i,j} = \tau_{i,k} \Leftrightarrow \sigma_{i,j}^{-1}\sigma_{i,k} \in G_i \Leftrightarrow \sigma_{i,j} = \sigma_{i,k}$$

and since any element of $\Gamma(\tilde{F}, \tilde{F}(\mu_i E) \to \Omega)$ may be extended to an element of G, it follows that

$$\{\tau_{i,j} | j = 1, \ldots, m_i\} = \Gamma(\tilde{F}, \tilde{F}(\mu_i E) \to \Omega)$$

This takes care of the first assertion.

Suppose next that $\mu \in \Gamma(F, E \to \Omega)$; then μ is equivalent to some μ_i, and there exists $\sigma \in G$ such that $\mu = \sigma \circ \mu_i$. Since we may write $\sigma = \sigma_{i,j}\rho_i$ with $\rho_i \in G_i$, we have $\mu = \sigma_{i,j} \circ \rho_i \circ \mu_i = \sigma_{i,j} \circ \mu_i$. Of course, every $\sigma_{i,j} \circ \mu_i$ belongs to $\Gamma(F, E \to \Omega)$, and it remains to show that they are distinct. This presents no difficulty; in fact,

$$\begin{aligned}
\sigma_{i,j}\mu_i = \sigma_{r,s}\mu_r &\Rightarrow \sigma_{r,s}^{-1}\sigma_{i,j}\mu_i = \mu_r \\
&\Rightarrow \mu_i \sim \mu_r \Rightarrow i = r \Rightarrow \sigma_{i,j}|(\mu_i E) = \sigma_{i,s}|(\mu_i E) \\
&\Rightarrow \sigma_{i,j}|\tilde{F}(\mu_i E) = \sigma_{i,s}|\tilde{F}(\mu_i E) \\
&\Rightarrow \tau_{i,j} = \tau_{i,s} \Rightarrow j = s
\end{aligned}$$

2-4-4. Proposition. Let $\tilde{F}/F$ be an arbitrary extension, and let Ω be an algebraic closure of $\tilde{F}$. Suppose that E/F is a finite extension with $E \cap \tilde{F} = F$. Let $\mu_1, \ldots, \mu_r$ be a full set of representatives for the equivalence classes of composition maps for $\{E, F, \tilde{F}, \Omega\}$, and put $[E:F] = n$, $[\tilde{F}(\mu_i E):\tilde{F}] = n_i$, $[E:F]_{\text{ins}}/[\tilde{F}(\mu_i E):\tilde{F}]_{\text{ins}} = g_i$. For any $\alpha \in E$, we have

(i) $$N_{E \to F}(\alpha) = \prod_{i=1}^{r} \{N_{\tilde{F}(\mu_i E) \to \tilde{F}}(\mu_i \alpha)\}^{g_i}$$

(ii) $$S_{E \to F}(\alpha) = \sum_{i=1}^{r} g_i S_{\tilde{F}(\mu_i E) \to \tilde{F}}(\mu_i \alpha)$$

(iii) $$f\left(\alpha, \frac{E}{F}\right) = \prod_{i=1}^{r} \left\{f\left(\mu_i \alpha, \frac{\tilde{F}(\mu_i E)}{\tilde{F}}\right)\right\}^{g_i}$$

(iv) $$n = \sum_{i=1}^{r} g_i n_i$$

Proof. The symbols N and S refer to the norm and trace function, respectively, while $f(\alpha, E/F)$ denotes the field polynomial of α for the extension E/F. We shall make use of (and assume that the reader is familiar with) the standard formulas which give the norm, trace, and field polynomial of an element in terms of its conjugates. To prove (i) we observe that

$$N_{E\to F}(\alpha) = \prod_{\mu} (\mu\alpha)^{[E:F]_{\text{ins}}} \text{ where } \mu \text{ runs over } \Gamma(F, E \to \Omega)$$

$$= \prod_{i,j} (\sigma_{i,j}\mu_i\alpha)^{[E:F]_{\text{ins}}} \qquad \text{(by (2-4-3))}$$

$$= \prod_{i=1}^{r} \left\{ \prod_{j=1}^{m_i} [\sigma_{i,j}(\mu_i\alpha)]^{[E:F]_{\text{ins}}} \right\}$$

$$= \prod_{i=1}^{r} \{N_{\tilde{F}(\mu_iE)\to\tilde{F}}(\mu_i\alpha)\}^{g_i}$$

since $N_{\tilde{F}(\mu_iE)\to\tilde{F}}(\mu_i\alpha) = \prod_{j=1}^{m_i} \sigma_{i,j}(\mu_i\alpha)^{[\tilde{F}(\mu_iE):\tilde{F}]_{\text{ins}}}$

The proofs of (ii) and (iii) proceed in an analogous fashion from

$$S_{E\to F}(\alpha) = [E:F]_{\text{ins}} \sum_{\mu} (\mu a) \text{ and } f(\alpha, E/F) = \prod_{\mu} (x - \mu\alpha)^{[E:F]_{\text{ins}}}$$

Since the degree of a field polynomial is equal to the degree of the extension, (iv) is an immediate consequence of (iii).

2-4-5. Proposition. Let the hypotheses and notation be as in (2-4-4), and suppose that E/F is a simple extension—say, $E = F(\alpha)$. Then there is a 1-1 correspondence between the equivalence classes of composition maps for $\{E, F, \tilde{F}, \Omega\}$ and the distinct irreducible factors of $f(\alpha, F)$ in $\tilde{F}[x]$.

Proof. For any $\beta \neq 0 \in E$ we know that $f(\beta, E/F) = f(\beta, F)^{[E:F(\beta)]}$, so that $E = F(\beta) \Leftrightarrow f(\beta, E/F) = f(\beta, F)$. Therefore, $f(\alpha, E/F) = f(\alpha, F)$ and from $\tilde{F}(\mu_iE) = \tilde{F}(\mu_i\alpha)$ it follows that $f(\mu_i\alpha, \tilde{F}(\mu_iE)/\tilde{F}) = f(\mu_i\alpha, \tilde{F})$ for $i = 1, \ldots, r$. According to (2-4-4)

$$f(\alpha, F) = \prod_{i=1}^{r} f(\mu_i\alpha, \tilde{F})^{g_i}$$

is the decomposition of $f(\alpha, F)$ into irreducible factors over $\tilde{F}$. Moreover, the irreducible factors $f(\mu_i\alpha, \tilde{F})$ are distinct. In fact, $f(\mu_i\alpha, \tilde{F}) = f(\mu_j\alpha, \tilde{F}) \Leftrightarrow \mu_i\alpha$ and $\mu_j\alpha$ are conjugates over $\tilde{F} \Leftrightarrow$ there exists an $\tilde{F}$ isomorphism ρ of $\tilde{F}(\mu_i\alpha)$ onto $\tilde{F}(\mu_j\alpha)$ such that $\rho(\mu_i\alpha) = \mu_j\alpha \Leftrightarrow$ there exists an $\tilde{F}$ isomorphism ρ of $\tilde{F}(\mu_iE)$ onto $\tilde{F}(\mu_jE)$ such that $\rho \circ \mu_i = \mu_j \Leftrightarrow \mu_i \sim \mu_j \Leftrightarrow i = j$.

The results about composite extensions (which do not mention prime divisors) may now be applied to the problem of counting the extensions of a prime divisor in an extension field of finite degree.

2-4-6. Theorem. Let F be a field with prime divisor P, let $\langle \tilde{F}, \tilde{P}, i \rangle$ be a completion of $\langle F, P \rangle$, and let Ω be an algebraic closure of $\tilde{F}$. If E/F is a finite extension, then the number of extensions of P to E is the same as the number of equivalence classes of composition maps for $\{E, F, \tilde{F}, \Omega\}$.

In fact, if $\mu_1, \ldots, \mu_r$ is a full set of representatives for the equivalence classes of composition maps for $\{E, F, \tilde{F}, \Omega\}$, then

$$\{Q_i = Q_{\mu_i} | i = 1, \ldots, r\}$$

is the set of all extensions of P to E.

Furthermore, if S_i is the unique extension of $\tilde{P}$ to $\tilde{E}_i = \tilde{F}(\mu_i E)$, then $\langle \tilde{E}_i, S_i, \mu_i \rangle$ is a completion of $\langle E, Q_i \rangle$.

Finally, if we put $[E:F] = n$, $[\tilde{E}_i:\tilde{F}] = n_i$, and $[E:F]_{\text{ins}}/[\tilde{E}_i:\tilde{F}]_{\text{ins}} = g_i$, then for any $\alpha \in E$

(i) $$N_{E\to F}(\alpha) = \prod_{i=1}^{r} \{N_{\tilde{E}_i\to\tilde{F}}(\mu_i\alpha)\}^{g_i}$$

(ii) $$S_{E\to F}(\alpha) = \sum_{i=1}^{r} g_i S_{\tilde{E}_i\to\tilde{F}}(\mu_i\alpha)$$

(iii) $$f\left(\alpha, \frac{E}{F}\right) = \prod_{i=1}^{r} \left\{f\left(\mu_i\alpha, \frac{\tilde{E}_i}{\tilde{F}}\right)\right\}^{g_i}$$

(iv) $$n = \sum_{i=1}^{r} g_i n_i$$

Proof. It is implicit in the statement of the theorem that $E \cap \tilde{F} = F$; this involves no loss of generality.

Let R denote the unique extension of $\tilde{P}$ to Ω. We recall that a composition map μ for E determines an extension $Q_\mu = \mu^*(R)$ of P to E. Let $S_\mu = i^*_{\tilde{F}(\mu E)\to\Omega}(R)$, so that S_μ is the unique extension of $\tilde{P}$ to $\tilde{F}(\mu E)$.

Assert. $\langle \tilde{F}(\mu E), S_\mu, \mu \rangle$ is a completion of $\langle E, Q_\mu \rangle$.

Proof. Since $\langle \tilde{F}(\mu E), S_\mu, i \rangle$ is a finite extension of $\langle \tilde{F}, \tilde{P} \rangle$ and $\tilde{F}$ is $\tilde{P}$-complete, it follows from (2-1-2) that $\tilde{F}(\mu E)$ is S_μ-complete. Clearly $\tilde{F}$ is the closure F^c of F in $\tilde{F}(\mu E)$, and $(\mu E)^c \supset \tilde{F}$ because $\mu E \supset F$. Furthermore, $(\mu E)^c$ is a field, and $(\mu E)^c \supset \tilde{F}(\mu E)$. It follows that $\tilde{F}(\mu E) = (\mu E)^c$ and that μE is dense in $\tilde{F}(\mu E)$. From $\mu = i_{\tilde{F}(\mu E)\to\Omega} \circ \mu$

we conclude that

$$\mu^*(S_\mu) = \mu^*[i^*_{\tilde{F}(\mu E)\to\Omega}(R)] = \mu^*(R) = Q_\mu$$

and the assertion is proved.

Assert. If μ_1, μ_2 are composition maps for E, then

$$\mu_1 \sim \mu_2 \Leftrightarrow Q_{\mu_1} = Q_{\mu_2}$$

Proof. $\Rightarrow$: By hypothesis, there exists $\sigma \in \mathcal{G}(\Omega/\tilde{F})$ such that $\sigma \circ \mu_1 = \mu_2$. Both $\langle \Omega, R, i\rangle$ and $\langle \Omega, \sigma^*(R), i\rangle$ are, therefore, extensions of $\langle \tilde{F}, \tilde{P}\rangle$. By uniqueness, $\sigma^*(R) = R$, and consequently

$$Q_{\mu_2} = \mu_2^*(R) = \mu_1^*\sigma^*(R) = \mu_1^*(R) = Q_{\mu_1}$$

$\Leftarrow$: For $i = 1, 2$, $\langle \tilde{F}(\mu_i E), S_{\mu_i}, \mu_i\rangle$ is a completion of $\langle E, Q_{\mu_1} = Q_{\mu_2}\rangle$. Hence, by uniqueness of the completion, there exists an isomorphism $\rho:\tilde{F}(\mu_1 E) \rightarrowtail\!\!\!\twoheadrightarrow \tilde{F}(\mu_2 E)$ such that $\rho^*(S_{\mu_2}) = S_{\mu_1}$ and $\rho \circ \mu_1 = \mu_2$. Now, ρ is a homeomorphism, and $\tilde{F}$ is the closure of F in both $\tilde{F}(\mu_1 E)$ and $\tilde{F}(\mu_2 E)$. Since $\rho|F$ is the identity, $\rho|\tilde{F}$ is also the identity map; hence $\mu_1 \sim \mu_2$.

It remains to show that, if Q is an extension of P to E, then $Q = Q_\mu$ for some composition map μ. Let $\langle \bar{E}, \bar{Q}, i\rangle$ be a Q completion of E. Let $\bar{F}$ denote the closure of F in $\bar{E}$, and let $\bar{P}$ be the projection of $\bar{Q}$ on $\bar{F}$. Thus, $\langle \bar{F}, \bar{P}, i\rangle$ is a completion of $\langle F, P\rangle$. By uniqueness of the completion, there exists an F isomorphism $\rho:\bar{F} \rightarrowtail\!\!\!\twoheadrightarrow \tilde{F}$ such that $\rho^*(\tilde{P}) = \bar{P}$.

Furthermore, we have $[E \cdot \bar{F}:\bar{F}] < \infty$ because $[E:F] < \infty$. Therefore, $E \cdot \bar{F}$ is complete in the topology of $i^*_{E\bar{F}\to\bar{E}}(\bar{Q})$. In particular, $E \cdot \bar{F}$ is a complete subset of $\bar{E}$ in the metric topology determined by $\bar{Q}$; hence, $E \cdot \bar{F}$ is closed in $\bar{E}$. On the other hand, E is dense in $\bar{E}$; so $E \cdot \bar{F}$ is dense in $\bar{E}$, and $E \cdot \bar{F} = \bar{E}$.

Since $[\bar{E}:\bar{F}] < \infty$, ρ may be extended to an F isomorphism $\tau:\bar{E} \to \Omega$. Putting $\mu = \tau|E$, we see that μ is a composition map for $\{E, F, \tilde{F}, \Omega\}$. Because $\tau(\bar{E}) = \tau(\bar{F} \cdot E) = \tilde{F}(\mu E)$, we may write $\tau^{-1}:\tilde{F}(\mu E) \rightarrowtail\!\!\!\twoheadrightarrow \bar{E}$. From $\tau^{-1} \circ i_{\tilde{F}\to\tilde{F}(\mu E)} = i_{\bar{F}\to\bar{E}} \circ \rho^{-1}$ it follows that

$$i^*(\tau^{-1})^*(\bar{Q}) = (\rho^{-1})^*i^*(\bar{Q}) = (\rho^{-1})^*(\bar{P}) = \tilde{P}$$

which means that $(\tau^{-1})^*(\bar{Q})$ is an extension of $\tilde{P}$ to $\tilde{F}(\mu E)$. By uniqueness, $(\tau^{-1})^*(\bar{Q}) = S_\mu$; and in virtue of $\mu = \tau \circ i_{E\to\bar{E}}$ we have

$$Q_\mu = \mu^*(S_\mu) = i^*_{E\to\bar{E}}\tau^*(S_\mu) = i^*_{E\to\bar{E}}(\bar{Q}) = Q$$

The remaining assertions of the theorem have already been proved.

The last result was stated for prime divisors, but it is clear that the applicable parts are valid for valuations. Thus, if E/F is a finite extension and φ is a valuation of F belonging to the prime divisor P of F, then the number of extensions of φ to E is equal to the number of equivalence classes of composition maps for $\{E, F, \tilde{F}, \Omega\}$. In particular, if $Q_{\mu_1}, \ldots, Q_{\mu_r}$ are the extensions of P to E, then φ has r extensions $\{\varphi_i \in Q_{\mu_i} | i = 1, \ldots, r\}$ to E—and these may be given explicitly. More precisely, if $\tilde{\varphi} \in \tilde{P}$ is the unique extension of φ to $\tilde{F}$, then $\tilde{\varphi}$ has a unique extension $\tilde{\varphi}_i$ to $\tilde{F}(\mu_i E) = \tilde{E}_i$ $(i = 1, \ldots, r)$ and the formula for $\tilde{\varphi}_i$ is known. When $\tilde{\varphi}_i$ is carried back to E via the monomorphism μ_i, we get a valuation $\varphi_i = \tilde{\varphi}_i \circ \mu_i$ which is an extension of φ and which is given by the formula

$$\varphi_i(\alpha) = [\tilde{\varphi}\{N_{\tilde{E}_i \to \tilde{F}}(\mu_i \alpha)\}]^{1/n_i} \qquad \alpha \in E$$

This formula holds in both the archimedean and nonarchimedean cases.

In practice, it is customary to suppress reference to the μ_i in the above formula, and also in (i), (ii), and (iii) of the theorem. One often writes F_P for $\tilde{F}$, E_{Q_i} for $\tilde{E}_i$ and $n(Q_i/P)$ for n_i; $n(Q_i/P)$ is called the ***local degree*** of E/F at Q_i. This is in conformity with a standard convention according to which objects associated with the complete layers E_{Q_i}/F_P are said to be "local" and objects associated with E/F are said to be "global." Thus, when $g_i = 1$ $(i = 1, \ldots, r)$ (which is true for algebraic number fields because E/F is always separable) (i), (ii), (iii), and (iv) of the theorem may be expressed as follows:

i. The global norm is the product of the local norms.

ii. The global trace is the sum of the local traces.

iii. The global field polynomial is the product of the local field polynomials.

iv. The global degree is the sum of the local degrees.

When we write F_P for $\tilde{F}$ and E_{Q_i} for $\tilde{E}_i = \tilde{F}(\mu_i E)$, the prime divisor of F_P will still be denoted by $\tilde{P}$, while the prime divisor S_{μ_i} of E_{Q_i} will be denoted by $\tilde{Q}_i$. Because μ_i is then an inclusion map, $\langle E_{Q_i}, \tilde{Q}_i, i\rangle$ is a completion of $\langle E, Q_i\rangle$.

2-4-7. Example. Suppose that we wish to find all the prime divisors of the gaussian field $\mathbf{Q}(\sqrt{-1}) = \mathbf{Q}(i)$. In view of (2-4-5) and (2-4-6) this involves examining the irreducible factorization of $f(\sqrt{-1}, \mathbf{Q}) = x^2 + 1$ over $\mathbf{Q}_p$ for all prime divisors p of $\mathbf{Q}$.

If $p = \infty$, then $\mathbf{Q}_\infty = \mathbf{R}$ and $x^2 + 1$ is irreducible over the reals. Hence, $\mathbf{Q}(\sqrt{-1})$ has a unique archimedean prime which contains an extension Φ_∞ of the absolute value on $\mathbf{Q}$; Φ_∞ is given by

$$\Phi_\infty(a + bi) = \sqrt{a^2 + b^2} \qquad a, b \in \mathbf{Q}$$

If $p = 2$, then $x^2 + 1$ is irreducible over $\mathbf{Q}_2$. This follows from an application of Exercise 1-6 to $(x + 1)^2 + 1 = x^2 + 2x + 2$ and will be discussed in more detail in Section 3-3. Thus, the normed valuation φ_2 of $\mathbf{Q}$ has a unique extension (as does the prime divisor to which it belongs) to $\mathbf{Q}(\sqrt{-1})$; it is given by

$$\Phi_2(a + bi) = \sqrt{\varphi_2(a^2 + b^2)} \qquad a, b \in \mathbf{Q}$$

If p is odd, then in virtue of Hensel's lemma we have

$$\begin{aligned} x^2 + 1 \text{ is irreducible over } \mathbf{Q}_p &\Leftrightarrow x^2 + 1 \text{ is irreducible over } \bar{\mathbf{Q}}_p \\ &\Leftrightarrow -1 \text{ is a quadratic nonresidue} \pmod p \\ &\Leftrightarrow p \equiv 3 \pmod 4 \end{aligned}$$

Therefore, for such p, φ_p has a unique extension given by

$$\Phi_p(a + bi) = \sqrt{\varphi_p(a^2 + b^2)}$$

On the other hand, for $p \equiv 1 \pmod 4$, $x^2 + 1$ has two distinct roots $\pm\zeta$ in $\mathbf{Q}_p$, and φ_p has two extensions $\Phi_p^{(1)}$ and $\Phi_p^{(2)}$ given by

$$\Phi_p^{(1)}(a + bi) = \tilde{\varphi}_p(a + b\zeta) \qquad \Phi_p^{(2)}(a + bi) = \tilde{\varphi}_p(a - b\zeta)$$

2-4-8. Exercise. Find all the prime divisors of $\mathbf{Q}(\sqrt{2})$ and $\mathbf{Q}(\sqrt{-3})$.

2-5. Consequences

This section is devoted to some easy and unrelated consequences of the results of the preceding section. The previous notation remains in force.

2-5-1. Proposition. Let E/F be a finite extension, let P be a prime divisor of F, and suppose that $Q_1, \ldots, Q_r$ are the extensions of P to E. Given $\varphi \in P$, we may choose $\varphi_{Q_i} \in Q_i$ so that

$$\prod_{i=1}^{r} \varphi_{Q_i}(\alpha) = \varphi\{N_{E\to F}(\alpha)\}$$

for all $\alpha \in E$.

Proof. For each $i = 1, \ldots, r$ we have an extension $\varphi_i \in Q_i$ of φ given by

$$\varphi_i(\alpha) = \tilde{\varphi}\{N_{\bar{E}_i\to\bar{F}}(\mu_i\alpha)\}^{1/n_i} \qquad \alpha \in E$$

Put $\varphi_{Q_i} = \varphi_i^{g_i n_i} \in Q_i$; then

$$\prod_1^r \varphi_{Q_i}(\alpha) = \tilde{\varphi}\left\{\prod_1^r N_{\tilde{E}_i \to \tilde{F}}(\mu_i \alpha)^{g_i}\right\}$$
$$= \tilde{\varphi}\{N_{E\to F}(\alpha)\}$$
$$= \varphi\{N_{E\to F}(\alpha)\}$$

The purpose of this result will appear in Chapter 5. Here we may note that, although such a φ_{Q_i} is not, as a rule, an extension of φ, it often takes a rather natural form. More precisely, suppose (as is the case when P is any nonarchimedean prime divisor of the rational field $\mathbf{Q}$) that P is discrete, that the residue class field is finite with $\mathfrak{N}P$ elements, and that all the g_i are 1. If we choose $\varphi \in P$ to be normalized [as defined in (1-9-7)], then each of the φ_{Q_i} is also normalized—that is,

$$\varphi_{Q_i}(\alpha) = (\mathfrak{N}Q_i)^{-\nu_{Q_i}(\alpha)} \qquad \alpha \in E$$

In fact,
$$\begin{aligned}\varphi_{Q_i}(\alpha) &= \tilde{\varphi}[N_{E_{Q_i}\to F_P}(\alpha)] \\ &= (\mathfrak{N}\tilde{P})^{-\nu_{\tilde{P}}(N_{E_{Q_i}\to F_P}(\alpha))} \qquad \text{(since } \tilde{\varphi} \in \tilde{P} \text{ is normalized)} \\ &= (\mathfrak{N}\tilde{P})^{-f(\tilde{Q}_i/\tilde{P})\nu_{\tilde{Q}_i}(\alpha)} = (\mathfrak{N}\tilde{Q}_i)^{-\nu_{\tilde{Q}_i}(\alpha)} \\ &= (\mathfrak{N}Q_i)^{-\nu_{Q_i}(\alpha)}\end{aligned}$$

When P is an archimedean prime divisor of F, we say, in virtue of the theorem of Ostrowski, that P is ***real*** or ***complex*** according as F may be viewed as the real field $\mathbf{R}$ or the complex field $\mathbf{C}$, respectively.

2-5-2. Proposition. Suppose that P is an archimedean prime divisor of F and that E/F is a finite extension of degree n.

If P is complex, then it has n extensions $Q_i = Q_{\mu_i}$ to E. Each Q_i is complex, and $n_i = n(Q_i/P) = 1$.

If P is real, then it has $r = r_1 + r_2$ extensions $Q_i = Q_{\mu_i}$ to E. These may be ordered so that $Q_1, \ldots, Q_{r_1}$ are real and $Q_{r_1+1}, \ldots, Q_{r_1+r_2}$ are complex. Moreover,

$$n_i = n\left(\frac{Q_i}{P}\right) = \begin{cases} 1 \text{ for } i = 1, \ldots, r_1 \\ 2 \text{ for } i = r_1 + 1, \ldots, r_1 + r_2 \end{cases}$$

and $r_1 + 2r_2 = n$.

Proof. F has characteristic 0; so E/F is separable, and all $g_i = 1$. In particular,

$$n = \sum_{Q \supset P} n\left(\frac{Q}{P}\right)$$

We may take Ω to be $\mathbf{C}$ with prime divisor determined by φ_∞. Since

E/F is separable, the number of elements in $\Gamma(F, E \to \Omega)$ is equal to $[E:F]$. We recall also that, for $\mu_1, \mu_2 \in \Gamma(F, E \to \Omega)$, $\mu_1 \sim \mu_2$ if and only if there exists $\sigma \in \mathcal{G}(\Omega/\tilde{F})$ with $\sigma \circ \mu_1 = \mu_2$.

If P is complex, then $\mathcal{G}(\Omega/\tilde{F}) = \{1\}$, so that $\mu_1 \sim \mu_2$ if and only if $\mu_1 = \mu_2$. The assertions about this case are now immediate.

If P is real, then $\mathcal{G}(\Omega/\tilde{F})$ is the cyclic group $\{1, \sigma\}$ of order 2 where the operation of σ is complex conjugation. Therefore,

$$\mu_1 \sim \mu_2 \Leftrightarrow \mu_1 = \mu_2 \text{ or } \sigma\mu_1 = \mu_2 \Leftrightarrow \mu_1 = \mu_2 \text{ or } \bar{\mu}_1 = \mu_2$$

Hence, we may choose the representatives $\mu_1, \ldots, \mu_r$ for the equivalence classes of composition maps for $\{E, F, \tilde{F}, \Omega\}$ so that $\mu_1, \ldots, \mu_r$ represent classes with one element, while the maps $\mu_{r_1+1}, \ldots, \mu_{r_1+r_2}$ represent classes which contain two elements. In other words, the equivalence classes are

$$\{\mu_1 = \bar{\mu}_1\}, \ldots, \{\mu_{r_1} = \bar{\mu}_{r_1}\}, \{\mu_{r_1+1}, \bar{\mu}_{r_1+1}\}, \ldots, \{\mu_{r_1+r_2}, \bar{\mu}_{r_1+r_2}\}$$

The assertions for this case now follow easily.

It may be noted that if P is a real archimedean prime divisor of F and $\varphi \in P$ is the absolute value (where F is viewed as a subfield of $\mathbf{R}$), then, for the complex primes $Q_{r_1+1}, \ldots, Q_{r_1+r_2}$ of E lying over P, each of the valuations $\varphi_{Q_i} \in Q_i$, as defined in (2-5-1), is the square of the absolute value (where E is viewed as a subfield of $\mathbf{C}$).

2-5-3. Example. Consider the field $E = \mathbf{Q}(\alpha)$, where $\alpha^4 = 2$. (One avoids confusion by viewing α as an abstract element not belonging to $\mathbf{C}$.) Since $x^4 - 2$ is irreducible over $\mathbf{Q}$ and its irreducible factorization over $\mathbf{Q}_\infty = \mathbf{R}$ is

$$x^4 - 2 = (x^2 + \sqrt{2})(x - \sqrt[4]{2})(x + \sqrt[4]{2})$$

we see that $[E:\mathbf{Q}] = 4$ and that the archimedean prime of $\mathbf{Q}$ has three extensions Q_1, Q_2, Q_3 to E with $n(Q_1/\infty) = 2$, $n(Q_2/\infty) = 1$ and $n(Q_3/\infty) = 1$. There are four composition maps described by

$$\mu_1:\alpha \to i\sqrt[4]{2} \qquad \mu_1':\alpha \to -i\sqrt[4]{2}$$
$$\mu_2:\alpha \to \sqrt[4]{2} \qquad \mu_3:\alpha \to -\sqrt[4]{2}$$

for some choice of $\sqrt[4]{2} \in \mathbf{R}$ (say, $\sqrt[4]{2} > 0$), and these fall into three equivalence classes, since $\mu_1 \sim \mu_1'$. For each $i = 1, 2, 3$ there is a $\Phi_i \in Q_i$ which extends the absolute value on $\mathbf{Q}$; a simple illustration of what these look like is given by

$$\begin{aligned} \Phi_1(1+\alpha) &= \sqrt{1+\sqrt{2}} \qquad \text{(positive square roots)} \\ \Phi_2(1+\alpha) &= |1+\sqrt[4]{2}| \\ \Phi_3(1+\alpha) &= |1-\sqrt[4]{2}| \end{aligned}$$

2-5-4. Proposition. Suppose that P is a nonarchimedean prime divisor of F and that E/F is a finite extension. If $Q_1, \ldots, Q_r$ are the extensions of P to E, then

$$O = \bigcap_{i=1}^{r} O_{Q_i}$$

is the integral closure of O_P in E.

Proof. Suppose that $\alpha \in E$ is integral over O_P; so α satisfies an equation of form

$$(*) \qquad -\alpha^m = a_1\alpha^{m-1} + \cdots + a_m$$

with all $a_i \in O_P$. Fix $\varphi \in P$, and let $\varphi_i \in Q_i$ denote its extensions to E. If any $\varphi_i(\alpha) > 1$, then applying φ_i to the equation $(*)$ leads to a contradiction. Therefore, $\varphi_i(\alpha) \leq 1$ for $i = 1, \ldots, r$ and $\alpha \in O$.

Suppose, on the other hand, that $\alpha \in O$. Now $\alpha \in O_{Q_i}$ implies that $\mu_i\alpha \in O_{\bar{Q}_i}$, and according to (2-2-10) this means that $\mu_i\alpha$ is integral over $O_{\bar{P}}$. In particular, $f(\mu_i\alpha, F_P)$ has coefficients in $O_{\bar{P}}$; and therefore the field polynomial $f(\mu_i\alpha, E_{Q_i}/F_P)$, which is a power of the minimum polynomial $f(\mu_i\alpha, F_P)$, belongs to $O_{\bar{P}}[x]$. Hence, the monic polynomial $f(\alpha, E/F) = \Pi f(\mu_i\alpha, E_{Q_i}/F_P)^{g_i}$ belongs to $O_{\bar{P}}[x] \cap F[x] = O_P[x]$, and α is integral over O_P.

2-5-5. Proposition. In the situation of (2-4-6) suppose that E/F is a Galois extension and that P is nonarchimedean. Then, for every $\sigma \in \mathcal{G} = \mathcal{G}(E/F)$, σ^* permutes the set $\{Q_1, \ldots, Q_r\}$, and if any Q_i and Q_j are given, there exists $\sigma \in \mathcal{G}$ such that $\sigma^* Q_j = Q_i$. Moreover, for all $i, j = 1, \ldots, r$ we have

$$e\left(\frac{Q_i}{P}\right) = e\left(\frac{Q_j}{P}\right) \qquad f\left(\frac{Q_i}{P}\right) = f\left(\frac{Q_j}{P}\right) \qquad n\left(\frac{Q_i}{P}\right) = n\left(\frac{Q_j}{P}\right)$$

Proof. Fix $\varphi_P \in P$, and for each i let φ_{Q_i} be the valuation in Q_i which extends φ_P. We observe immediately that for any $\sigma \in \mathcal{G}$ we have $\sigma^* Q_i \in \{Q_1, \ldots, Q_r\}$ and

$$(*) \qquad \varphi_{Q_i} \circ \sigma = \varphi_{\sigma^* Q_i}$$

Thus, $\sigma^*Q_i = \sigma^*Q_j \Rightarrow \varphi_{Q_i} \circ \sigma = \varphi_{Q_j} \circ \sigma \Rightarrow \varphi_{Q_i} = \varphi_{Q_j} \Rightarrow i = j$, and σ^* is indeed a permutation of $\{Q_1, \ldots, Q_r\}$.

We may also note that for any $\alpha \in E$ and any i,

$$\varphi_P[N_{E\to F}(\alpha)] = \prod_{\sigma \in \mathcal{G}} \varphi_{\sigma^* Q_i}(\alpha)$$

In fact,

$$\begin{aligned} \varphi_P[N_{E\to F}(\alpha)] &= \varphi_{Q_i}[N_{E\to F}(\alpha)] \\ &= \prod_{\sigma \in \mathcal{G}} \varphi_{Q_i}(\sigma\alpha) \\ &= \prod_{\sigma \in \mathcal{G}} \varphi_{\sigma^* Q_i}(\alpha) \end{aligned}$$

Suppose now that $i \neq j$ are such that $Q_i \notin \{\sigma^*Q_j | \sigma \in \mathcal{G}\}$, and consider the relation

$$(**) \qquad \prod_{\sigma \in \mathcal{G}} \varphi_{\sigma^* Q_i}(\alpha) = \prod_{\sigma \in \mathcal{G}} \varphi_{\sigma^* Q_j}(\alpha) \qquad \alpha \in E$$

By the approximation theorem, there exists an element $\beta \in E$ such that $\varphi_{Q_i}(\beta) > 1$, $\varphi_{Q_t}(\beta) < 1$ for $t \neq i$ and $\Pi\varphi_{Q_t}(\beta) > 1$. This β makes the left side of (**) greater than 1 and the right side less than 1, a contradiction.

The remaining assertions are now immediate. In fact, if $\sigma^*Q_j = Q_i$, then $\varphi_{Q_i}(E^*) = \varphi_{Q_j}(\sigma E^*) = \varphi_{Q_j}(E^*)$ and σ maps O_{Q_i} and $\mathfrak{Q}_i$ isomorphically onto O_{Q_j} and $\mathfrak{Q}_j$, respectively.

2-5-6. Exercise. The assertions of (2-5-5) remain valid (in so far as they have meaning) when P is archimedean and also when E/F is an infinite Galois extension.

2-5-7. Proposition. Suppose that P is a nonarchimedean prime divisor of F and that E/F is a finite extension of degree n. Then

$$\sum_{Q \supset P} g\left(\frac{Q}{P}\right) e\left(\frac{Q}{P}\right) f\left(\frac{Q}{P}\right) \leq \sum_{Q \supset P} g\left(\frac{Q}{P}\right) n\left(\frac{Q}{P}\right) = n$$

In particular,

$$\sum_{Q \supset P} e\left(\frac{Q}{P}\right) f\left(\frac{Q}{P}\right) = n \Leftrightarrow \begin{cases} g\left(\frac{Q}{P}\right) = 1 & \forall Q \supset P \\ e\left(\frac{Q}{P}\right) f\left(\frac{Q}{P}\right) = n\left(\frac{Q}{P}\right) & \forall Q \supset P \end{cases}$$

Proof. Let Q denote any extension of P to E. We know that $n(Q/P) = [E_Q:F_P]$, that $\langle F_P, \tilde{P}, i\rangle$ is a completion of $\langle F, P\rangle$, and that

$\langle E_Q, \tilde{Q}, i\rangle$ is a completion of $\langle E, Q\rangle$. Since $e(\tilde{P}/P) = f(\tilde{P}/P) = 1$ and $e(\tilde{Q}/Q) = f(\tilde{Q}/Q) = 1$, it follows from the multiplicative character of e and f that

$$e\left(\frac{Q}{P}\right) = e\left(\frac{\tilde{Q}}{\tilde{P}}\right) \quad \text{and} \quad f\left(\frac{Q}{P}\right) = f\left(\frac{\tilde{Q}}{\tilde{P}}\right)$$

Therefore, $n(Q/P) \geq e(\tilde{Q}/\tilde{P})f(\tilde{Q}/\tilde{P}) = e(Q/P)f(Q/P)$, and our assertions hold.

2-5-8. Proposition. Suppose that P is a discrete prime divisor of F and that E/F is a finite extension of degree n. Then

(i) Every extension of P to E is discrete

(ii) $$n = \sum_{Q \supset P} g\left(\frac{Q}{P}\right) e\left(\frac{Q}{P}\right) f\left(\frac{Q}{P}\right)$$

Furthermore, the normalized exponential valuations are related by

(iii) $$\nu_Q(a) = e\left(\frac{Q}{P}\right) \nu_P(a) \qquad a \in F$$

(iv) $$\nu_P[N_{E\to F}(\alpha)] = \sum_{Q \supset P} g\left(\frac{Q}{P}\right) f\left(\frac{Q}{P}\right) \nu_Q(\alpha) \qquad \alpha \in E$$

Proof. i. Let Q be any extension of P to E; then P discrete on $F \Rightarrow \tilde{P}$ discrete on $F_P \Rightarrow \tilde{Q}$ discrete on $E_Q \Rightarrow Q$ discrete on E.

ii. This is immediate from (2-5-7) because $n(Q/P) = [E_Q:F_P] = e(\tilde{Q}/\tilde{P})f(\tilde{Q}/\tilde{P}) = e(Q/P)f(Q/P)$.

iii. $\nu_Q(a) = \nu_{\tilde{Q}}(a) = e(\tilde{Q}/\tilde{P})\nu_{\tilde{P}}(a) = e(Q/P)\nu_P(a)$.

iv. Identifying the various $\mu_i\alpha$ with α, we have

$$\begin{aligned}
\nu_P[N_{E\to F}(\alpha)] &= \nu_{\tilde{P}}[N_{E\to F}(\alpha)] \\
&= \nu_{\tilde{P}}\left[\prod_{Q\supset P}\{(N_{E_Q\to F_P}(\alpha))^{g(Q/P)}\}\right] \\
&= \sum_{Q\supset P} g\left(\frac{Q}{P}\right) \nu_{\tilde{P}}[N_{E_Q\to F_P}(\alpha)] \\
&= \sum_{Q\supset P} g\left(\frac{Q}{P}\right) f\left(\frac{\tilde{Q}}{\tilde{P}}\right) \nu_{\tilde{Q}}(\alpha) \\
&= \sum_{Q\supset P} g\left(\frac{Q}{P}\right) f\left(\frac{Q}{P}\right) \nu_Q(\alpha)
\end{aligned}$$

2-5-9. Corollary. Suppose that P is a nonarchimedean prime divisor of the algebraic number field F and that E/F is a finite extension of degree n. If $Q_1, \ldots, Q_r$ are the extensions of P to E, then

$$\text{(i)} \qquad \sum_1^r e\left(\frac{Q_i}{P}\right) f\left(\frac{Q_i}{P}\right) = n$$

$$\text{(ii)} \qquad \nu_{Q_i}(a) = e\left(\frac{Q_i}{P}\right) \nu_P(a) \qquad a \in F$$

$$\text{(iii)} \qquad \nu_P[N_{E \to F}(\alpha)] = \sum_1^r f\left(\frac{Q_i}{P}\right) \nu_{Q_i}(\alpha) \qquad \alpha \in E$$

2-5-10. Exercise. Suppose that $f(x) \in \mathbf{Z}[x]$ is a monic polynomial of degree n which is irreducible over $\mathbf{Q}$. Let α be a root of $f(x)$, and consider the algebraic number field $E = \mathbf{Q}(\alpha)$ of degree n over $\mathbf{Q}$. Consider a prime divisor p of $\mathbf{Q}$ and its extensions $Q_1, \ldots, Q_r$ to E.

i. If $p = \infty$, then the $n(Q_i/p)$ are the degrees of the irreducible factors of $f(x)$ over $\mathbf{Q}_\infty = \mathbf{R}$.

ii. If $p \neq \infty$ and all the irreducible factors of $\bar{f}(x)$ over $\bar{\mathbf{Q}}_p$ are distinct, then their degrees are the $f(Q_i/p)$, so that all $e(Q_i/p) = 1$ and $\Sigma f(Q_i/p) = n$.

iii. Find the $n(Q_i/p)$, $e(Q_i/p)$, and $f(Q_i/p)$ for $f(x) = x^3 - 2$ and $p = \infty$, 5, 7, 11, 13, 17, 19, 23, 29, 31. In general, if $p > 3$, then all e's $= 1$. What about $p = 2$ or 3? A good deal of information about questions of this type will be developed in Chapter 3.

2-5-11. Remark. Let us place ourselves in the situation of (2-4-6) and consider the tensor product $\tilde{F} \otimes_F E$. [The standard facts about tensor products may be found in N. Bourbaki's "Algèbre multilinèaire," Hermann & Cie, Paris, 1948.] The mapping $a \to a \otimes 1$ for $a \in \tilde{F}$ is an isomorphism of $\tilde{F}$ into $\tilde{F} \otimes E$. We identify $\tilde{F}$ with its image, and in the same way we identify E with its image under the isomorphism $\alpha \to 1 \otimes \alpha$ for $\alpha \in E$. Therefore, $\tilde{F}E = \tilde{F} \otimes E$, $\tilde{F} \cap E = F$, and $\tilde{F} \otimes E$ is a commutative algebra with 1 over $\tilde{F}$. We recall that $[\tilde{F} \otimes E:\tilde{F}] = [E:F] = n$—in fact, if $\omega_1, \ldots, \omega_n$ is a basis for E/F, then, under our identifications, $\omega_1, \ldots, \omega_n$ is a basis for $\tilde{F} \otimes E$ over $\tilde{F}$. Note that for $\tilde{F} \otimes E$ its ideals as a ring are identical with its ideals as an algebra. Moreover, in $\tilde{F} \otimes E$ every proper prime ideal is maximal. In fact, if $\mathfrak{P}$ is a proper prime ideal, then $(\tilde{F} \otimes E)/\mathfrak{P}$ becomes a finite-dimensional commutative algebra (over $\tilde{F}$) with 1 and with no zero divisors. It follows easily that $(\tilde{F} \otimes E)/\mathfrak{P}$ is a field, so that $\mathfrak{P}$ is maximal.

Suppose that μ is a composition map for $\{E, F, \tilde{F}, \Omega\}$. The map $1 \times \mu : \tilde{F} \times E \to \Omega$ given by $(1 \times \mu)(a, \alpha) = a\mu(\alpha)$ is bilinear over F; hence, it determines a unique F-linear map (which we denote by μ) $\mu = 1 \otimes \mu : \tilde{F} \otimes E \to \Omega$ such that $\mu(a \otimes \alpha) = a\mu(\alpha)$. This map μ is an algebra homomorphism, and its image is clearly $\tilde{F}(\mu E)$; consequently, the kernel is a prime (maximal) ideal $\mathfrak{P}_\mu$ and $(\tilde{F} \otimes E)/\mathfrak{P}_\mu \approx \tilde{F}(\mu E)$. If $\mu_1 \sim \mu_2$, then there exists $\rho : \tilde{F}(\mu_1 E) \longmapsto\!\!\!\!\to \tilde{F}(\mu_2 E)$ such that $\rho | \tilde{F}$ is the identity and $\rho \circ \mu_1 = \mu_2$. It is immediate that $\mathfrak{P}_{\mu_1} = \mathfrak{P}_{\mu_2}$. Thus, an equivalence class of composition maps determines a maximal ideal of $\tilde{F} \otimes E$.

Suppose next that $\mathfrak{P}$ is a maximal ideal of $\tilde{F} \otimes E$. Then $\tilde{F} \cap \mathfrak{P} = E \cap \mathfrak{P} = \{0\}$ and $(\tilde{F} \otimes E)/\mathfrak{P}$ is a finite extension field of $\tilde{F}$. We may view $(\tilde{F} \otimes E)/\mathfrak{P}$ as a subfield of Ω. The map $\mu : E \to \Omega$ given by $\mu(\alpha) = (1 \otimes \alpha) + \mathfrak{P}$ is clearly an F isomorphism, so that $\mu E \subset \tilde{F} \otimes E$, μ is a composition map, $(\tilde{F} \otimes E)/\mathfrak{P} = \tilde{F}(\mu E)$, and $\mathfrak{P}_\mu = \mathfrak{P}$. We have shown that the maximal (prime) ideals of $\tilde{F} \otimes E$ are in 1-1 correspondence with the equivalence classes of composition maps for $\{E, F, \tilde{F}, \Omega\}$.

Let $\{\mathfrak{P}_i = \mathfrak{P}_{\mu_i} | i = 1, \ldots, r\}$ be the set of all maximal ideals of $\tilde{F} \otimes E$. We have then isomorphic $\tilde{F}$ algebras

$$\frac{\tilde{F} \otimes E}{\overset{r}{\underset{1}{\bigcap}} \mathfrak{P}_i} \approx \frac{\tilde{F} \otimes E}{\mathfrak{P}_1} \oplus \cdots \oplus \frac{\tilde{F} \otimes E}{\mathfrak{P}_r} \approx E_{Q_1} \oplus \cdots \oplus E_{Q_r}$$

Observe that $\overset{r}{\underset{1}{\bigcap}} \mathfrak{P}_i$ is the radical N of $\tilde{F} \otimes E$. We conclude that

$$g_i = 1 \text{ for } i = 1, \ldots, r \Leftrightarrow n = \sum_1^r n_i \Leftrightarrow N = (0)$$
$$\Leftrightarrow \tilde{F} \otimes E \text{ is semisimple}$$

EXERCISES

2-1. Let P be a nonarchimedean prime divisor of F; then O_P is integrally closed in its quotient field F. In fact, for any collection of nonarchimedean prime divisors P, $O = \cap O_P$ is integrally closed in F.

2-2. Suppose that the existence of an extension in the complete nonarchimedean case has been proved, as was done in (2-2-10). Give a direct proof of uniqueness which does not use any results from Section 2-1. Of course, this method does not give information about the relation between the topologies of the extension field and the base field.

2-3. Let F be complete with respect to the nonarchimedean prime divisor P, let E/F be an extension of degree n, and let p be the characteristic of $\tilde{F}$.

a. If q is a prime number, then $q|e \Longrightarrow q|n$.

b. For any $\alpha \in O_E$, $\deg f(\bar{\alpha}, \bar{F})$ divides $\deg f(\alpha, F)$.

c. If q is a prime number, then $q|f \Longrightarrow q|n$.

d. If E/F is purely inseparable, then $n = efp^\delta$ for some integer $\delta \geq 0$.

e. Suppose that n is a prime q not equal to p; then E/F is separable, and we may write $E = F(\alpha)$, where $\alpha \in O_E$ and $f(\alpha, F) = x^q + a_1x^{q-1} + \cdots + a_q \in O_F[x]$ has $a_1 = 0$. Furthermore, $ef = q$.

f. In the general case, $n = efp^\delta$ for some integer $\delta \geq 0$. This is known as ***Ostrowski's defect theorem.*** (*Hint:* Use Galois theory and elementary facts about Sylow subgroups.)

2-4. Suppose that we are in the situation of (2-3-2).

a. What is the O_E ideal generated by $\mathcal{P}_F$—that is, what is $\mathcal{P}_FO_E$? What is $\mathcal{P}_FO_E \cap O_F$? What happens if $\mathcal{P}_F$ is replaced by an arbitrary O_F ideal $\mathfrak{A}$?

b. If $\mathfrak{A}$ is an arbitrary O_E ideal, what can be said about $\mathfrak{A} \cap O_F$ and $(\mathfrak{A} \cap O_F)O_E$?

c. If $\mathfrak{A}$ is an O_E ideal, what is the O_F ideal generated by $\{N_{E\to F}(\alpha)|\alpha \in \mathfrak{A}\}$? Call this ideal $N_{E\to F}(\mathfrak{A})$. If we write $\mathfrak{A} = \alpha O_E$, then $N_{E\to F}(\mathfrak{A}) = (N_{E\to F}\alpha)O_F$.

2-5. We know that in $\mathbf{Q}_5$ there are two square roots of -1 [see (2-2-8)]; find the first six terms in their power series expansions.

2-6. Let E be a finite extension of $F = \mathbf{Q}_p$, and let ν be the unique extension of $\nu_F = \nu_{\bar{p}}$ to E.

a. The series

$$\exp x = 1 + x + \frac{x^2}{2!} + \cdots + \frac{x^n}{n!} + \cdots$$

converges for all $x \in E$ such that $\nu(x) > 1/(p-1)$. [*Hint:* Express $\nu(n!)$ in terms of the coefficients in the p-adic expansion of n.] When this is so, $\nu(x^n/n!) > \nu(x)$ for $n \geq 2$, so that $\nu(x) = \nu(\exp x - 1)$.

b. If both $\nu(x)$ and $\nu(y)$ are $>1/(p-1)$, then

$$\exp(x + y) = (\exp x)(\exp y)$$

c. The series

$$\log(1 + x) = x - \frac{x^2}{2} + \frac{x^3}{3} + \cdots + \frac{(-1)^{n+1}x^n}{n} + \cdots$$

converges for all $x \in E$ such that $\nu(x) > 0$. If $\nu(x) > 1/(p-1)$, then $\nu(x^n/n) > \nu(x)$ for $n \geq 2$, so that $\nu(x) = \nu[\log(1 + x)]$.

d. For $x \equiv 1 \pmod{\mathcal{P}_E}$ we put $\log x = \log(1 + (x - 1))$. If both x and y are $\equiv 1 \pmod{\mathcal{P}_E}$, then

$$\log(xy) = \log x + \log y$$

e. If $\nu(x) > 1/(p-1)$, then

$$\exp \log(1 + x) = 1 + x$$
$$\log \exp x = x$$

In particular, for $r \geq \left[\frac{e}{p-1}\right] + 1$, log and exp provide inverse isomorphisms between $U_r = 1 + \mathcal{P}^r$ and $\mathcal{P}^r$. Of course, we know [from (1-5-3)] that, for $r \geq 1$, $U_r/U_{r+1} \approx \mathcal{P}^r/\mathcal{P}^{r+1}$.

f. The kernel of the homomorphism $x \to \log x$ for $x \equiv 1 \pmod{\mathfrak{P}}$ consists of the p^sth roots of unity in E (where $s = 0, 1, 2, \ldots$).

2-7. Give a direct proof of (2-4-5)—that is, without making use of (2-4-4).

2-8. In the situation of (2-4-6) we have

$$[E:F]_{\text{sep}} \leq \sum_{Q \supset P} n\left(\frac{Q}{P}\right) \leq [E:F]$$

2-9. Suppose that E/F is normal; then, in the situation of (2-4-6), all the g_i are equal.

2-10. Let $k = \{0, 1\}$ be the two-element field, and put $K = k(x)$, where x is an indeterminate. Let P be the prime divisor of K corresponding to the irreducible polynomial x. Denote the completion of $\langle K, P\rangle$ by $\langle \tilde{K}, \tilde{P}\rangle$; so $\tilde{K} = K_P = k\langle x\rangle$ is the field of formal power series in x over k. There exists in $\tilde{K}$ an element of form

$$u = \sum_{0}^{\infty} a_r x^{2r} \qquad a_r \in k$$

which is transcendental over K. Let $F = k(x, u) = K(u)$, and denote the restriction of $\tilde{P}$ to F by P_F. If we put $E = F(\sqrt{u})$, then $[E:F] = 2$ and P_F has a unique extension to E with $e = 1, f = 1, g = 2$. In particular, this is an example in which not all g's equal 1.

3

Local Fields

One way to get information about algebraic number fields is to study their various completions; this is a technique which will be applied repeatedly, and in keeping with it, the present chapter is directed toward the examination of the complete fields which arise in this way. Thus, because the facts in the archimedean case are well known, our concern is, in particular, with a field of characteristic 0, complete with respect to a discrete prime divisor and having finite residue class field, and also with the structure of finite extensions of such fields. However, because many of the results one wants carry over to the case of an arbitrary field complete with respect to a non-archimedean prime divisor, there is no need for such restrictive hypotheses.

Let us fix the basic hypotheses and notation that will remain in force throughout this chapter. Let F denote a field complete with respect to the discrete prime divisor P. Thus our hypotheses are sufficiently general to include the complete fields that occur in algebraic function theory. Choose φ and $\nu \in P$, and fix an algebraic closure Ω of F. All algebraic extensions of F will be taken to be subfields of Ω. Since there is no danger of confusion, the unique extensions of φ and ν to Ω or to any intermediate field E will also be denoted by φ and ν, respectively. It will be convenient to index objects associated with such a field E by using the subscript E—thus it is clear what is meant by O_E, $\mathcal{P}_E$, or U_E. If $[E:F] < \infty$, the normalized exponential valuation of E will be denoted by ν_E. In virtue of the standard identification, the same symbol ψ (or $^-$) may be used for the residue class map $O_E \rightarrow O_E/\mathcal{P}_E = \bar{E}$ for all intermediate fields E. Naturally, we put $e(E/F) = (\varphi(E^*):\varphi(F^*)) = (\nu(E^*):\nu(F^*))$ and $f(E/F) = [\bar{E}:\bar{F}]$.

3-1. Newton's Method

Consider a polynomial $f(x) = a_0 + a_1x + \cdots + a_nx^n \in F[x]$, and assume, for simplicity, that $a_0a_n \neq 0$. One way to get information about the factorization of $f(x)$ over F is to make it primitive and then make use of Hensel's lemma. An alternative and sometimes useful procedure (which we are about to describe) is based on Newton's polygon.

Let us associate a polygon in euclidean 2-space with the polynomial $f(x)$. To each term of $f(x)$ we assign a point in $\mathbf{R} \times \mathbf{R}$ in the following manner:

If $a_ix^i \neq 0$, take the point $(i, \nu(a_i))$
If $a_ix^i = 0$, take the nonexistent point $(i, \infty) = (i, \nu(a_i))$

Now, form the lower convex envelope of the set of points

$$\{(i, \nu(a_i)) | i = 0, \ldots, n\}$$

The polygon thus determined is called the ***Newton polygon of*** $f(x)$ (with respect to ν). One may give a precise definition of *lower convex envelope* in terms of the slopes of the line segments connecting the various points $(i, \nu(a_i))$, but rather than give the details, let us merely observe that a typical Newton's polygon might look as follows.

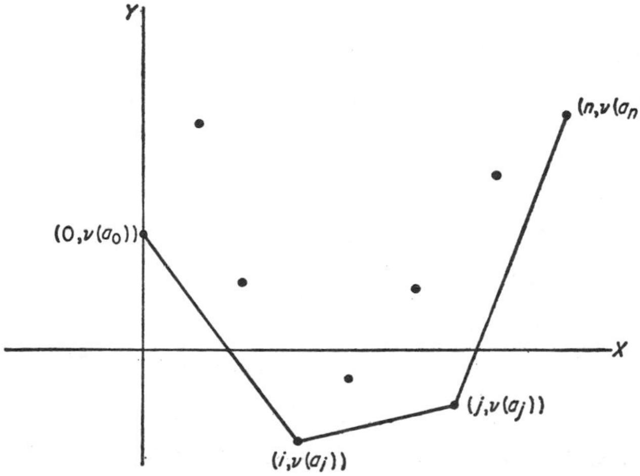

Note that as we go from left to right the slopes of the segments of the Newton polygon increase. The requirement that $a_0a_n \neq 0$ means that the polygon starts from a finite point on the Y axis and ends at a finite point.

3-1-1. Proposition. Let $f(x) = a_0 + a_1x + \cdots + a_nx^n$ with $a_0a_n \neq 0$ be a polynomial in $F[x]$. Suppose that $(r, \nu(a_r)) \leftrightarrow (s, \nu(a_s))$ with $s > r$ is any segment of the Newton polygon of $f(x)$ and that its slope is $-m$. Then $f(x)$ has exactly $s - r$ roots $\alpha_1, \ldots, \alpha_{s-r}$ with

$$\nu(\alpha_1) = \cdots = \nu(\alpha_{s-r}) = m$$

Moreover,

$$f_m(x) = \prod_{i=1}^{s-r} (x - \alpha_i)$$

is in $F[x]$, and $f_m(x)$ divides $f(x)$.

Proof. Dividing through by a_n moves the Newton polygon but does not affect any of the statements or conclusions of the theorem; hence, we may assume that $a_n = 1$.

Let $\alpha_1, \ldots, \alpha_n \in \Omega$ be the roots of $f(x)$, and index them so that

$$\begin{aligned}
\nu(\alpha_1) &= \cdots = \nu(\alpha_{s_1}) = m_1 \\
\nu(\alpha_{s_1+1}) &= \cdots = \nu(\alpha_{s_2}) = m_2 \\
&\cdots\cdots\cdots\cdots \\
\nu(\alpha_{s_t+1}) &= \cdots = \nu(\alpha_n) = m_{t+1}
\end{aligned}$$

with $m_1 < m_2 < \cdots < m_{t+1}$. We have immediately

$$\begin{aligned}
\nu(a_n) &= \nu(1) = 0 \\
\nu(a_{n-1}) &\geq \min_i \{\nu(\alpha_i)\} = m_1 \\
\nu(a_{n-2}) &\geq \min_{i_1,i_2} \{\nu(\alpha_{i_1}\alpha_{i_2})\} = 2m_1 \\
&\cdots\cdots\cdots\cdots \\
\nu(a_{n-s_1}) &= \min_{i_1,\ldots,i_{s_1}} \{\nu(\alpha_{i_1} \cdots \alpha_{i_{s_1}})\} = s_1m_1
\end{aligned}$$

(because there is exactly one term with minimal ν, namely, $\alpha_1 \cdots \alpha_{s_1}$)

$$\begin{aligned}
\nu(a_{n-s_1-1}) &\geq \min_{i_1,\ldots,i_{s_1+1}} \{\nu(\alpha_{i_1} \cdots \alpha_{i_{s_1+1}})\} = s_1m_1 + m_2 \\
\nu(a_{n-s_1-2}) &\geq \min_{1,\ldots,i_{s_1+2}} \{\nu(\alpha_{i_1} \cdots \alpha_{i_{s_1+2}})\} = s_1m_1 + 2m_2 \\
&\cdots\cdots\cdots\cdots \\
\nu(a_{n-s_2}) &= \min_{i_1,\ldots,i_{s_2}} \{\nu(\alpha_{i_1} \cdots \alpha_{i_{s_2}})\} = s_1m_1 + (s_2 - s_1)m_2
\end{aligned}$$

and this process continues in the obvious way.

What does the Newton polygon of $f(x)$ look like? It is not hard to see that its vertices (going from right to left) are

$$(n, 0),\ (n - s_1, s_1m_1),\ (n - s_2, s_1m_1 + (s_2 - s_1)m_2),\ \ldots$$

and that the slope of the segment farthest to the right is

$$\frac{0 - s_1m_1}{n - (n - s_1)} = -m_1$$

while the remaining slopes are

$$\frac{[s_1m_1 + \cdots + (s_j - s_{j-1})m_j] - [s_1m_1 + \cdots + (s_{j+1} - s_j)m_{j+1}]}{(n - s_j) - (n - s_{j+1})} = -m_{j+1}$$

for $j = 1, \ldots, t$. This proves the first assertion.

The remaining part is proved by induction on n. The case $n = 1$ is trivial, and the case $n = 2$ is quite easy because the Newton polygon of an irreducible polynomial is a single line segment. Suppose that the desired conclusion holds for all polynomials of degree $<n$ and that $f(x)$ has degree n. If we put $s_0 = 0$ and

$$f_{s_j}(x) = \prod_{i=s_j+1}^{s_{j+1}} (x - \alpha_i) \qquad j = 0, 1, \ldots, t$$

then

$$f(x) = \prod_{j=0}^{t} f_{s_j}(x)$$

Consider

$$g(x) = \frac{f(x)}{f(\alpha_1, F)} \in F[x]$$

It is clear that

$$g(x) = g_0(x) \prod_{j=1}^{t} f_{s_j}(x)$$

where $g_0(x) = f_{s_0}(x)/f(\alpha_1, F)$. From the induction hypothesis applied to $g(x)$ it follows that $g_0(x)$ and $f_{s_j}(x)$ $(j = 1, \ldots, t)$ are in $F[x]$. It is now clear that $f_{s_0}(x) \in F[x]$, and the proof is complete.

Newton's polygon gives information about the "size" of the roots of a polynomial but not about their actual location. The problem of finding good approximations to the roots may sometimes be handled on the basis of our next result.

3-1-2. Newton's Method. Suppose that $f(x) \in O_F[x]$ and that $a_0 \in O = O_F$ satisfies the relation

$$\varphi\left\{\frac{f(a_0)}{f'(a_0)^2}\right\} = r < 1 \qquad r \neq 0$$

Then a_0 can be refined to a root of $f(x)$. More precisely, the sequence

$$a_{i+1} = a_i - \frac{f(a_i)}{f'(a_i)} \qquad i = 0, 1, 2, \ldots$$

converges to a root $\alpha \in O$ of $f(x)$. Moreover,

$$\varphi(\alpha - a_0) \leq \varphi\left\{\frac{f(a_0)}{f'(a_0)}\right\} \leq r < 1$$

Proof. We show first (inductively) that

(i) $$\varphi(a_i) \leq 1$$

(ii) $$\varphi(a_{i+1} - a_i) = \varphi\left\{\frac{f(a_i)}{f'(a_i)}\right\} \leq \varphi\left\{\frac{f(a_i)}{f'(a_i)^2}\right\} \leq r^{2^i}$$

(iii) $$\varphi(a_i - a_0) \leq \varphi\left\{\frac{f(a_0)}{f'(a_0)}\right\} \leq r$$

In the course of the proof it will also appear that

$$\varphi\{f'(a_{i+1})\} = \varphi\{f'(a_i)\}$$

We have at the start $a_0 \in O$, $f(a_0) \in O$, $f'(a_0) \in O$, $\varphi\{f'(a_0)\} \leq 1$, $\varphi\{f(a_0)\} \leq r$ and $\varphi\{f(a_0)/f'(a_0)\} \leq r < 1$. Thus for $i = 0$ the conditions are trivially satisfied. Suppose then that the conditions hold for i. Since $\varphi(a_i) \leq 1$, it follows from (ii) that $\varphi(a_{i+1}) \leq 1$—so that (i) holds. Now from the Taylor expansion and the fact that both a_i and $f(a_i)/f'(a_i)$ are in O it follows that

$$\begin{aligned}\varphi\{f(a_{i+1})\} &= \varphi\left\{f\left(a_i - \frac{f(a_i)}{f'(a_i)}\right)\right\} \\ &= \varphi\left\{f(a_i) - f'(a_i)\frac{f(a_i)}{f'(a_i)} + (\text{in } O)\left(\frac{f(a_i)}{f'(a_i)}\right)^2\right\} \\ &\leq \varphi\left\{\frac{f(a_i)}{f'(a_i)}\right\}^2\end{aligned}$$

Again by Taylor's theorem we have

$$f'(a_{i+1}) = f'\left(a_i - \frac{f(a_i)}{f'(a_i)}\right) = f'(a_i) + (\text{in } O)\left(\frac{f(a_i)}{f'(a_i)}\right)$$

Therefore,
$$\frac{f'(a_{i+1})}{f'(a_i)} = 1 + (\text{in } O)\left(\frac{f(a_i)}{f'(a_i)^2}\right)$$

and from (ii) we have $\varphi\{f'(a_{i+1})/f'(a_i)\} = 1$; that is,

$$\varphi\{f'(a_{i+1})\} = \varphi\{f'(a_i)\}$$

This shows, in particular, that $f'(a_{i+1}) \neq 0$, so that a_{i+2} is well defined. Combining the various relations yields

$$\varphi\left\{\frac{f(a_{i+1})}{f'(a_{i+1})^2}\right\} \leq \frac{\varphi\{f(a_i)/f'(a_i)\}^2}{\varphi\{f'(a_i)\}^2} = \varphi\left\{\frac{f(a_i)}{f'(a_i)^2}\right\}^2 \leq r^{2^{i+1}}$$

This proves (ii). The last equation also implies that

$$\varphi\left\{\frac{f(a_{i+1})}{f'(a_{i+1})^2}\right\} \leq \varphi\left\{\frac{f(a_i)}{f'(a_i)^2}\right\}$$

and hence $\varphi\{f(a_{i+1})/f'(a_{i+1})\} \leq \varphi\{f(a_i)/f'(a_i)\}$. Now (iii) follows easily, and the construction is finished. Note that any a_i can be taken as a starting point (in the statement of Newton's method) with the given r.

According to (ii), $\{a_i\}$ is a Cauchy sequence, and by (i) $a_i \to \alpha \in O$. Since $\varphi\{f(a_i)\} \leq r^{2^i}$ we get by continuity $\varphi\{f(a_i)\} \to \varphi\{f(\alpha)\}$. The fact that $\varphi(\alpha - a_0) \leq \varphi\{f(a_0)/f'(a_0)\}$ follows from (iii). This completes the proof.

3-1-3. Example. Consider the polynomial

$$f(x) = x^3 - x^2 - 2x - 8 \in \mathbf{Q}[x]$$

By testing ± 1, ± 2, ± 4, ± 8 one sees that there is no linear factor; so $f(x)$ is irreducible over $\mathbf{Q}$. Now, view $f(x) \in \mathbf{Q}_2[x]$. The Newton polygon of $f(x) = -8 - 2x - x^2 + x^3$ taken with respect to the normalized exponential valuation $\nu = \nu_2$ is determined from the points (0, 3), (1, 1), (2, 0), (3, 0). Therefore, the Newton polygon has three distinct sides with slopes -2, -1, 0, respectively, and $f(x)$ has three distinct roots $\alpha_1, \alpha_2, \alpha_3$ with $\nu(\alpha_1) = 2$, $\nu(\alpha_2) = 1$, $\nu(\alpha_3) = 0$. We have then $f(x) = (x - \alpha_1)(x - \alpha_2)(x - \alpha_3)$ with $\alpha_1, \alpha_2, \alpha_3 \in O = O_{\frac{2}{2}}$ (the ring of integers in $\mathbf{Q}_2$). Since $\alpha_1, \alpha_2, \alpha_3 \notin \mathbf{Q}$, we can do no better than approximate the roots by rational numbers; this is done via Newton's method.

According to Newton's method, if $\nu\{f(a)\} > 2\nu\{f'(a)\}$ for some $a \in O$, then a can be refined to an element $c \in O$ such that $f(c) = 0$ and $\nu(c - a) \geq \nu\{f(a)/f'(a)\}$.

Suppose we start with $a = 0$; then $\nu\{f(0)\} = 3 > 2\nu\{f'(0)\} = 2$. Hence $a = 0$ is an admissible starting point, and the root which it leads to satisfies $\nu(c - 0) \geq \nu\{f(0)/f'(0)\} = 2$; consequently $c = \alpha_1$.

In this situation, Newton's method gives

$$\begin{aligned} a_0 &= 0 \qquad \nu(\alpha_1 - 0) \geq 2 \\ a_1 &= a_0 - \frac{f(a_0)}{f'(a_0)} = -4 \qquad \nu(\alpha_1 + 4) \geq 3 \\ a_2 &= a_1 - \frac{f(a_1)}{f'(a_1)} = -\frac{68}{27} \end{aligned}$$

and the computations now become burdensome. Thus far we know that

$$\begin{aligned} -4 &= 0 \cdot 2^0 + 0 \cdot 2^1 + 1 \cdot 2^2 + 1 \cdot 2^3 + \cdots \\ &= (\cdots 0{\downarrow}0011 \cdots 1 \cdots) \end{aligned}$$

is an approximation to α_1 good to three or more places starting from the arrow (which is the analogue of the decimal point of a real number). Hence any number whose expansion starts with

$$0 \cdot 2^0 + 0 \cdot 2^1 + 1 \cdot 2^2 + \cdots = (\cdots 0{\downarrow}001 \cdots)$$

is a likely candidate for a fresh start. Let us try

$$4 = (\cdots 0{\downarrow}00100 \cdots)$$

4 is indeed admissible and gives $\nu(\alpha_1 - 4) \geq 4$. Taking $a_0 = 4$ gives $a_1 = \frac{60}{19}$, which is not particularly good for computations. Since 4 is an approximation to α_1 good to four or more places, it is natural to try to replace 4 by $20 = (\cdots 0{\downarrow}001010 \cdots 0 \cdots)$, which does turn out to be admissible with

$$\nu(\alpha_1 - 20) \geq \nu \left\{ \frac{2^7 \cdot 59}{2 \cdot 579} \right\} = 6$$

Thus 20 as an approximation to α_1 is correct to six or more places—not a bad approximation, at that.

As for the other roots; starting with 1 as an approximation to α_3, one arrives eventually at 87, which is good to ≥ 7 places. The reader may also show that 22 as an approximation to α_2 is good to ≥ 6 places.

Now, let us give some interesting applications of Newton's method.

3-1-4. Proposition. Suppose that η is a unit in $\mathbf{Q}_p$; then, for $p \neq 2$,

$$\eta \text{ is a square in } \mathbf{Q}_p \Leftrightarrow \bar{\eta} \text{ is a quadratic residue (mod } p)$$

while, for $p = 2$

$$\eta \text{ is a square in } \mathbf{Q}_2 \Leftrightarrow \nu_2(\eta - 1) \geq 3$$

Proof. Suppose that $p \neq 2$. If $\eta \in \mathbf{Q}_p^{*^2}$, then $\eta = \varepsilon^2$ for some unit ε and $\bar{\eta} = \bar{\varepsilon}^2$. This says that $\bar{\eta}$ is a square in the p-element field $\mathbf{Z}/(p)$—that is, $\bar{\eta}$ is a quadratic residue (mod p). For the converse, consider $x^2 - \eta \in O_{\bar{p}}[x]$. By hypothesis, there exists an element $\gamma \in \mathbf{Z}/(p)$ such that

$$x^2 - \bar{\eta} = (x - \gamma)(x + \gamma)$$

Since $-\gamma \neq \gamma$, (2-2-6) says that η is a square in $O_{\bar{p}}$.

Suppose next that $p = 2$. If ε is any unit, then, from the expansion of elements of $\mathbf{Q}_p$, it is clear that $\nu_2(\varepsilon - 1) \geq 1$ and $\nu_2(\varepsilon + 1) \geq 1$. Therefore, $\nu_2(\varepsilon^2 - 1) \geq 3$ unless $\nu_2(\varepsilon - 1) = \nu_2(\varepsilon + 1) = 1$. But this case may be excluded because then (from the expansion of $\varepsilon - 1$) $\nu_2(\varepsilon - 3) \geq 2$ and $\nu_2(\varepsilon - 3) = \nu_2((\varepsilon + 1) - 4) = 1$, a contradiction. This shows that, if η is a square, then $\nu_2(\eta - 1) \geq 3$, or, to put it another way, $\eta \equiv 1 \pmod 8$. For the converse, consider

$$f(x) = x^2 - \eta \in O_{\bar{2}}[x]$$

If we take $a_0 = 1$, then

$$\nu_2[f(a_0)] \geq 3 > 2 = 2\nu_2[f'(a_0)]$$

so that, according to Newton's method, $a_0 = 1$ can be refined to a root of $f(x)$. Thus η is a square in $\mathbf{Q}_p$.

One may note that, in the proof for the case $p \neq 2$, Newton's method could be used in place of Hensel's lemma.

3-1-5. Corollary. Suppose that b is a rational integer not divisible by the prime p. If $p \neq 2$, then $b \in \mathbf{Q}_p^{*^2}$ if and only if b is a quadratic residue (mod p). If $p = 2$, then $b \in \mathbf{Q}_2^{*^2}$, if and only if $b \equiv 1 \pmod 8$.

3-1-6. Proposition. Suppose that F has characteristic p and that m is a natural number with $(m, p) = 1$ (no restriction on m if $p = 0$). If $\nu_F(m) = s$, then for $i \geq s + 1$ the map $a \to a^m$ is an isomorphism of U_i onto U_{i+s}.

Proof. With $\nu = \nu_F$ we have $U_i = 1 + \mathcal{P}^i = \{a | \nu(a - 1) \geq i\}$. It is quite easy to check that $a \to a^m$ is a homomorphism into. Furthermore, the kernel consists of all the mth roots of unity which belong to U_i. Suppose that $\zeta \neq 1$ is in the kernel; then $\zeta \equiv 1 \pmod{\mathcal{P}^i}$, and even more $\zeta^j \equiv 1 \pmod{\mathcal{P}^i}$ for $j = 1, \ldots, m - 1$. Since $\zeta^{m-1} + \zeta^{m-2} + \cdots + \zeta + 1 = 0$, we have $m \equiv 0 \pmod{\mathcal{P}^i}$. This means that $\nu(m) \geq i > s$, a contradiction, so that our map is a monomorphism.

To show that the map is onto, consider any $b \in U_{i+s}$; then take $f(x) = x^m - b$, and apply Newton's method with $a_0 = 1$. The conclusion is that there exists $a \in O$ with $f(a) = a^m - b = 0$ and $\nu(a-1) \geq \nu \left\{\frac{f(a_0)}{f'(a_0)}\right\} = \nu\left[\frac{(1-b)}{m}\right] \geq i$; thus, $a \in U_i$.

3-2. Unramified Extensions

In this section we initiate the study of the connections between a finite extension E/F and its residue class field extension $\bar{E}/\bar{F}$.

3-2-1. Proposition. $\bar{\Omega}$ is an algebraic closure of $\bar{F}$.

Proof. Any element of $\bar{\Omega}$ is of form $\bar{\alpha}$ for some $\alpha \in O_\Omega$. Because $\bar{\alpha} \in \overline{F(\alpha)}$ and $[\overline{F(\alpha)}:\bar{F}] = f(F(\alpha)/F) \leq [F(\alpha):F] < \infty$, it is clear that $\bar{\alpha}$ is algebraic over $\bar{F}$. Thus, $\bar{\Omega}$ is algebraic over $\bar{F}$.

On the other hand, any monic polynomial in $\bar{\Omega}[x]$ may be written as $\bar{f}(x)$ for some monic polynomial $f(x) \in O_\Omega[x]$. Since Ω is algebraically closed, we have a factorization of form

$$f(x) = (x - \alpha_1) \cdots (x - \alpha_n) \qquad \alpha_i \in \Omega$$

Now, each α_i is integral over O_Ω; and because O_Ω is integrally closed in Ω, we have $\alpha_i \in O_\Omega$. Consequently, $\bar{f}(x) = (x - \bar{\alpha}_1) \cdots (x - \bar{\alpha}_n)$ with $\bar{\alpha}_i \in \bar{\Omega}$, and $\bar{\Omega}$ is algebraically closed.

Henceforth, whenever we deal with algebraic extensions of $\bar{F}$, it is to be understood that they are contained in $\bar{\Omega}$.

3-2-2. Lemma. An arbitrary element of $\bar{\Omega}$ is of form $\bar{\alpha}$ for some $\alpha \in O_\Omega$; and α may be chosen in such a way that

(i) $\overline{F(\alpha)} = \bar{F}(\bar{\alpha})$

(ii) $f\left(\frac{F(\alpha)}{F}\right) = [F(\alpha):F] \qquad e\left(\frac{F(\alpha)}{F}\right) = 1$

(iii) If $f(x) = f(\alpha, F)$, then $\bar{f}(x) = f(\bar{\alpha}, \bar{F})$

Proof. We describe a procedure for making the choice of α. Suppose that $[\bar{F}(\bar{\alpha}):\bar{F}] = n$; then $f(\bar{\alpha}, \bar{F})$ is a polynomial of degree n which may be carried back to a monic polynomial $f(x)$ of degree n, $f(x) \in O_F[x]$. Thus, $\bar{f}(x) = f(\bar{\alpha}, \bar{F})$. Note that there are many possible choices for $f(x)$ and that we do not assert at this time that the one we have fixed is irreducible.

From the factorization

$$f(x) = (x - \alpha_1) \cdots (x - \alpha_n) \qquad \alpha_i \in O_\Omega$$

it follows that

$$\bar{f}(x) = (x - \bar{\alpha}_1) \cdots (x - \bar{\alpha}_n)$$

and that one of the $\bar{\alpha}_i$ must be $\bar{\alpha}$. Fix the notation so that $\alpha = \alpha_1$. From the relations

$$\begin{aligned} n = [\bar{F}(\bar{\alpha}):\bar{F}] \leq [\overline{F(\alpha)}:\bar{F}] &= f\left(\frac{F(\alpha)}{F}\right) \\ &\leq e\left(\frac{F(\alpha)}{F}\right) f\left(\frac{F(\alpha)}{F}\right) = [F(\alpha):F] \\ &\leq n \end{aligned}$$

and the fact that $[F(\alpha):F] = n$ implies that $f(x) = f(\alpha, F)$, it is immediate that this α will do.

3-2-3. Proposition. The map $E \rightarrow \bar{E}$ takes the set of all finite extensions E of F for which $f(E/F) = [E:F]$ onto the set of all finite extensions of $\bar{F}$.

Proof. If a finite extension of $\bar{F}$ is simple, then (3-2-2) shows that it is of form $\overline{F(\alpha)} = \bar{F}(\bar{\alpha})$. Since any finite extension of $\bar{F}$ arises by the adjunction of a finite number of elements from $\bar{\Omega}$, and since f is multiplicative, the proof consists of repeated applications of (3-2-2). For example, in the case of a two-story tower, we get $\overline{F(\alpha)} = \bar{F}(\bar{\alpha})$ at the first stage, and when the second element is adjoined,

$$\bar{F}(\bar{\alpha}, \bar{\beta}) = \overline{F(\alpha)}(\bar{\beta}) = \overline{F(\alpha, \beta)}$$

Moreover, $f(F(\alpha)/F) = [F(\alpha):F]$ and $f(F(\alpha, \beta)/F(\alpha)) = [F(\alpha, \beta):F(\alpha)]$ imply that $f(F(\alpha, \beta)/F) = [F(\alpha, \beta):F]$.

Is it possible to select a class of finite extensions E/F for which the mapping $E \rightarrow \bar{E}$ is 1-1? For this it is convenient to make a definition. A finite extension E/F is said to be ***unramified*** when

(i) $$f\left(\frac{E}{F}\right) = [E:F]$$

(ii) $$\frac{\bar{E}}{\bar{F}} \text{ is separable}$$

Of course, the condition $f(E/F) = [E:F]$ is equivalent to $e(E/F) = 1$; for certain generalizations, where ef need not be equal to $[E:F]$, it is necessary to state the condition as we have done. It will become clear from the discussion why the separability of the residue class field extension (which holds for the local fields of algebraic number theory) is essential.

Since separability is transitive, and since field degree and residue class degree are both multiplicative, we have the following proposition.

3-2-4. Proposition. Suppose that $F \subset E \subset K \subset \Omega$ and $[K:F] < \infty$; then K/F is unramified $\Leftrightarrow$ both K/E and E/F are unramified.

A somewhat curious result, which we shall use in order to characterize unramified extensions, is the following lemma.

3-2-5. Lemma (Krasner). Consider $\alpha \in \Omega$, and put

$$r = \min_{\sigma} \{\varphi[\sigma(\alpha) - \alpha]\} \qquad \sigma \neq 1 \in \Gamma(F, F(\alpha) \to \Omega)$$

If $\beta \in \Omega$ is such that $\varphi(\beta - \alpha) < r$, then

$$\Gamma(F(\beta), F(\beta, \alpha) \to \Omega) = \{1\}$$

If, in addition, α is separable over F, then $F(\alpha) \subset F(\beta)$.

Proof. The elements $\sigma(\alpha)$ for $\sigma \neq 1 \in \Gamma(F, F(\alpha) \to \Omega)$ are the conjugates of α (other than itself) over F—so $r > 0$. Suppose that $\tau \in \Gamma(F(\beta), F(\beta, \alpha) \to \Omega)$. Since elements of Ω which are conjugate over F have the same value, so that in particular $\varphi \circ \tau = \varphi$, we have

$$\varphi(\beta - \alpha) = \varphi[\tau(\beta - \alpha)] = \varphi[\beta - \tau(\alpha)]$$

Therefore,

$$\varphi[\alpha - \tau(\alpha)] = \varphi[\alpha - \beta + \beta - \tau(\alpha)] \leq \max \{\varphi(\alpha - \beta), \varphi[\beta - \tau(\alpha)]\} < r$$

and τ must be the identity map.

If, in addition, α is separable over F, then $F(\beta, \alpha)$ is separable over $F(\beta)$, so that $[F(\beta, \alpha):F(\beta)]$ is equal to the number of elements in $\Gamma(F(\beta), F(\beta, \alpha) \to \Omega)$. Therefore, $F(\beta, \alpha) = F(\beta)$, and $F(\alpha) \subset F(\beta)$.

3-2-6. Theorem. A finite extension E/F is unramified $\Leftrightarrow E = F(\alpha)$, where α is a root of a monic polynomial $f(x) \in O_F[x]$ such that $\bar{\alpha}$ is a simple root of $\bar{f}(x)$. Moreover, for such an α we have

(i) $\bar{E} = \bar{F}(\bar{\alpha})$

(ii) If $[E:F] = n$, then $\{1, \alpha, \alpha^2, \ldots, \alpha^{n-1}\}$ is an integral basis for E/F—that is, $O_E = O_F \oplus O_F\alpha \oplus \cdots \oplus O_F\alpha^{n-1} = O_F[\alpha]$.

Proof. $\Rightarrow$: Since the extension $\bar{E}/\bar{F}$ is separable, $\bar{E}$ is generated over $\bar{F}$ by a single element from $\bar{\Omega}$. Carrying over the results and notation of (3-2-2) bodily, we observe that $\bar{E} = \bar{F}(\bar{\alpha}) = \overline{F(\alpha)}$, that $F(\alpha)/F$ is unramified, and that

$$[F(\alpha):F] = f\left(\frac{F(\alpha)}{F}\right) = [\overline{F(\alpha)}:\bar{F}] = [\bar{E}:\bar{F}] = [E:F]$$

We wish to show that $E = F(\alpha)$. There exists $\beta \in O_E$ such that $\bar{\beta} = \bar{\alpha}$. This means that $\varphi(\beta - \alpha) < 1$. Since $\bar{\alpha}$ is separable over $\bar{F}$, we know from (3-2-2) that $\bar{\alpha}_i \neq \bar{\alpha}_j$ for $i \neq j$. Therefore, $\alpha_i - \alpha_j \in O_\Omega$, and $\overline{\alpha_i - \alpha_j} \neq 0$, so that $\varphi(\alpha_i - \alpha_j) = 1$ for $i \neq j$. An application of (3-2-5) to β and the separable element $\alpha = \alpha_1$ yields

$$F(\alpha) \subset F(\beta) \subset E$$

Consequently, $E = F(\alpha)$, α is a root of the irreducible polynomial $f(x) = f(\alpha, F) \in O_F[x]$, and $\bar{\alpha}$ is a simple root of $\bar{f}(x)$.

Of course, (i) has already been proved. As for (ii), we know that $\{1, \bar{\alpha}, \ldots, \bar{\alpha}^{n-1}\}$ constitutes a basis for $\bar{E}/\bar{F}$. Since $e = 1$, the assertion is immediate from the proof of (2-3-2) as soon as we take $\pi_0 = 1$ in that proof. It may be noted that $\{1, \beta, \ldots, \beta^{n-1}\}$ is also an integral basis for E/F.

$\Leftarrow$: It is immediate that $f(\alpha, F)$ is a monic polynomial in $O_F[x]$ and $\bar{\alpha}$ is a simple root of $\overline{f(\alpha, F)}$. Therefore, we may assume at the start that $f(x)$ is irreducible over F.

From (2-2-5) it follows that $\bar{f}(x)$ is a power of an irreducible polynomial in $\bar{F}[x]$. But $\bar{\alpha}$ is a simple root of $\bar{f}(x)$; so $\bar{f}(x)$ is irreducible. In particular, $\bar{F}(\bar{\alpha})/\bar{F}$ is separable, and since $\deg f(x) = \deg \bar{f}(x)$, we have

$$[E:F] = [F(\alpha):F] = [\bar{F}(\bar{\alpha}):\bar{F}] \leq [\overline{F(\alpha)}:\bar{F}] = [\bar{E}:\bar{F}] = f\left(\frac{E}{F}\right) \leq [E:F]$$

The consequent equalities imply that E/F is unramified. This completes the proof.

In virtue of this result, it is easy to construct unramified extensions. For example, consider any $\mathbf{Q}_p$, and let ζ be a primitive nth root of unity where $(n, p) = 1$; then the extension $\mathbf{Q}_p(\zeta)/\mathbf{Q}_p$ is unramified.

3-2-7. Corollary. An unramified extension E/F is separable.

Proof. In the first part of the preceding proof we had $\overline{\alpha_i - \alpha_j} \neq 0$ for $i \neq j$. Hence, $\alpha_i \neq \alpha_j$, and $E = F(\alpha)$ is separable over F.

3-2-8. Corollary. Suppose that E/F is unramified and that K/F is a finite extension; then KE/K is unramified. If, in addition, K/F is unramified, then KE/F is unramified.

Proof. According to (3-2-6), $E = F(\alpha)$, where α is a root of a monic polynomial $f(x) \in O_F[x]$ and $\bar{\alpha}$ is a simple root of $\bar{f}(x)$. Now $KE = K(\alpha)$ is a finite extension of K, and (3-2-6) applied over the complete field K shows that KE/K is unramified. [We do not know that $f(x)$ is irreducible over K; in fact, this is why the statement of (3-2-6) says nothing about the irreducibility of $f(x)$.] The remaining part is now a corollary of (3-2-4).

3-2-9. Corollary. Suppose that E/F is unramified, and let K/F be a finite extension; then

$$E \subset K \Leftrightarrow \bar{E} \subset \bar{K}$$

Proof. $\Rightarrow$: Trivially true in general.

$\Leftarrow$: This, too, is substantially included in the proof of the first part of (3-2-6). Namely, we have $E = F(\alpha)$ in the canonical way, and choosing $\beta \in O_K$ such that $\bar{\beta} = \bar{\alpha}$ leads to

$$E = F(\alpha) \subset F(\beta) \subset K$$

3-2-10. Theorem. Any finite extension E/F has a unique maximal unramified subfield T. It contains every unramified subfield and is called the ***inertia field*** (*Trägheitskörper*) of E/F.

Proof. $\bar{E}/\bar{F}$ is a finite extension, and from (3-2-3) it follows that the separable part of $\bar{E}$ over $\bar{F}$ is of form $\bar{T}$, where T/F is unramified. According to (3-2-9), $T \subset E$. Furthermore, if T'/F is any unramified extension with $T' \subset E$, then $\overline{T'} \subset \bar{E}$; and since $\overline{T'}/\bar{F}$ is separable, $\overline{T'} \subset \bar{T}$. Thus, another application of (3-2-9) gives $T' \subset T$.

3-2-11. Theorem. The mapping $E \rightarrow \bar{E}$ is a lattice isomorphism from the lattice of all unramified extensions of F onto the lattice of all finite separable extensions of $\bar{F}$.

Proof. By the term ***lattice*** we mean a partially ordered set in which any two elements have a join and a meet. Here, of course, the join (meet) of two fields is their compositum (intersection). A lattice isomorphism is a map which is 1-1 onto and preserves join and meet.

From (3-2-8), (3-2-4), and properties of separability, it follows easily that the sets in question are indeed lattices. We already know that the map is onto, and (3-2-9) says that it is 1-1. In addition, we know that the map $E \rightarrow \bar{E}$ and its inverse are both order (that is, inclusion) preserving—this is the full force of (3-2-9). These facts guarantee that $E \rightarrow \bar{E}$ preserves joins and meets; symbolically

$$\overline{E_1E_2} = \overline{E_1} \cdot \overline{E_2} \qquad \text{and} \qquad \overline{E_1 \cap E_2} = \overline{E_1} \cap \overline{E_2}$$

This completes the proof.

The reader who wishes may verify that the preceding results are valid when P is an arbitrary nonarchimedean prime divisor.

3-2-12. Proposition. Suppose that $\bar{F}$ has $q = p^r$ elements; then:

i. The lattice of unramified extensions of F is isomorphic to the lattice of all finite extensions of $\bar{F}$.

ii. For every positive integer f there exists a unique unramified extension T/F of degree f. It is of form $T = F(\zeta)$, where ζ is any primitive $(q^f - 1)$st root of unity.

iii. If m is a positive integer such that $(m, p) = 1$ and $\zeta \in \Omega$ is a primitive mth root of unity, then $T = F(\zeta)$ is an unramified extension of F. Furthermore, $[T{:}F] = f(T/F)$ is the smallest positive integer f such that $q^f \equiv 1 \pmod{m}$, and $\{1, \zeta, \ldots, \zeta^{f-1}\}$ is an integral basis for T/F.

Proof. Since $\bar{F}$ is a perfect field, (i) is simply a restatement of (3-2-11). Furthermore, for each integer $f > 0$, $\bar{F}$ has a unique extension of degree f; hence, F has a unique unramified extension T/F of degree f.

Suppose that $m > 0$ is such that $(m, p) = 1$; in particular, the remarks which follow apply for $m = q^f - 1$. Since F has characteristic 0 or p, there exist primitive mth roots of unity in Ω; choose any one, and call it ζ. Applying (3-2-6) for ζ and the polynomial $f(x) = x^m - 1 \in O_F[x]$, we see [since $(p, m) = 1$ guarantees that $\bar{\zeta}$ is a simple root of $\bar{f}(x)$] that $T = F(\zeta)$ is an unramified extension of F and that $\{1, \zeta, \ldots, \zeta^{f(T/F)-1}\}$ is an integral basis for T/F. It may also be noted that the equations

$$f(x) = x^m - 1 = \prod_0^{m-1} (x - \zeta^\nu)$$
$$\bar{f}(x) = x^m - 1 = \prod_0^{m-1} (x - \bar{\zeta}^\nu)$$

say that $1, \bar{\zeta}, \ldots, \bar{\zeta}^{m-1}$ are the m distinct mth roots of unity in $\bar{\Omega}$; so $\bar{\zeta}$ is a primitive mth root of unity.

In the case where $m = q^f - 1$, it is clear that $\bar{T} = \overline{F(\zeta)} = \bar{F}(\bar{\zeta})$ is the field with q^f elements, so that $f(T/F) = [\bar{T}{:}\bar{F}] = f$. This proves (ii). It may be remarked that this case could also be handled by starting with the extension of $\bar{F}$ of degree f, choosing a primitive $(q^f - 1)$st root of unity in $\bar{\Omega}$, and working backward to get T.

Now let us turn to the proof of (iii). It remains to determine $[T{:}F] = f(T/F) = [\bar{T}{:}\bar{F}]$. We recall that $\bar{T}/\bar{F}$ is a cyclic extension whose Galois group $\mathcal{G}(\bar{T}/\bar{F})$ is generated by the automorphism

$$\rho{:}\xi \to \xi^q \qquad \forall \xi \in \bar{T}$$

Thus $\rho(\bar{\zeta}) = \bar{\zeta}^q$, and because $\bar{T} = \bar{F}(\bar{\zeta})$, an element of $\mathcal{G}(\bar{T}/\bar{F})$ is completely determined as soon as its action on $\bar{\zeta}$ is known. Since the order of $\rho = \#(\mathcal{G}(\bar{T}/\bar{F})) = f(T/F)$, $f(T/F)$ is the smallest positive integer f such that $\rho^f = 1 \in \mathcal{G}(\bar{T}/\bar{F})$—that is, such that $\bar{\zeta}^{q^f} = \bar{\zeta}$. This means that $\bar{\zeta}^{q^f-1} = 1$, so that f is the smallest integer such that $m|(q^f - 1)$.

3-3. Totally Ramified Extensions

At the other end of the spectrum from the unramified extensions is the class of finite extensions E/F which have the property that $e(E/F) = [E:F]$. Such extensions are said to be ***totally ramified,*** or ***fully ramified,*** or ***purely ramified.*** No hypothesis on the separability of the residue class field extension is necessary, since E/F is totally ramified $\Leftrightarrow f(E/F) = 1 \Leftrightarrow \bar{E} = \bar{F}$. It is also immediate that a tower of extensions is totally ramified $\Leftrightarrow$ each of the steps is totally ramified.

The characterization of totally ramified extensions is not particularly difficult, and it is conveniently expressed in terms of the following definition: A polynomial $f(x) = x^n + a_1x^{n-1} + \cdots + a_n \in O_F[x]$ is said to be an ***Eisenstein polynomial*** when

$$\nu_F(a_n) = 1 \qquad \text{and} \qquad \nu_F(a_i) \geq 1 \text{ for } i = 1, \ldots, n-1$$

It has already been pointed out (see Exercise 1-6) that the notion of an Eisenstein polynomial may be defined for the case of an arbitrary nonarchimedean prime divisor and that such a polynomial is irreducible. We do not need this fact, because the irreducibility of an Eisenstein polynomial when the prime divisor is discrete is a byproduct of the next theorem.

3-3-1. Theorem. i. Suppose that E/F is a totally ramified extension. Let ω be any prime element of E, and let $f(x)$ be its minimum polynomial over F; then $f(x)$ is Eisenstein and $E = F(\omega)$.

ii. Suppose that $f(x) \in F[x]$ is an Eisenstein polynomial. Let $\omega \in \Omega$ be a root of $f(x)$, and put $E = F(\omega)$; then E/F is totally ramified, ω is a prime element of E, and $f(x)$ is irreducible.

Proof. i. Let us write

$$f(x) = x^m + a_1x^{m-1} + \cdots + a_{m-1}x + a_m$$

Clearly $m \leq [E:F] = e(E/F) = e$. Now $\nu_E(\omega) = 1$ implies $\omega \in O_E$. Thus ω is integral over O_F, which implies that each $a_i \in O_F$—that is,

$\nu_F(a_i) \geq 0$. We have then

$$(*) \qquad \omega^m + a_1\omega^{m-1} + \cdots + a_{m-1}\omega + a_m = 0 \qquad a_i \in O_F$$

According to (1-3-4) at least two of the terms on the left side of (*) have the same minimal value for ν_E. By making use of the fact that for $a \in F$ we have $\nu_E(a) = e\nu_F(a)$, the values of the terms in (*) are

$$\{m, e\nu_F(a_1) + m - 1, \ldots, e\nu_F(a_{m-1}) + 1, e\nu_F(a_m)\}$$

Since the residues of these values (mod e) are $\{m, m - 1, \ldots, 1, 0\}$ and since $e \geq m > 1$, it follows that the minimum must be achieved at the end points and nowhere else—that is, for $i = 1, \ldots, m - 1$,

$$e\nu_F(a_i) + m - i > m = e\nu_F(a_m)$$

In particular, $e \leq m$, so that $m = e = [E:F]$ and $E = F(\omega)$. Furthermore, $\nu_F(a_m) = 1$, and $\nu_F(a_i) \geq 1$ for $i = 1, \ldots, m - 1$; that is, $f(x)$ is Eisenstein.

ii. We write $f(x) = x^n + a_1x^{n-1} + \cdots + a_{n-1}x + a_n$ with $\nu_F(a_i) \geq 1$ for $i = 1, \ldots, n - 1$ and $\nu_F(a_n) = 1$. Consider the equation

$$\omega^n + a_1\omega^{n-1} + \cdots + a_{n-1}\omega + a_n = 0$$

with $\nu_E(\omega) \geq 0$. The set of values ν_E for the terms of the left side is

$$\{n\nu_E(\omega), e\nu_F(a_1) + (n - 1)\nu_E(\omega), \ldots, e\nu_F(a_{n-1}) + \nu_E(\omega), e\}$$

Excluding the first term $n\nu_E(\omega)$, it is clear that e is the minimum of the remaining terms. Since these $n + 1$ terms cannot have a unique minimum, it follows that $n\nu_E(\omega) \neq 0$, and so $\nu_E(\omega) \geq 1$. The non-uniqueness of the minimum now implies that $n\nu_E(\omega) = e \leq [E:F] \leq n$. Therefore, $\nu_E(\omega) = 1$, $e = n$, and $f(x)$ is irreducible.

3-3-2. Corollary. Suppose that E/F is a totally ramified extension of degree n and that ω is a prime element of E; then the set of elements $\{1, \omega, \omega^2, \ldots, \omega^{n-1}\}$ is an integral basis for E/F.

Proof. Immediate from the proof of (2-3-2).

3-3-3. Proposition. Suppose that E/F is a finite extension and that $\bar{E}/\bar{F}$ is separable; then E/F consists of an unramified extension followed by a totally ramified extension. Moreover, if $[E:F] = n$, there exists an element $\beta \in U_E$ such that $\{1, \beta, \beta^2, \ldots, \beta^{n-1}\}$ is an integral basis for E/F.

Proof. Since $\bar{E}/\bar{F}$ is separable, we have $\bar{T} = \bar{E}$, where T is the unique maximal unramified subfield in the extension E/F, and $\bar{E} = \bar{T}$ says that E/T is totally ramified. Thus, $e(E/T) = e(E/F)$, $f(E/T) = 1$, $e(T/F) = 1$, and $f(T/F) = f(E/F)$.

Now let us turn to the question of an integral basis. According to the results of Section 3-2 there exists an element $\alpha \in U_T$ such that $T = F(\alpha)$, $\bar{T} = \bar{F}(\bar{\alpha}) = \bar{E}$, $f(x) = f(\alpha, F) \in O_F[x]$, $\bar{f}(x) = f(\bar{\alpha}, \bar{F})$, $\bar{\alpha}$ is a simple root of $\bar{f}(x)$ [since $\bar{f}' = \overline{f'}$ and $\overline{f'(\alpha)} = \overline{f'}(\bar{\alpha})$, this is equivalent to saying that $\nu_T[f'(\alpha)] = 0$], and $\{1, \alpha, \ldots, \alpha^{f(T/F)-1}\}$ is an integral basis for T/F. Note that it will not be necessary to consider the cases $T = F$ (in which case there is no harm in taking $\alpha = 1$) and $T = E$ (in which case $\beta = \alpha$ will do) separately.

Take any prime element π_E of E, and put

$$\beta = \alpha + \pi_E$$

Since $\nu_E(\alpha) = 0$, we have $\nu_E(\beta) = 0$ and $\beta \in U_E$. Because $\bar{\beta} = \bar{\alpha}$ and $f(T/F) = f(E/F)$, we know that the elements $1, \bar{\beta}, \ldots, \bar{\beta}^{f(E/F)-1}$ form a basis for $\bar{E}$ over $\bar{F}$. Furthermore, from $f(\alpha) = 0$, $\nu_E[f'(\alpha)] = 0$, and the Taylor expansion

$$f(\beta) = f(\alpha) + \pi_E f'(\alpha) + \pi_E^2 \text{ (in } O_E)$$

it follows that $\nu_E[f(\beta)] = 1$.

The proof of (2-3-2) implies that the set

$$\{\beta^i f(\beta)^j | i = 0, 1, \ldots, f(E/F) - 1, j = 0, 1, \ldots, e(E/F) - 1\}$$

is an integral basis for E/F. Therefore,

$$O_E = \sum_{i,j} O_F \beta^i f(\beta)^j \subset O_F[\beta] \subset O_F + O_F\beta + \cdots + O_F\beta^{n-1} \subset O_E$$

We conclude that $\{1, \beta, \ldots, \beta^{n-1}\}$ spans E/F and is indeed an integral basis for E/F.

3-3-4. Exercise. Suppose that β is the element constructed in the proof of (3-3-3). If $\theta \in E$ is such that $\nu_E(\theta - \beta) \geq 2$, then the elements $1, \theta, \ldots, \theta^{n-1}$ form an integral basis for E/F. Does this assertion carry over for an arbitrary $\beta \in U_E$ for which $\{1, \beta, \ldots, \beta^{n-1}\}$ is an integral basis?

3-4. Tamely Ramified Extensions

In this section, let p denote the characteristic of $\bar{F}$. The finite extension E/F is said to be ***tamely ramified*** when

(i) $$\frac{\bar{E}}{\bar{F}} \text{ is separable}$$

(ii) $$p \nmid e\left(\frac{E}{F}\right) = e$$

If condition (ii) is replaced by $p|e$, the extension is said to be ***wildly ramified.*** It is understood, of course, that, when $p = 0$, condition (ii) is trivially satisfied.

3-4-1. Proposition. i. An unramified extension is tamely ramified.

ii. A tower of extensions is tamely ramified if and only if each of the steps is tamely ramified.

iii. E/F is both fully and tamely ramified if and only if $p \nmid e = [E:F]$.

Proof. Immediate from the definitions.

We have already seen that an arbitrary finite extension E/F for which $\bar{E}/\bar{F}$ is separable consists of an unramified extension T/F followed by the fully ramified extension E/T; moreover, each of the component extensions has been characterized in terms of a single generator. We shall see that the fully ramified part E/T may be split into a tamely ramified extension V/T followed by a wildly ramified extension E/V. Moreover, in order to characterize V/T in terms of a single generator, it is not necessary to know anything about totally ramified extensions.

3-4-2. Proposition. Suppose that π is any prime element of F and that $(m, p) - 1$. If ζ is any root of tho polynomial

$$f(x) = x^m - \pi$$

and we put $E = F(\zeta)$, then E/F is a totally and tamely ramified extension of degree m.

Proof. From $\nu_E(\zeta) = e/m$ it follows that $m \leq e \leq [E:F] \leq m$, and the proof is complete. The proof of the converse, however, is considerably more complicated.

3-4-3. Proposition. Suppose that E/F is both fully and tamely ramified and that $[E:F] = e = e(E/F)$; then

$$E = F(\sqrt[e]{\pi})$$

where $\sqrt[e]{\pi}$ is an eth root of some prime element π of F.

Proof. Fix π_E and π_F; so there exists $\mu \in U_E$ such that

$$\pi_E^e = \pi_F \mu$$

Since $\bar{E} = \bar{F}$, there exists $u \in U_F$ such that $\bar{u} = \bar{\mu}$—that is, $\nu_E(\mu - u) > 0$. We may therefore write

$$\pi_E^e = \pi_F u + \pi_F(\mu - u)$$

Suppose that we can prove the following general result.

3-4-4. Lemma. If $a \in F^*$ and $(m, p) = 1$, then the polynomial

$$f(x) = x^m - a$$

has distinct roots $\alpha_1, \ldots, \alpha_m \in \Omega$, and

$$\nu(\alpha_i - \alpha_j) = \nu(\alpha_1) \qquad i \neq j, \quad i, j = 1, \ldots, m$$

Suppose further that $\alpha \in E$ (where E is any finite extension of F) is such that $\nu_E(\alpha) > \nu_E(a)$. If the polynomial

$$g(x) = x^m - a - \alpha$$

has a root $\xi \in E$, then there exists a root $\zeta \in \{\alpha_1, \ldots, \alpha_m\}$ of $f(x)$ such that

$$F(\zeta) \subset F(\xi) \subset E \qquad \text{and} \qquad \nu_E(\zeta) = \nu_E(\xi)$$

This lemma may then be used to show that $\pi = \pi_F u$ is the element we seek; more precisely, let us apply the lemma with $m = e$, $a = \pi = \pi_F u$, $\alpha = \pi_F(\mu - u)$, $\xi = \pi_E$, and ζ = one of the choices of $\pi^{1/e}$. Then, from $\nu_E(\zeta) = 1$, it follows that $e \leq e(F(\zeta)/F)$. Combining this with $e(F(\zeta)/F) \leq [F(\zeta):F] \leq [E:F] = e$, we conclude that $F(\zeta) = F(\xi) = E$, and the proof is complete.

Proof of Lemma. $f(x)$ is separable over F, since $f'(x) = mx^{m-1}$ and $(m, p) = 1$; so its roots $\alpha_1, \ldots, \alpha_m$ are distinct. The relations $\alpha_i^m - a = 0$ imply that $\nu(\alpha_i) = (1/m)\nu(a)$ and

$$\nu(\alpha_i - \alpha_j) \geq \min\{\nu(\alpha_i), \nu(\alpha_j)\} = \nu(\alpha_1)$$

Furthermore, it follows from

$$f'(\alpha_1) = m\alpha_1^{m-1} = (\alpha_1 - \alpha_2)(\alpha_1 - \alpha_3) \cdots (\alpha_1 - \alpha_m)$$

and $\nu(m) = 0$ that

$$(m - 1)\nu(\alpha_1) = \sum_{j=2}^{m} \nu(\alpha_1 - \alpha_j) \geq (m - 1)\nu(\alpha_1)$$

Clearly, this implies that $\nu(\alpha_i - \alpha_j) = \nu(\alpha_1)$ for all $i \neq j$.

For the remaining part, observe that

$$f(\xi) = \xi^m - a = \alpha = \prod_{i=1}^{m} (\xi - \alpha_i)$$

so that, for $j = 1, \ldots, m$,

$$\sum_{1}^{m} \nu(\xi - \alpha_i) = \nu(\alpha) > \nu(a) = m\nu(\alpha_j)$$

Therefore, there exists an i such that, for $j = 1, \ldots, m$,

$$\nu(\xi - \alpha_i) > \nu(\alpha_j) = \nu(\alpha_i - \alpha_j) = \nu(\alpha_i)$$

Put this $\alpha_i = \zeta$. By (3-2-5) we have $F(\zeta) \subset F(\xi) \subset E$, and finally $\nu(\xi) = \nu[(\xi - \alpha_i) + \alpha_i] = \nu(\alpha_i) = \nu(\zeta)$.

Now, let us generalize (3-4-2) and show how tamely ramified extensions with a single generator may be constructed.

3-4-5. Proposition. Suppose that $a \in F^*$ and that $(m, p) = 1$. If α is any root of the polynomial

$$f(x) = x^m - a$$

and we put $E = F(\alpha)$, then E/F is tamely ramified.

Proof. Since $m\nu(\alpha) = \nu(a)$, we have $d|m$, where d is the order of the class of $\nu(\alpha)$ in the factor group $\nu(E^*)/\nu(F^*)$. Furthermore, there exists $b \in F^*$ such that $d\nu(\alpha) = \nu(b)$. It follows that $\beta - \alpha^d/b \in U_E$ and that β is a root of

$$g(x) = x^{m/d} - \frac{a}{b^{m/d}} \in O_F[x]$$

Making use of the fact that $(m, p) = 1$, we see that $\bar{\beta}$ is a simple root of $\bar{g}(x)$. According to (3-2-6), $F(\beta)/F$ is unramified.

Now, α is a root of the polynomial

$$h(x) = x^d - b\beta \in F(\beta)[x]$$

Thus, it follows that $E \supset F(\beta)$ and

$$d \leq e\left(\frac{E}{F(\beta)}\right) \leq [E:F(\beta)] \leq d$$

This implies that $E/F(\beta)$ is a fully ramified extension of degree d; and because $p \nmid d$, it is also tamely ramified. In particular, $h(x)$ is irreducible.

In view of (3-4-1), E/F is tamely ramified; it consists of an unramified extension followed by a fully and tamely ramified extension.

3-4-6. Proposition. Suppose that E/F is tamely ramified and that K/F is a finite extension; then KE/K is tamely ramified. If, in addition, K/F is tamely ramified, then KE/F is tamely ramified.

Proof. Let T be the inertia field of E/F. By (3-2-8), KT/K is unramified. According to (3-4-3), we can write $E = T(\sqrt[e]{\pi_T})$, where π_T is some prime element of T, $e = e(E/T) = e(E/F)$, and $p \nmid e$. Since $KE = KT(\sqrt[e]{\pi_T})$, (3-4-5) implies that KE/KT is tamely ramified. Therefore, KE/K is tamely ramified. The remaining part is now trivial.

3-4-7. Theorem. Any finite extension E/F contains a unique maximal tamely ramified subfield V. It contains every tamely ramified subfield and is called the ***ramification field*** (*Verzweigungskörper*) of E/F. The extension V/T is totally and tamely ramified, and if we write $e(E/F) = e = e_0p^r$ with $(e_0, p) = 1$ (so that $e_0 = e$ when $p = 0$), then $[V:T] = e(V/T) = e_0$.

Proof. The set of all tamely ramified subextensions of E/F is nonempty; for example, it includes T. Let V be one of maximal degree. In virtue of (3-4-6), it is immediate that V is the unique maximal tamely ramified subfield. In particular, $T \subset V$, $\bar{T} = \bar{V}$, and V/T is fully and tamely ramified. It remains to show that $[V:T] = e_0$; and for this it suffices to construct a field W between T and E such that W/T is tamely ramified and $[W:T] = e_0$.

Suppose that $\bar{E}/\bar{F}$ is separable, so that $\bar{E} = \bar{T}$ and E/T is totally ramified of degree e. If the proof of (3-4-3) is applied to E/T with e replaced by e_0, π_E by $\pi_E^{p^r}$, and F by T, then one may conclude (with the remaining notation as before) that $\nu_E(\zeta) = \nu_E(\xi)$, $T(\zeta) \subset T(\xi) \subset E$, and $e_0 \leq e(T(\zeta)/T) \leq [T(\zeta):T] \leq e_0$. Thus $W = T(\zeta) = T(\pi^{1/e_0})$ is the desired field.

The case where $\bar{E}/\bar{F}$ is inseparable may be treated by a slight variation on the technique above; note that, in this situation, $[E:V] = p^r[\bar{E}:\bar{F}]_{\text{ins}}$. The details are left to the reader.

The reader who is so inclined may also show that the basic results of this section carry over when the prime divisor is not discrete.

3-4-8. Exercise. Consider any $\mathbf{Q}_p$, and let ζ be a primitive nth root of unity, where n is a power of p, say, $n = p^s$. Show that $\mathbf{Q}_p(\zeta)/\mathbf{Q}_p$ is a totally ramified extension of degree $\phi(p^s)$ and that $(1 - \zeta)$ is a prime element. When this extension is wildly ramified, what is V?

3-5. Inertia Group

Consider the finite extension E/F and any element $\sigma \in \Gamma(F, E \to \Omega)$— so σE denotes the image field of E under σ. From extension theory, it is clear that σ provides an isomorphism of O_E onto $O_{\sigma E}$ and also of $\mathcal{P}_E$ onto $\mathcal{P}_{\sigma E}$. Consequently, σ determines an isomorphism $\bar{\sigma}$ of $O_E/\mathcal{P}_E$ onto $O_{\sigma E}/\mathcal{P}_{\sigma E}$; more precisely, we have an isomorphism

$\bar{\sigma}:\bar{E} \rightarrowtail\!\!\!\twoheadrightarrow \overline{\sigma E} \subset \bar{\Omega}$, where $\bar{\sigma}$ is defined by

$$\bar{\sigma}(\bar{\alpha}) = \overline{\sigma\alpha} \qquad \forall \alpha \in O_E$$

Of course, $\bar{\sigma}$ leaves $\bar{F}$ pointwise fixed, so that $\sigma \rightarrow \bar{\sigma}$ is a mapping of $\Gamma(F, E \rightarrow \Omega) \rightarrow \Gamma(\bar{F}, \bar{E} \rightarrow \bar{\Omega})$. This mapping is easily seen to be a homomorphism on the category of isomorphisms—which means that if $\tau \in \Gamma(F, E \rightarrow \Omega)$ and $\sigma \in \Gamma(F, \tau E \rightarrow \Omega)$ then $\overline{\sigma \circ \tau} = \bar{\sigma} \circ \bar{\tau}$. Thus, if E/F is normal [so that $\Gamma(F, E \rightarrow \Omega) = \mathcal{G}(E/F)$], then we have

$$\bar{\sigma}\bar{E} = \overline{\sigma E} = \bar{E}$$

for every $\sigma \in \mathcal{G}(E/F)$, and the mapping $\sigma \rightarrow \bar{\sigma}$ is a homomorphism of $\mathcal{G}(E/F)$ into $\mathcal{G}(\bar{E}/\bar{F})$. The purpose of this section is to study the map $\sigma \rightarrow \bar{\sigma}$.

3-5-1. Proposition. Suppose that E/F is unramified; then the mapping $\sigma \rightarrow \bar{\sigma}$ of $\Gamma(F, E \rightarrow \Omega) \rightarrow \Gamma(\bar{F}, \bar{E} \rightarrow \bar{\Omega})$ is both 1-1 and onto. Furthermore,

$$\frac{E}{F} \text{ is Galois} \Leftrightarrow \frac{\bar{E}}{\bar{F}} \text{ is Galois}$$

and in this situation

$$\mathcal{G}\left(\frac{E}{F}\right) \approx \mathcal{G}\left(\frac{\bar{E}}{\bar{F}}\right)$$

Proof. According to (3-2-6) we may write $E = F(\alpha)$, $\bar{E} = \bar{F}(\bar{\alpha})$,

$$f(x) = f(\alpha, F) = (x - \alpha_1) \cdots (x - \alpha_n)$$

$$\bar{f}(x) = f(\bar{\alpha}, \bar{F}) = (x - \bar{\alpha}_1) \cdots (x - \bar{\alpha}_n)$$

where $\alpha_i \neq \alpha_j$ and $\bar{\alpha}_i \neq \bar{\alpha}_j$ for $j \neq i$ and where $\alpha = \alpha_1$. For each $i = 1, \ldots, n$ we have a distinct element $\sigma_i \in \Gamma(F, E \rightarrow \Omega)$—namely, the F isomorphism for which $\sigma_i\alpha = \alpha_i$—and it is clear that

$$\Gamma(F, E \rightarrow \Omega) = \{\sigma_1, \sigma_2, \ldots, \sigma_n\}$$

Since $\bar{\sigma}_i(\bar{\alpha}) = \bar{\alpha}_i$, it follows immediately that

$$\Gamma(\bar{F}, \bar{E} \rightarrow \bar{\Omega}) = \{\bar{\sigma}_1, \bar{\sigma}_2, \ldots, \bar{\sigma}_n\}$$

and the first part of the proof is complete.

Since both E/F and $\bar{E}/\bar{F}$ are separable, it remains to show that

$$\frac{E}{F} \text{ is normal} \Leftrightarrow \frac{\bar{E}}{\bar{F}} \text{ is normal}$$

Suppose that E/F is normal. From above we know that any element of $\Gamma(\bar{F}, \bar{E} \to \bar{\Omega})$ is of form $\bar{\sigma}$ for some $\sigma \in \Gamma(F, E \to \Omega)$. Because $\bar{\sigma}\bar{E} = \overline{\sigma E} = \bar{E}$, we conclude that $\bar{E}/\bar{F}$ is normal. Suppose conversely that $\bar{E}/\bar{F}$ is normal, so that $\overline{\sigma E} = \bar{\sigma}\bar{E} = \bar{E}$ for every $\sigma \in \Gamma(F, E \to \Omega)$. Since both E/F and $\sigma E/F$ are unramified, we conclude [by (3-2-11)] that $\sigma E = E$; hence, E/F is normal.

3-5-2. Proposition. Suppose that E/F is a finite extension with inertia field T; then the mapping $\sigma \to \bar{\sigma}$ of $\Gamma(F, E \to \Omega) \to \Gamma(\bar{F}, \bar{E} \to \bar{\Omega})$ is onto. Furthermore, for $\sigma, \tau \in \Gamma(F, E \to \Omega)$ we have

$$\bar{\sigma} = \bar{\tau} \Leftrightarrow \sigma|T = \tau|T$$

and, in particular,

$$\bar{\sigma} = 1 \Leftrightarrow \sigma \in \Gamma(T, E \to \Omega)$$

Proof. Let $\mu \in \Gamma(\bar{F}, \bar{E} \to \bar{\Omega})$ be given, and put $\mu' = \mu|\bar{T} \in \Gamma(\bar{F}, \bar{T} \to \bar{\Omega})$. Now $\bar{T}/\bar{F}$ is the separable part of $\bar{E}/\bar{F}$, and $\bar{E}/\bar{T}$ is purely inseparable; therefore, μ is the unique extension of μ' to $\bar{E}$. By (3-5-1) there exists a unique $\sigma' \in \Gamma(F, T \to \Omega)$ such that $\overline{\sigma'} = \mu'$. Let $\sigma \in \Gamma(F, E \to \Omega)$ be any extension of σ' to E. It is clear that $\bar{\sigma} \in \Gamma(\bar{F}, \bar{E} \to \bar{\Omega})$ is an extension of $\overline{\sigma'} = \mu'$; hence, $\bar{\sigma} = \mu$.

Furthermore,

$$\begin{aligned} \bar{\sigma} = \bar{\tau} &\Leftrightarrow \bar{\sigma}|\bar{T} = \bar{\tau}|\bar{T} && \left(\frac{\bar{E}}{\bar{T}} \text{ is purely inseparable}\right) \\ &\Leftrightarrow \overline{\sigma|T} = \overline{\tau|T} && (\bar{\sigma}|\bar{T} = \overline{\sigma|T}) \\ &\Leftrightarrow \sigma|T = \tau|T && (\text{by (3-5-1)}) \end{aligned}$$

In particular, $\Gamma(T, E \to \Omega)$ is the "kernel."

3-5-3. Theorem. Suppose that E/F is a finite normal extension with inertia field T; then $\bar{E}/\bar{F}$ is normal, and both T/F and $\bar{T}/\bar{F}$ are Galois extensions. The mapping $\sigma \to \bar{\sigma}$ of $\mathcal{G}(E/F) \to \mathcal{G}(\bar{E}/\bar{F})$ is a homomorphism onto with kernel $\mathfrak{J} = \mathcal{G}(E/T)$ [$\mathfrak{J}$ is called the ***inertia group*** (*Trägheitsgruppe*) of E/F] so that

$$\mathcal{G}\left(\frac{T}{F}\right) \approx \frac{\mathcal{G}(E/F)}{\mathfrak{J}} \approx \mathcal{G}\left(\frac{\bar{E}}{\bar{F}}\right) \approx \mathcal{G}\left(\frac{\bar{T}}{\bar{F}}\right)$$

Proof. Since $\sigma \to \bar{\sigma}$ maps $\mathcal{G}(E/F) = \Gamma(F, E \to \Omega)$ onto $\Gamma(\bar{F}, \bar{E} \to \bar{\Omega})$, the simple argument used in the proof of (3-5-1) shows that $\bar{E}/\bar{F}$ is normal. Now, $\bar{T}/\bar{F}$ is a Galois extension, because it is the separable part of the normal extension $\bar{E}/\bar{F}$. According to (3-5-1), T/F is a Galois extension. The remaining part is immediate from (3-5-2).

3-5-4. Corollary. If E/F is a finite Galois extension, then the inertia field T and the inertia group $\mathfrak{J}$ correspond to each other; that is, $\mathfrak{J} = \mathcal{G}(E/T)$, and T is the fixed field of $\mathfrak{J}$.

3-5-5. Remark. Suppose that E/F is unramified and that $\bar{F}$ has $q = p^r$ elements. Then $\bar{E}/\bar{F}$ is a cyclic Galois extension, and, by (3-5-1), so is E/F. Now $\mathcal{G}(\bar{E}/\bar{F})$ has a canonical generator ρ—its action consists of raising to the qth power. Since $\mathcal{G}(E/F) \approx \mathcal{G}(\bar{E}/\bar{F})$, it is possible to single out a canonical generator of the cyclic group $\mathcal{G}(E/F)$—namely, the element corresponding to ρ. It is known as the ***Frobenius automorphism*** of E/F, and we shall denote it by $[E/F]$. Thus the Frobenius automorphism is characterized by the property that, if $\sigma \in \mathcal{G}(E/F)$, then

$$\sigma = [E/F] = \left[\frac{E}{F}\right] \Leftrightarrow \sigma\alpha \equiv \alpha^q \pmod{\mathcal{P}_E} \quad \forall \alpha \in O_E$$

If H is any intermediate field, then both E/H and H/F are unramified and it follows easily that

$$\left[\frac{E}{H}\right] = \left[\frac{E}{F}\right]^{[H:F]} \qquad \text{and} \qquad \left[\frac{E}{F}\right] \Big| H = \left[\frac{H}{F}\right]$$

Furthermore, if K/F is a finite extension, then KE/K is unramified and one checks that

$$\left[\frac{KE}{K}\right] \Big| E = \left[\frac{E}{F}\right]^{f(K/F)}$$

3-5-6. Exercise. We sketch an alternative approach to some of the preceding results; the methods are group-theoretic, and no use is made of any results from this chapter.

1. Let K/F be a finite Galois extension, and put $\mathcal{G} = \mathcal{G}(K/F)$. Then $\sigma \in \mathcal{G}$ determines, as before, an automorphism $\bar{\sigma}$ of $\bar{K}$; denote the set of all of these by $\bar{\mathcal{G}}$. If $\bar{K}/\bar{F}$ is separable, then the set of elements of $\bar{K}$ fixed under all $\bar{\sigma} \in \bar{\mathcal{G}}$ is precisely $\bar{F}$; moreover, $\bar{K}/\bar{F}$ is then a Galois extension, and the map $\sigma \to \bar{\sigma}$ is a homomorphism of $\mathcal{G}(K/F)$ onto $\mathcal{G}(\bar{K}/\bar{F})$. Denote the kernel of this map by $\mathfrak{J}$, and call it the ***inertia group*** of K/F. Thus $\mathfrak{J}$ is a normal subgroup of $\mathcal{G}$, $\mathcal{G}/\mathfrak{J} \approx \bar{\mathcal{G}}$, and

$$\mathfrak{J} = \{\sigma \in \mathcal{G} | \sigma\alpha \equiv \alpha \pmod{\mathcal{P}_K} \quad \forall \alpha \in O_K\}$$

2. Now consider the general case. Let E/F be a finite separable extension, and imbed it in a finite Galois extension K/F. Assume further that $\bar{K}/\bar{F}$ is separable, or make the blanket assumption that $\bar{F}$

is perfect (as is the case in number theory). We have then $\mathcal{G} = \mathcal{G}(K/F)$ and the inertia group $\mathfrak{J}$ of K/F. Put $\mathcal{H} = \mathcal{G}(K/E)$; so E is the fixed field of $\mathcal{H}$. Then $\bar{\mathcal{G}} = \mathcal{G}(\bar{K}/\bar{F})$, $\bar{\mathcal{H}} = \mathcal{G}(\bar{K}/\bar{E})$, and $\mathcal{H}\mathfrak{J}$ is the complete inverse image of $\bar{\mathcal{H}}$ under the map $\sigma \to \bar{\sigma}$ of $\mathcal{G} \to \bar{\mathcal{G}}$. Thus,

$$\frac{\mathcal{G}}{\mathfrak{J}} \approx \bar{\mathcal{G}} = \mathcal{G}\left(\frac{\bar{K}}{\bar{F}}\right)$$

and

$$\frac{\mathcal{H}\mathfrak{J}}{\mathfrak{J}} \approx \bar{\mathcal{H}} = \mathcal{G}\left(\frac{\bar{K}}{\bar{E}}\right)$$

Furthermore, we have

$$\frac{\mathcal{H}}{\mathcal{H} \cap \mathfrak{J}} \approx \bar{\mathcal{H}}$$

which says that $\mathcal{H} \cap \mathfrak{J}$ is the inertia group of K/E.

Suppose that $n = [E:F] = (\mathcal{G}:\mathcal{H})$; then

$$f(E/F) = [\bar{E}:\bar{F}] = (\bar{\mathcal{G}}:\bar{\mathcal{H}}) = (\mathcal{G}:\mathcal{H}\mathfrak{J})$$

and

$$e(E/F) = \frac{n}{f(E/F)} = \frac{(\mathcal{G}:\mathcal{H})}{(\mathcal{G}:\mathcal{H}\mathfrak{J})} = (\mathcal{H}\mathfrak{J}:\mathcal{H}) = (\mathfrak{J}:\mathcal{H} \cap \mathfrak{J})$$

Let $T \subset E$ be the fixed field of $\mathcal{H}\mathfrak{J}$, so that $\mathcal{H}\mathfrak{J} = \mathcal{G}(K/T)$; T is called the ***inertia field*** of E/F.

3. Suppose that E'/F is any subextension of K/F, and put $\mathcal{H}' = \mathcal{G}(K/E')$. We have then

$$f\left(\frac{E'}{F}\right) = (\mathcal{G}:\mathcal{H}'\mathfrak{J}) \qquad \text{and} \qquad e\left(\frac{E'}{F}\right) = (\mathfrak{J}:\mathcal{H}' \cap \mathfrak{J})$$

and therefore

$$\begin{aligned} \frac{E'}{F} \text{ is unramified} &\Leftrightarrow f\left(\frac{E'}{F}\right) = [E':F] \\ &\Leftrightarrow (\mathcal{G}:\mathcal{H}'\mathfrak{J}) = (\mathcal{G}:\mathcal{H}') \\ &\Leftrightarrow \mathcal{H}'\mathfrak{J} = \mathcal{H}' \\ &\Leftrightarrow \mathcal{H}' \supset \mathfrak{J} \end{aligned}$$

Of course, $E' \subset E \Leftrightarrow \mathcal{H}' \supset \mathcal{H}$; so, for $E' \subset E$, E'/F is unramified $\Leftrightarrow$ $\mathcal{H}' \supset \mathcal{H}\mathfrak{J}$. It follows that T/F is unramified, that it is the unique maximal unramified extension in E/F, and that it contains every unramified extension. In particular, the composite of two unramified extensions is unramified. Furthermore, T is uniquely characterized by the properties:

(i) $F \subset T \subset E$

(ii) $\bar{T} = \bar{E}$

(iii) T/F is unramified

4. There are natural 1-1 correspondences between:

i. The finite extensions of $\bar{F}$ in $\bar{K}$
ii. The subgroups of $\bar{\mathcal{G}} = \mathcal{G}(\bar{K}/\bar{F})$
iii. The groups between $\mathcal{G}$ and $\mathfrak{J}$
iv. The unramified extensions of F in K

Moreover, joins and meets are preserved. By dropping the restriction to subfields of K and $\bar{K}$, we get a lattice isomorphism between the lattice of unramified extensions of F and the lattice of finite (separable) extensions of $\bar{F}$.

3-6. Ramification Groups

In order to keep the hypotheses fairly uniform, we assume in this section that E/F is a finite Galois extension of degree n, that $\bar{E}/\bar{F}$ is separable, and that $\bar{F}$ has characteristic p (p may be 0). In particular, $\bar{E}/\bar{F}$ is a Galois extension [by (3-5-3)], and the map $\sigma \to \bar{\sigma}$ is a homomorphism of $\mathcal{G} = \mathcal{G}(E/F)$ onto $\mathcal{G}(\bar{E}/\bar{F})$ with kernel $\mathfrak{J}$. Thus, $\sigma \in \mathfrak{J} \Leftrightarrow \bar{\sigma} = 1 \Leftrightarrow \sigma\alpha \equiv \alpha \pmod{\mathcal{P}_E}$ for all $\alpha \in O_E \Leftrightarrow (\sigma - 1)O_E \subset \mathcal{P}_E$. One may note that the map $(\sigma - 1): E \to E$ is a continuous endomorphism of the topological vector space E (over the field F) and that the elements of F are in the kernel.

We generalize $\mathfrak{J}$ by putting, for $r = 0, 1, 2, \ldots$,

$$\mathcal{V}_r = \{\sigma \in \mathcal{G} | (\sigma - 1)O_E \subset \mathcal{P}_E^{r+1}\}$$

Note that $\mathcal{V}_0 = \mathfrak{J}$ and that, if we allow $r = -1$, then $\mathcal{V}_{-1} = \mathcal{G}$.

3-6-1. Proposition. For all $r \geq 0$ we have:

i. $\mathcal{V}_r$ is a normal subgroup of $\mathcal{G}$; it is called the ***rth ramification group*** of E/F.

ii. $\mathcal{V}_r \supset \mathcal{V}_{r+1}$.

Moreover, there exists an integer s such that $\mathcal{V}_s = \{1\}$.

Proof. i. If $\sigma, \tau \in \mathcal{V}_r$, then

$$\begin{aligned}(\sigma\tau - 1)O_E &\subset (\sigma - 1)(\tau - 1)O_E + (\sigma - 1)O_E + (\tau - 1)O_E \\ &\subset (\sigma - 1)\mathcal{P}_E^{r+1} + \mathcal{P}_E^{r+1} + \mathcal{P}_E^{r+1} \\ &\subset \mathcal{P}_E^{r+1}\end{aligned}$$

and also
$$(\sigma^{-1} - 1)O_E = (\sigma^{-1} - 1)\sigma O_E = (1 - \sigma)O_E$$
$$= (\sigma - 1)O_E \subset \mathcal{P}_E^{r+1}$$

Thus, $\mathcal{V}_r$ is a group. Furthermore, for $\tau \in \mathcal{G}$ and $\sigma \in \mathcal{V}_r$ we have

$$(\tau\sigma\tau^{-1} - 1)O_E = (\tau\sigma - \tau)\tau^{-1}O_E = \tau(\sigma - 1)O_E$$
$$\subset \tau\mathcal{P}_E^{r+1} = \mathcal{P}_E^{r+1}$$

so that $\mathcal{V}_r$ is indeed a normal subgroup of $\mathcal{G}$.

Of course (ii) is trivial, and, for the remaining part, let us note first that $\bigcap_{r=0}^{\infty} \mathcal{V}_r = \{1\}$; in fact,

$$\sigma \in \bigcap_0^\infty \mathcal{V}_r \Rightarrow (\sigma - 1)O_E \subset \bigcap_r \mathcal{P}_E^{r+1} = \{0\} \Rightarrow \sigma = 1$$

Since each $\mathcal{V}_r$ is a finite group, we conclude from (ii) that some $\mathcal{V}_s = \{1\}$.

Before trying to analyze the chain of groups

$$(*) \quad \mathcal{G} = \mathcal{V}_{-1} \supset \mathfrak{J} = \mathcal{V}_0 \supset \cdots \supset \mathcal{V}_r \supset \mathcal{V}_{r+1} \supset \cdots \supset \mathcal{V}_s = \{1\}$$

we give some alternative characterizations for $\mathcal{V}_r$.

3-6-2. Proposition. Fix a prime element π_E of E and an element $\beta \in O_E$ [whose existence is guaranteed by (3-3-3)] such that the set $\{1, \beta, \ldots, \beta^{n-1}\}$ is an integral basis for E/F; then for $\sigma \in \mathcal{G}$ and $r = 0, 1, 2, \ldots$ the following conditions are equivalent:

(i) $\sigma \in \mathcal{V}_r$

(ii) $\sigma\alpha \equiv \alpha \pmod{\mathcal{P}_E^{r+1}}$ for all $\alpha \in O_E$

(iii) $\nu_E[(\sigma - 1)\alpha] \geq r + 1$ for all $\alpha \in O_E$

(iv) $\nu_E[(\sigma - 1)\beta] \geq r + 1$

(v) $\sigma\beta \equiv \beta \pmod{\mathcal{P}_E^{r+1}}$

(vi) $\sigma \in \mathfrak{J}$ and $\sigma\pi_E \equiv \pi_E \pmod{\mathcal{P}_E^{r+1}}$

(vii) $\sigma \in \mathfrak{J}$ and $\sigma\pi \equiv \pi \pmod{\mathcal{P}_E^{r+1}}$ for all primes π of E

(viii) $\sigma \in \mathfrak{J}$ and $\nu_E\left(\frac{\sigma\pi_E}{\pi_E} - 1\right) \geq r$

(ix) $\sigma \in \mathfrak{J}$ and $\nu_E\left(\frac{\sigma\pi}{\pi} - 1\right) \geq r$ for all primes π of E

(x) $\sigma \in \mathfrak{J}$ and $\nu_E\left(\frac{\sigma\alpha}{\alpha} - 1\right) \geq r$ for all $\alpha \in E^*$

Proof. We have at the start the trivial equivalences: (i) $\Leftrightarrow$ (ii) $\Leftrightarrow$ (iii), (iv) $\Leftrightarrow$ (v), (vi) $\Leftrightarrow$ (viii), (vii) $\Leftrightarrow$ (ix). An element $\alpha \in O_E$ may be put in the form $\alpha = a_0 + a_1\beta + \cdots + a_{n-1}\beta^{n-1}$ with $a_i \in O_F$. Then

$$\begin{aligned}\sigma\alpha - \alpha &= a_1(\sigma\beta - \beta) + \cdots + a_{n-1}(\sigma\beta^{n-1} - \beta^{n-1}) \\ &= (\sigma\beta - \beta)(\text{in } O_E)\end{aligned}$$

so that $\nu_E(\sigma\alpha - \alpha) \geq \nu_E(\sigma\beta - \beta)$. We conclude that

$$\nu_E(\sigma\beta - \beta) = \min_{\alpha \in O_E} \{\nu_E(\sigma\alpha - \alpha)\}$$

which gives (iii) $\Leftrightarrow$ (iv).

Next let us observe that $(\sigma - 1)O_E \subset \mathfrak{P}_E^{r+1} \Rightarrow \sigma \in \mathfrak{J}$ and $\nu_E(\sigma\pi_E - \pi_E) \geq r + 1$. For the converse, note that $\mathfrak{J}$ is the Galois group of the Galois extension E/T and that the powers of π_E are an integral basis for E/T. Thus the same argument used to show that (iii) and (iv) are equivalent enables us to conclude that (i) $\Leftrightarrow$ (vi). The same proof gives (i) $\Leftrightarrow$ (vii).

It is clear that (x) $\Rightarrow$ (ix). On the other hand, the primes π of E serve as a set of generators for the multiplicative group E^*. Moreover, for $\gamma, \gamma' \in E^*$ and $r \geq 1$, we have

$$\frac{\sigma\gamma}{\gamma}, \frac{\sigma\gamma'}{\gamma'} \in 1 + \mathfrak{P}_E^r \Rightarrow \frac{\sigma(\gamma\gamma')}{\gamma\gamma'} = \left(\frac{\sigma\gamma}{\gamma}\right)\left(\frac{\sigma\gamma'}{\gamma'}\right) \in 1 + \mathfrak{P}_E^r$$

and

$$\frac{\sigma\gamma}{\gamma} \in 1 + \mathfrak{P}_E^r \Rightarrow \frac{\sigma\gamma^{-1}}{\gamma^{-1}} = \frac{\gamma}{\sigma\gamma} = \left(\frac{\sigma\gamma}{\gamma}\right)^{-1} \in 1 + \mathfrak{P}_E^r$$

This implies that (ix) and (x) are equivalent when $r \geq 1$. Finally, we note that $\nu_E[(\sigma\alpha/\alpha) - 1] \geq 0$ for all $\alpha \in E^*$, since $\sigma\alpha/\alpha \in U_E$, so that (ix) $\Rightarrow$ (x) when $r = 0$. This completes the proof.

3-6-3. Corollary. For $r \geq 1$ we have

$$\mathfrak{V}_r = \left\{\sigma \in \mathfrak{G} \,\middle|\, \nu_E\left(\frac{\sigma\alpha}{\alpha} - 1\right) \geq r \quad \forall \alpha \in E^*\right\}$$

Proof. By the arguments used in the proof of (3-6-2), we have

$$\begin{aligned}\sigma \in \mathfrak{V}_r &\Rightarrow \nu_E(\sigma\pi - \pi) \geq r + 1 && \forall \text{ primes } \pi \text{ of } E \\ &\Rightarrow \nu_E\left(\frac{\sigma\pi}{\pi} - 1\right) \geq r && \forall \text{ primes } \pi \text{ of } E \\ &\Rightarrow \nu_E\left(\frac{\sigma\alpha}{\alpha} - 1\right) \geq r && \forall \alpha \in E^* \\ &\Rightarrow \nu_E(\sigma\alpha - \alpha) \geq r && \forall \alpha \in O_E \\ &\Rightarrow \sigma \in \mathfrak{V}_r\end{aligned}$$

Now let us examine the factor groups of the chain of ramification groups (*). We recall that $\mathcal{G}/\mathfrak{J} \approx \mathcal{G}(\bar{E}/\bar{F})$, so that $\mathcal{G}/\mathfrak{J}$ is a group of order $f = f(E/F)$. Thus our interest will center on $\mathfrak{J}/\mathcal{V}_1$ and on $\mathcal{V}_r/\mathcal{V}_{r+1}$ for $r \geq 1$; when $\mathcal{V}_r \neq \mathcal{V}_{r+1}$, r is said to be a ***discontinuity*** in the sequence of ramification groups.

3-6-4. Proposition. The group $\mathfrak{J}/\mathcal{V}_1$ is isomorphic to a (multiplicative) subgroup of $\bar{E}^*$. In particular, $\mathfrak{J}/\mathcal{V}_1$ is a cyclic group whose order is prime to p.

Proof. Fix a prime element $\pi = \pi_E$ of E. Since $\nu_E(\sigma\pi) = \nu_E(\pi)$ for every $\sigma \in \mathcal{G}$, we may put

$$u_\sigma = \frac{\sigma\pi}{\pi} \in U_E$$

Thus, for $\sigma, \tau \in \mathcal{G}$ we have

$$u_{\sigma\tau} = \frac{\sigma\tau\pi}{\pi} = \left(\frac{\sigma\tau\pi}{\sigma\pi}\right)\left(\frac{\sigma\pi}{\pi}\right) = (\sigma u_\tau)(u_\sigma)$$

and then $$\bar{u}_{\sigma\tau} = (\bar{\sigma}(\bar{u}_\tau))(\bar{u}_\sigma)$$

If $\sigma \in \mathfrak{J}$, then $\bar{\sigma} = 1$, so that the mapping $\sigma \rightarrow \bar{u}_\sigma$ is a homomorphism of $\mathfrak{J} \rightarrow \bar{E}^*$. Furthermore, by (3-6-2)

$$\bar{u}_\sigma = 1 \Leftrightarrow \frac{\sigma\pi}{\pi} \equiv 1 \pmod{\mathcal{P}_E} \Leftrightarrow \sigma \in \mathcal{V}_1$$

and therefore $\mathfrak{J}/\mathcal{V}_1$ is isomorphic to a subgroup of $\bar{E}^*$. Finally, any finite multiplicative group in a field is cyclic, and if $p \neq 0$ divides its order, there exists an element of multiplicative order p, which is impossible.

3-6-5. Proposition. For $r \geq 1$, the group $\mathcal{V}_r/\mathcal{V}_{r+1}$ is isomorphic to an additive subgroup of $\bar{E}$. If $p = 0$, then $\mathcal{V}_1 = \{1\}$, and if $p \neq 0$, then, at a discontinuity in the sequence of ramification groups, $\mathcal{V}_r/\mathcal{V}_{r+1}$ is an abelian group of type $(p, p, \ldots, p)$.

Proof. For $\pi = \pi_E$ and $\sigma \in \mathcal{V}_r$ we may write

$$\sigma\pi \equiv \pi + \lambda_\sigma \pi^{r+1} \pmod{\mathcal{P}_E^{r+2}}$$

with $\lambda_\sigma \in O_E$; and since $\bar{T} = \bar{E}$, we may take $\lambda_\sigma \in O_T$. It follows that for $\sigma, \tau \in \mathcal{V}_r$ we have (mod $\mathcal{P}_E^{r+2}$) the congruences

$$\begin{aligned} \pi + \lambda_{\sigma\tau}\pi^{r+1} &\equiv \sigma\tau\pi \\ &\equiv \sigma[\pi + \lambda_\tau \pi^{r+1}] \\ &\equiv \pi + \lambda_\sigma\pi^{r+1} + \lambda_\tau(\sigma\pi^{r+1}) \\ &\equiv \pi + \lambda_\sigma\pi^{r+1} + \lambda_\tau\pi^{r+1} \end{aligned}$$

and this implies that

$$\lambda_{\sigma\tau} \equiv \lambda_\sigma + \lambda_\tau \pmod{\mathcal{P}_E}$$

Therefore, the mapping $\sigma \to \bar{\lambda}_\sigma$ is a homomorphism of $\mathcal{V}_r$ into the additive group of $\bar{E}$. The kernel is given by

$$\bar{\lambda}_\sigma = 0 \Leftrightarrow \lambda_\sigma \in \mathcal{P}_T \subset \mathcal{P}_E \Leftrightarrow \sigma\pi \equiv \pi \pmod{\mathcal{P}_E^{r+2}} \Leftrightarrow \sigma \in \mathcal{V}_{r+1}$$

If $p = 0$, the additive group of $\bar{E}$ contains no finite subgroups other than $\{0\}$; hence, $\mathcal{V}_r = \mathcal{V}_{r+1}$ for all $r \geq 1$, and all these groups consist of a single element. If $p \neq 0$ and $\mathcal{V}_r \neq \mathcal{V}_{r+1}$, then every nontrivial element of the finite abelian group $\mathcal{V}_r/\mathcal{V}_{r+1}$ has order p.

3-6-6. Proposition. If $\bar{F}$ is finite, then $\mathcal{G}$ is a solvable group.

Proof. Immediate from (3-6-4), (3-6-5), and the fact that here $\mathcal{G}/\mathcal{J}$ is a cyclic group of order f.

3-6-7. Proposition. If we write $e(E/F) = e = e_0p^r$ with $(e_0, p) = 1$ (so that $e_0 = e$ when $p = 0$), then $(\mathcal{J}:\mathcal{V}_1) = e_0$.

Proof. Consider the equation

$$n = (\mathcal{G}:\mathcal{J})(\mathcal{J}:\mathcal{V}_1)\prod_1^\infty (\mathcal{V}_r:\mathcal{V}_{r+1})$$

We know that $(\mathcal{G}:\mathcal{J}) = f$. If $p = 0$, then $\mathcal{V}_1 = \{1\}$, so that $(\mathcal{J}:\mathcal{V}_1) = e = e_0$. If $p \neq 0$, then $(\mathcal{J}:\mathcal{V}_1)$ is prime to p and $\Pi(\mathcal{V}_r:\mathcal{V}_{r+1})$ is a power of p; therefore, $(\mathcal{J}:\mathcal{V}_1) = e_0$, and

$$\prod_1^\infty (\mathcal{V}_r:\mathcal{V}_{r+1}) = p^r$$

3-6-8. Proposition. The ramification field V and the first ramification group $\mathcal{V}_1$ correspond to each other; in other words, V is the fixed field of $\mathcal{V}_1$, and $\mathcal{V}_1$ is the Galois group of E/V.

Proof. From (3-6-7) it follows that the order of $\mathcal{V}_1$ is the p part of e, so that in view of (3-4-7) we have $(\mathcal{V}_1:\{1\}) = [E:V]$. Since $\mathcal{G}(E/V)$ has order $[E:V]$, it suffices to show that $\mathcal{V}_1 \subset \mathcal{G}(E/V)$.

According to (3-4-3) we may write $V = T(\pi_V)$, where π_V is a prime element of V such that $\pi_V^{e_0} \in T$. By (3-6-2x), $\sigma \in \mathcal{V}_1 \subset \mathcal{J}$ implies that

$$\nu_E(\sigma\pi_V - \pi_V) > \nu_E(\pi_V)$$

Applying the first part of (3-4-4) with $m = e_0$, $a = \pi_V^{e_0}$, and $F = T$, we conclude that $\sigma\pi_V = \pi_V$. Therefore, σ is the identity map on V, and $\sigma \in \mathcal{G}(E/V)$. This completes the proof.

If for $r \geq 0$ we denote the fixed field of $\mathcal{V}_r$ by V_r (and call it the ***rth ramification field*** of E/F), then corresponding to the descending chain of groups (*) there exists the ascending chain of Galois extensions of F,

$$F \subset T = V_0 \subset V = V_1 \subset V_2 \subset \cdots \subset V_s = E$$

The extensions T/F, V/T, and E/T have been treated already. Since $\mathcal{V}_1 = \{1\} \Leftrightarrow V = V_1 = E \Leftrightarrow E/F$ is tamely ramified, it is only in the wildly ramified case that nontrivial $\mathcal{V}_r$ and V_r can occur for $r \geq 1$.

If H is any intermediate field and $\mathcal{H}$ is the corresponding subgroup of $\mathcal{G}(E/F)$, then it is clear that the ramification groups of the extension E/H are the groups $\mathcal{V}_i \cap \mathcal{H}$. If, furthermore, H/F is a Galois extension, then the ramification groups of this extension can be found, but the situation is more complicated (the groups are the $\mathcal{V}_i\mathcal{H}/\mathcal{H}$, but the indexing must be adjusted).

3-6-9. Example. Let $F = \mathbf{Q}_p$ and $E = \mathbf{Q}_p(\sqrt{p})$. Then $\sqrt{p}$ is a prime element for the totally ramified quadratic extension E/F, and $\{1, \sqrt{p}\}$ is an integral basis. Let σ be the nontrivial element of $\mathcal{G}(E/F)$; so $\sigma(\sqrt{p}) = -\sqrt{p}$. Since $\mathfrak{J} = \mathcal{G}$ and $\nu_E(\sigma\sqrt{p} - \sqrt{p}) = \nu_E(-2\sqrt{p})$ it follows from (3-6-2) that, if $p \neq 2$, then $\mathcal{V}_0 = \mathcal{G}$ and $\mathcal{V}_1 = \mathcal{V}_2 = \cdots = \{1\}$ and, if $p = 2$, then $\mathcal{V}_0 = \mathcal{V}_1 = \mathcal{V}_2 = \mathcal{G}$ and $\mathcal{V}_3 = \mathcal{V}_4 = \cdots = \{1\}$.

3-6-10. Exercise. Let E be the splitting field of $x^4 - 2$ over $F = \mathbf{Q}_2$. Find the Galois group, the ramification groups, and the ramification fields.

3-7. Different and Discriminant

It is convenient to begin this section with preparations of an algebraic nature in which valuations play no role. The treatment is somewhat sketchy because the results are elementary and fairly standard.

Suppose that V is a vector space of dimension n over an arbitrary field F. By a ***bilinear form*** on V we mean a mapping $B: V \times V \to F$ which is linear in each variable; in other words, for each $w \in V$, the mappings $v \to B(v, w)$ and $v \to B(w, v)$ are linear functionals on V. The bilinear form B is said to be ***degenerate*** when there exists a $v \neq 0 \in V$ such that $B(v, w) = 0$ for all $w \in V$. If $v_1, \ldots, v_n$ are elements of V, then det $[B(v_i, v_j)]$ [that is, the determinant of the $n \times n$ matrix whose (i, j) component is $B(v_i, v_j)$] is called the ***discriminant*** of $v_1, \ldots, v_n$ with respect to B; it is denoted by $\Delta(v_1, \ldots, v_n)$.

3-7-1. Proposition. Let $\{v_1, \ldots, v_n\}$ be a basis of V, and let $w_1, \ldots, w_n$ be arbitrary elements of V. If for $i = 1, \ldots, n$ we write

$$w_i = \sum_{j=1}^{n} a_{ij}v_j$$

with $a_{ij} \in F$, then with $A = (a_{ij})$ we have

$$\Delta(w_1, \ldots, w_n) = (\det A)^2 \, \Delta(v_1, \ldots, v_n)$$

Proof. If we write $w_k = \sum_{l=1}^{n} a_{kl}v_l$ for $k = 1, \ldots, n$, then

$$B(w_i, w_k) = \sum_{j,l} a_{ij}B(v_j, v_l)a_{kl}$$

3-7-2. Corollary. If $w_1, \ldots, w_n$ are linearly dependent, then

$$\Delta(w_1, \ldots, w_n) = 0$$

3-7-3. Proposition. Let $\{v_1, \ldots, v_n\}$ be any basis of V; then B is degenerate $\Leftrightarrow \Delta(v_1, \ldots, v_n) = 0$.

Proof. $\Delta(v_1, \ldots, v_n) = 0 \Leftrightarrow$ the rows of the matrix $[B(v_i, v_j)]$ are linearly dependent $\Leftrightarrow$ there exist $c_1, \ldots, c_n \in F$, not all 0, such that $\sum_{i=1}^{n} c_iB(v_i, v_j) = 0$ for $j = 1, \ldots, n \Leftrightarrow$ there exists an element $v \neq 0 \in V$ (namely, $v = \sum_{1}^{n} c_iv_i$) such that $B(v, v_j) = 0$ for $j = 1, \ldots, n$ $\Leftrightarrow B$ is degenerate.

Suppose that $\{v_1, \ldots, v_n\}$ and $\{w_1, \ldots, w_n\}$ are bases of V; they are said to be ***complementary*** when $B(v_i, w_j) = \delta_{ij}$ for all $i, j = 1, \ldots, n$ (δ_{ij} is the Kronecker delta).

3-7-4. Proposition. If $\{v_1, \ldots, v_n\}$ and $\{w_1, \ldots, w_n\}$ are complementary bases, then

$$\Delta(v_1, \ldots, v_n) \, \Delta(w_1, \ldots, w_n) = 1$$

Proof. This follows from (3-7-1) since

$$B(w_i, w_k) = \sum_{j=1}^{n} a_{ij}B(v_j, w_k) = a_{ik}$$

implies that $\Delta(w_1, \ldots, w_n) = \det A \neq 0$.

3-7-5. Proposition. B is nondegenerate $\Leftrightarrow$ every basis of V has a complementary basis. Moreover, the complementary basis is then unique.

Proof. $\Leftarrow$: Trivial.

$\Rightarrow$: For $v \in V$ let $B_v \in V^*$ (V^* is the dual space of V) be given by $B_v : w \to B(w, v)$. The map $v \to B_v$ is then an isomorphism of V onto V^*. The complementary basis is essentially the dual basis in V^*.

Suppose that E is an extension field of F of degree n, and let $S = S_{E\to F}$ be the trace function. For $\alpha, \beta \in E$ we put $B(\alpha, \beta) = S(\alpha\beta)$. Clearly, B is a bilinear form on E; and because $B(\alpha, \beta) = B(\beta, \alpha)$, we say that B is ***symmetric.*** (Henceforth, we shall deal exclusively with this bilinear form.) By definition, we have, for $\alpha_1, \ldots, \alpha_n \in E$,

$$\Delta(\alpha_1, \ldots, \alpha_n) = \det \{S(\alpha_i\alpha_j)\} \in F$$

3-7-6. Proposition. B is degenerate $\Leftrightarrow E/F$ is inseparable.

Proof. B is degenerate $\Leftrightarrow$ there exists $\alpha \neq 0 \in E$ such that $S(\alpha\beta) = 0$ for all $\beta \in E \Leftrightarrow S(E) = 0 \Leftrightarrow E/F$ is inseparable.

In view of this, our interest will center on the separable case. From (3-7-6), (3-7-5), and (3-7-4) we see that the following corollary holds.

3-7-7. Corollary. If E/F is separable, then for any basis $\{\alpha_1, \ldots, \alpha_n\}$ of E/F we have $\Delta(\alpha_1, \ldots, \alpha_n) \neq 0$.

3-7-8. Proposition. Suppose that E/F is separable, and let

$$\Gamma(F, E \to \Omega) = \{\sigma_1 = 1, \sigma_2, \ldots, \sigma_n\}$$

Then, for any $\alpha_1, \ldots, \alpha_n \in E$,

$$\Delta(\alpha_1, \ldots, \alpha_n) = \begin{vmatrix} \sigma_1\alpha_1 & \sigma_1\alpha_2 & \cdots & \sigma_1\alpha_n \\ \sigma_2\alpha_1 & \sigma_2\alpha_2 & \cdots & \sigma_2\alpha_n \\ \cdot & \cdot & \cdots & \cdot \\ \sigma_n\alpha_1 & \sigma_n\alpha_2 & \cdots & \sigma_n\alpha_n \end{vmatrix}^2 = [\det (\sigma_i\alpha_j)]^2$$

Proof. If we multiply the matrix $(\sigma_i\alpha_j)$ by its transpose, the resulting matrix has in its (j, k) place the element

$$\sum_i (\sigma_i\alpha_j)(\sigma_i\alpha_k) = \sum_i \sigma_i(\alpha_j\alpha_k) = S(\alpha_j\alpha_k)$$

3-7-9. Corollary. Suppose that α is a primitive element for the separable extension E/F, and for $i = 1, \ldots, n$ let $\alpha_i = \sigma_i\alpha$; then

$$\Delta(1, \alpha, \ldots, \alpha^{n-1}) = \begin{vmatrix} 1 & \alpha_1 & \cdots & \alpha_1^{n-1} \\ 1 & \alpha_2 & \cdots & \alpha_2^{n-1} \\ \cdot & \cdot & \cdots & \cdot \\ 1 & \alpha_n & \cdots & \alpha_n^{n-1} \end{vmatrix}^2 = \prod_{i<j} (\alpha_i - \alpha_j)^2$$

Of course, if α is not primitive, then $\Delta(1, \alpha, \ldots, \alpha^{n-1}) = 0$. It should be noted that, for primitive α,

$$\Delta(1, \alpha, \ldots, \alpha^{n-1}) = \prod_{i<j} (\alpha_i - \alpha_j)^2$$

is the discriminant (according to the usual definition) of the minimum polynomial $f(\alpha, F)$, and that $\Delta(1, \alpha, \ldots, \alpha^{n-1})$ is often called the ***discriminant of*** α.

A straightforward and fairly efficient procedure for finding the discriminant of a concrete (separable, irreducible) polynomial

$$f(x) = f(\alpha, F) = x^n + a_1x^{n-1} + \cdots + a_{n-1}x + a_n \in F[x]$$

is based on the observation that [with the notation of (3-7-9)] column-by-column multiplication gives

$$\begin{vmatrix} 1 & \alpha_1 & \cdots & \alpha_1^{n-1} \\ 1 & \alpha_2 & \cdots & \alpha_2^{n-1} \\ \cdot & \cdot & \cdots & \cdot \\ 1 & \alpha_n & \cdots & \alpha_n^{n-1} \end{vmatrix}^2 = \begin{vmatrix} s_0 & s_1 & \cdots & s_{n-1} \\ s_1 & s_2 & \cdots & s_n \\ \cdot & \cdot & \cdots & \cdot \\ s_{n-1} & s_n & \cdots & s_{2n-2} \end{vmatrix}$$

where $s_i = \alpha_1^i + \alpha_2^i + \cdots + \alpha_n^i$ for $i = 0, 1, \ldots, 2n - 2$. Now $s_0 = n$, and the s_i are related to the coefficients of $f(x)$ by the equations

$$\begin{aligned} s_i + a_1s_{i-1} + a_2s_{i-2} + \cdots + a_{i-1}s_1 + ia_i &= 0 \text{ for } i = 1, 2, \ldots, n \\ s_i + a_1s_{i-1} + a_2s_{i-2} + \cdots + a_ns_{i-n} &= 0 \text{ for } i > n \end{aligned}$$

We have then $s_1 = -a_1$, $s_2 = a_1^2 - 2a_2$, $s_3 = -a_1^3 + 3a_1a_2 - 3a_3$; and, in general, if $s_1, \ldots, s_{i-1}$ are known, then the ith equation may be used to solve for s_i. Thus, it is quite easy to find all the s_i and the discriminant of $f(x)$. One may check, for example, that when

$$f(x) = x^2 + a_1x + a_2$$

we have

$$\Delta(1, \alpha) = a_1^2 - 4a_2$$

and, if

$$f(x) = x^3 + a_1x^2 + a_2x + a_3$$

then

$$\Delta(1, \alpha, \alpha^2) = a_1^2a_2^2 - 4a_2^3 - 4a_1^3a_3 - 27a_3^2 + 18a_1a_2a_3$$

3-7-10. Proposition. Suppose that $F \subset E \subset K$. If $\{\alpha_1, \ldots, \alpha_n\}$ is a basis for E/F and $\{\beta_1, \ldots, \beta_n\}$ is a basis for K/E, then

$$\Delta_{K/F}(\{\alpha_i\beta_j\}) = \pm[\Delta_{E/F}(\{\alpha_i\})]^{[K:E]}N_{E\to F}[\Delta_{K/E}(\{\beta_j\})]$$

Proof. This involves a straightforward verification; the details are left to the reader.

Suppose that α belongs to the separable extension E/F, and let $f(x) = f(\alpha, E/F)$. The element $f'(\alpha) \in E$ is called the ***different of*** α (for E/F) and is denoted by $\mathfrak{D}_{E/F}(\alpha)$ or $\mathfrak{D}(\alpha)$. Because $f(\alpha, E/F)$ is a power of $f(\alpha, F)$, it is immediate that $f'(\alpha) \neq 0 \Leftrightarrow \alpha$ is a primitive element for E/F. Therefore, the different (and the discriminant too) will be considered *only* for a primitive element of a separable extension. In this case, we have

$$f(x) = f(\alpha, F) = f\left(\alpha, \frac{E}{F}\right) = \prod_1^n (x - \sigma_i\alpha) = \prod_1^n (x - \alpha_i)$$

Consequently,

$$f'(\alpha) = \prod_{j=2}^{n} (\alpha - \alpha_j) \qquad f'(\alpha_i) = \prod_{j\neq i} (\alpha_i - \alpha_j) \qquad \sigma_i f'(\alpha) = f'(\alpha_i)$$

The relation between different and discriminant is now clear—namely, we have the following result.

3-7-11. Proposition. Let α be a primitive element for the separable extension E/F of degree n; then

$$N_{E\to F}[\mathfrak{D}(\alpha)] = (-1)^{[n(n-1)]/2}\,\Delta(1, \alpha, \ldots, \alpha^{n-1})$$

3-7-12. Proposition (Euler). Suppose that α is a primitive element for the separable extension E/F of degree n. If

$$f(x) = f(\alpha, F) = a_0 + a_1x + \cdots + a_{n-1}x^{n-1} + x^n \in F[x]$$

and we put $f(x)/(x - \alpha) = \beta_0 + \beta_1x + \cdots + \beta_{n-1}x^{n-1} \in E[x]$, then $\{1, \alpha, \ldots, \alpha^{n-1}\}$ and $\{\beta_0/f'(\alpha), \beta_1/f'(\alpha), \ldots, \beta_{n-1}/f'(\alpha)\}$ are complementary bases. Moreover,

$$\begin{aligned}
\beta_{n-1} &= 1\\
\beta_{n-2} &= \alpha + a_{n-1}\\
\beta_{n-3} &= \alpha^2 + a_{n-1}\alpha + a_{n-2}\\
&\cdots\cdots\cdots\cdots\\
\beta_0 &= \alpha^{n-1} + a_{n-1}\alpha^{n-2} + \cdots + a_2\alpha + a_1
\end{aligned}$$

Proof. Let $\Gamma(F, E \to \Omega) = \{\sigma_1 = 1, \sigma_2, \ldots, \sigma_n\}$ and put $\sigma_i\alpha = \alpha_i$; however, for arbitrary $\beta \in E$ we shall write $\sigma_i\beta = \beta^{(i)}$. Because $f(x) = (x - \alpha_1) \cdots (x - \alpha_n)$ it follows that

$$\frac{f(x)}{(x - \alpha_i)} = \beta_0^{(i)} + \beta_1^{(i)}x + \cdots + \beta_{n-1}^{(i)}x^{n-1}$$

and

$$f'(\alpha_i) = \prod_{j\neq i} (\alpha_i - \alpha_j) = \beta_0^{(i)} + \beta_1^{(i)}\alpha_i + \cdots + \beta_{n-1}^{(i)}\alpha_i^{n-1}$$

We observe that, for each $r = 0, 1, \ldots, n-1$,

$$\sum_{i=1}^{n} \left[\frac{\beta_0^{(i)} + \beta_1^{(i)}x + \cdots + \beta_{n-1}^{(i)}x^{n-1}}{f'(\alpha_i)} \alpha_i^r \right] = x^r$$

To see this, it suffices to evaluate both sides (they are polynomials of degree $n-1$) at the n distinct points $\alpha_1, \ldots, \alpha_n$. By comparing coefficients we see that

$$\sum_{i=1}^{n} \left(\frac{\beta_j^{(i)}}{f'(\alpha_i)} \alpha_i^r \right) = \begin{cases} 0 \text{ if } j \neq r \\ 1 \text{ if } j = r \end{cases}$$

In other words,

$$S\left(\alpha^r \frac{\beta_j}{f'(\alpha)}\right) = \delta_{rj} \qquad j, r = 0, 1, \ldots, n-1$$

so that we have complementary bases.

The values of the β's may be read off from

$$\begin{aligned} \beta_0 + \beta_1 x + \cdots + \beta_{n-1}x^{n-1} &= \frac{f(x)}{x-\alpha} = \sum_1^n a_i \left(\frac{x^i - \alpha^i}{x - \alpha} \right) \qquad (a_n = 1) \\ &= a_1 + a_2(x+\alpha) + a_3(x^2 + \alpha x + \alpha^2) + \cdots \\ &\qquad + 1(x^{n-1} + \alpha x^{n-2} + \cdots + \alpha^{n-1}) \end{aligned}$$

The remaining part of our preliminary discussion will deal with the following situation: Let R be a domain which is integrally closed in its quotient field F, let E/F be a separable extension of degree n, and let S be the integral closure of R in E.

For any subset M of E, we put

$$M^* = \{\alpha \in E | S_{E \to F}(\alpha M) \subset R\}$$

and call it the ***complementary set*** (of M with respect to R). It is clear that $M_1 \subset M_2 \Rightarrow M_2^* \subset M_1^*$ and that, if M is an S module, then so is M^*.

3-7-13. Proposition. Suppose that $\{\alpha_1, \ldots, \alpha_n\}$ is a basis for E/F, and let $\{\alpha_1^*, \ldots, \alpha_n^*\}$ be the complementary basis. If we put $M = R\alpha_1 + \cdots + R\alpha_n$, then $M^* = R\alpha_1^* + \cdots + R\alpha_n^*$ and $(M^*)^* = M$.

Proof. Write $\beta \in E$ in the form $\beta = \Sigma a_i \alpha_i^*$ with $a_i \in F$. Then

$$\begin{aligned} \beta \in M^* &\Leftrightarrow S(\beta M) \subset R \Leftrightarrow S(\beta \alpha_i) \in R \text{ for } i = 1, \ldots, n \\ &\Leftrightarrow a_i \in R, \; i = 1, \ldots, n \end{aligned}$$

3-7-14. Proposition. Suppose that $E = F(\alpha)$ with $\alpha \in S$ and $f(x) = f(\alpha, F) = f(\alpha, E/F)$. If $M = R \cdot 1 + R\alpha + \cdots + R\alpha^{n-1}$, then

$$M^* = \frac{M}{f'(\alpha)}$$

Moreover,

$$\begin{aligned} R[\alpha] &= R \cdot 1 + R\alpha + \cdots + R\alpha^{n-1} \\ &\subset S \subset S^* \\ &\subset (R[\alpha])^* = \frac{R[\alpha]}{f'(\alpha)} \subset \frac{S}{f'(\alpha)} \end{aligned}$$

Proof. Note that such an α can always be found; in fact, any primitive element for E/F can be multiplied by a suitable element of R to yield a primitive element belonging to S.

From (3-7-12) and (3-7-13) we see that

$$M^* = R\frac{\beta_0}{f'(\alpha)} + \cdots + R\frac{\beta_{n-1}}{f'(\alpha)}$$

The formulas for the β's (plus the fact that $a_0, \ldots, a_{n-1} \in R$) imply that

$$M^* = \frac{R \cdot 1 + R\alpha + \cdots + R\alpha^{n-1}}{f'(\alpha)}$$

Since $f(\alpha, F) \in R[x]$, it follows that $R[\alpha] = R + R\alpha + \cdots + R\alpha^{n-1}$, and the formula for $S_{E\to F}$ (as a coefficient of the minimum polynomial) implies that $S \subset S^*$.

Suppose that $\{\alpha_1, \ldots, \alpha_n\}$ is a basis for E/F, and consider the R module $\mathfrak{A} = R\alpha_1 + \cdots + R\alpha_n$. We put $\Delta(\mathfrak{A}) = \Delta(\alpha_1, \ldots, \alpha_n)$ and call it the ***discriminant of*** $\mathfrak{A}$. Of course, $\Delta(\mathfrak{A})$ is *not* uniquely determined; however, if $\{\alpha_1', \ldots, \alpha_n'\}$ is another basis for $\mathfrak{A}$, then, by (3-7-1), $\Delta(\alpha_1', \ldots, \alpha_n')$ and $\Delta(\alpha_1, \ldots, \alpha_n)$ differ multiplicatively by the square of a unit from R. One often eliminates the ambiguity by dealing with the R ideal [see (1-5-1) for the definition]

$$\boldsymbol{\Delta}(\mathfrak{A}) = \Delta(\mathfrak{A})R$$

In general, any definition or notation related to Δ may be transferred to $\boldsymbol{\Delta}$. We may note that, if $\mathfrak{A} = R\alpha_1 + \cdots + R\alpha_n \subset S$, then $\Delta(\mathfrak{A}) = \Delta(\alpha_1, \ldots, \alpha_n) \in R$. In the case where there exists an integral basis for S over R, we call $\Delta(S)$ the ***discriminant of the extension*** E/F (or of S over R). The discriminant gives a measure for the size of a module; in fact, the following proposition holds.

3-7-15. Proposition. Suppose that $\{\alpha_1, \ldots, \alpha_n\}$ and $\{\beta_1, \ldots, \beta_n\}$ are bases for E/F, and let

$$\mathfrak{A} = R\alpha_1 + \cdots + R\alpha_n \qquad \mathfrak{B} = R\beta_1 + \cdots + R\beta_n$$

If $\mathfrak{A} \subset \mathfrak{B}$, then

$$\frac{\Delta(\mathfrak{A})}{\Delta(\mathfrak{B})} = \frac{\Delta(\alpha_1, \ldots, \alpha_n)}{\Delta(\beta_1, \ldots, \beta_n)}$$

is a square in R; and then $\mathfrak{A} = \mathfrak{B}$ if and only if $\Delta(\mathfrak{A})/\Delta(\mathfrak{B})$ is a unit of R.

Proof. Of course, questions of divisibility are not affected by the choice of representatives for $\Delta(\mathfrak{A})$ and $\Delta(\mathfrak{B})$. If $\mathfrak{A} \subset \mathfrak{B}$, then writing

$$\alpha_i = \sum_{j=1}^{n} a_{ij}\beta_j \qquad a_{ij} \in R,\ j = 1, \ldots, n$$

leads to $\Delta(\mathfrak{A}) = [\det (a_{ij})]^2\, \Delta(\mathfrak{B})$. Moreover, if $\Delta(\mathfrak{A})/\Delta(\mathfrak{B})$ is a unit of R, then the inverse matrix of (a_{ij}) has all its entries from R, and it follows that $\mathfrak{A} = \mathfrak{B}$.

Rather than pursue the general formulation further, let us return to the standard situation of this chapter (namely, that F is complete with respect to a discrete prime divisor and that E/F is a finite extension of degree n) and add the hypothesis that E/F is ***separable.*** Since O_E is the integral closure of O_F in E [see (2-2-10)] and O_F is integrally closed in its quotient field F [see item 7 of (2-2-9)], all the previous results of this section apply. We shall also make use of the fact [see (1-5-1)] that the O_F ideals of F form a group—as do the O_E ideals of E.

3-7-16. Proposition. Suppose that $\mathfrak{A}$ is an O_E ideal and that $\mathfrak{A}^* = \{\beta \in E | S(\beta\mathfrak{A}) \subset O_F\}$ is the complementary set; then

(i) $S(\mathfrak{A})$ is an O_F ideal

(ii) O_E^* is an O_E ideal containing O_E

(iii) $S(\mathfrak{A}) \subset O_F \Leftrightarrow \mathfrak{A} \subset O_E^*$

(iv) $(O_E^*)^{-1}$ is an integral O_E ideal

(v) $\mathfrak{A}^* = O_E^*\mathfrak{A}^{-1}$; that is, $\mathfrak{A}\mathfrak{A}^* = O_E^*$

(vi) $\mathfrak{A}^*$ is an O_E ideal, and $\mathfrak{A}^{-1} \subset \mathfrak{A}^*$

(vii) $\mathfrak{A}$ is an integral O_E ideal $\Leftrightarrow (\mathfrak{A}^*)^{-1} \subset (O_E^*)^{-1}$

(viii) $(\mathfrak{A}^*)^* = \mathfrak{A}$

(ix) $S(O_E^*) = O_F$

(x) $O_E^* = S_{E\to F}^{-1}(O_F)$

Proof. i. It is clear that $S(\mathfrak{A})$ is an O_F module. To see that $S(\mathfrak{A}) \neq (0)$, note that there exists $\alpha \neq 0 \in E$ with $S(\alpha) \neq 0$ [in fact, $S(E) = F$]; then it follows from the properties of ideals and valuations that there exists $a \neq 0 \in O_F$ with $a\alpha \in \mathfrak{A}$. Thus, $S(a\alpha) = aS(\alpha) \neq 0$. Finally, there exists $c \neq 0 \in F$ such that $c\mathfrak{A} \subset O_E$, so that $cS(\mathfrak{A}) \subset S(O_E) \subset O_F$.

ii. It is clear that O_E^* is an O_E module and that $O_E^* \supset O_E$. Now let us write $E = F(\alpha)$ with $\alpha \in O_E$. According to (3-7-14) we have $O_E^* \subset O_E/f'(\alpha)$, where $f(x) = f(\alpha, F) = f(\alpha, E/F)$. This implies that O_E^* is indeed an O_E ideal.

iii. $S(\mathfrak{A}) \subset O_F \Leftrightarrow S(\mathfrak{A}O_E) \subset O_F \Leftrightarrow \mathfrak{A} \subset O_E^*$.

iv. Immediate from (ii).

v. $\beta \in \mathfrak{A}^* \Leftrightarrow S(\beta\mathfrak{A}) \subset O_F \Leftrightarrow S(\beta\mathfrak{A}O_E) \subset O_F \Leftrightarrow \beta\mathfrak{A} \subset O_E^* \Leftrightarrow \beta \in O_E^*\mathfrak{A}^{-1}$.

vi. From (ii), (iv), and (v) we see that $\mathfrak{A}^*$ is an O_E ideal, and $\mathfrak{A}^{-1} = \mathfrak{A}^*(O_E^*)^{-1} \subset \mathfrak{A}^*O_E \subset \mathfrak{A}^*$.

vii. We have $\mathfrak{A} \subset O_E \Leftrightarrow \mathfrak{A} \subset \mathfrak{A}^*(\mathfrak{A}^*)^{-1} \Leftrightarrow \mathfrak{A}\mathfrak{A}^* \subset \mathfrak{A}^* \Leftrightarrow O_E^* \subset \mathfrak{A}^* \Leftrightarrow (\mathfrak{A}^*)^{-1} \subset (O_E^*)^{-1}$.

viii. According to (v) we have $(O_E^*\mathfrak{A}^{-1})(\mathfrak{A}^*)^* = \mathfrak{A}^*(\mathfrak{A}^*)^* = O_E^*$; so $\mathfrak{A}^{-1}(\mathfrak{A}^*)^* = O_E$, and $\mathfrak{A} = (\mathfrak{A}^*)^*$.

ix. Suppose that $S(O_E^*) = \mathfrak{B} < O_F$; so $\mathfrak{B}^{-1} > O_F$. We have $S(\mathfrak{B}^{-1}O_E^*) = \mathfrak{B}^{-1}S(O_E^*) = \mathfrak{B}^{-1}\mathfrak{B} = O_F$, so that $\mathfrak{B}^{-1}O_E^* \subset O_E^*$, and $\mathfrak{B}^{-1}O_E \subset O_E$. This means that $\mathfrak{B}^{-1} \subset O_E \cap F = O_F$, a contradiction.

x. Immediate from (iii) and (ix).

For an arbitrary O_E ideal $\mathfrak{A}$, the O_E ideal $\mathfrak{D}(\mathfrak{A}) = (\mathfrak{A}^*)^{-1}$ is called the ***different of*** $\mathfrak{A}$, and the O_E ideal

$$\mathfrak{D}_{E/F} = \mathfrak{D}(O_E) = (O_E^*)^{-1} = \{\alpha \in E | \alpha O_E^* \subset O_E\}$$

is known as the ***different of*** E/F.

Translating some of the assertions of (3-7-16) to the language of differents, we have the following corollary.

3-7-17. Corollary. Let $\mathfrak{A}$ be an E ideal; then

(i) $\mathfrak{D}_{E/F}$ is an integral E ideal

(ii) $\mathfrak{D}(\mathfrak{A}) = \mathfrak{A}\mathfrak{D}_{E/F}$; in particular, $\mathfrak{D}(\mathfrak{A}) \subset \mathfrak{A}$

(iii) $\mathfrak{A}$ is integral $\Leftrightarrow \mathfrak{D}(\mathfrak{A}) \subset \mathfrak{D}_{E/F}$

(iv) $S(\mathfrak{D}_{E/F}^{-1}) = O_F$ and $\mathfrak{D}_{E/F}^{-1} = S_{E\to F}^{-1}(O_F)$

(v) $\mathfrak{A} \subset \mathfrak{D}_{E/F}^{-1} \Leftrightarrow S(\mathfrak{A})$ is an integral F ideal

A simple but important property of the different is its transitivity; more precisely, the following proposition applies.

3-7-18. Proposition. Suppose that $F \subset E \subset K$ and that K/F is separable; then

$$\mathfrak{D}_{K/F} = \mathfrak{D}_{K/E} \cdot \mathfrak{D}_{E/F}$$

Proof. The right-hand side may be interpreted either as the usual product of two additive groups in a ring or as the ideal product of the K ideals $\mathfrak{D}_{K/E}$ and $\mathfrak{D}_{E/F}O_K$.

It is easy to see that $(\mathfrak{D}_{E/F}O_K)^{-1} = \mathfrak{D}_{E/F}^{-1}O_K$, and therefore $\mathfrak{D}_{K/F} = \mathfrak{D}_{K/E} \cdot \mathfrak{D}_{E/F} \Leftrightarrow \mathfrak{D}_{K/F}^{-1} = \mathfrak{D}_{K/E}^{-1} \cdot \mathfrak{D}_{E/F}^{-1}$. Now, for $\alpha \in K$ we have

$$\begin{aligned}
\alpha \in \mathfrak{D}_{K/F}^{-1} &\Leftrightarrow S_{K\to F}(\alpha O_K) \subset O_F \\
&\Leftrightarrow S_{E\to F}[S_{K\to E}(\alpha O_K)] \subset O_F \\
&\Leftrightarrow S_{K\to E}(\alpha O_K) \subset \mathfrak{D}_{E/F}^{-1} \\
&\Leftrightarrow [S_{K\to E}(\alpha O_K)]\mathfrak{D}_{E/F} \subset O_E \\
&\Leftrightarrow S_{K\to E}(\alpha \mathfrak{D}_{E/F} O_K) \subset O_E \\
&\Leftrightarrow \alpha \mathfrak{D}_{E/F} \subset \mathfrak{D}_{K/E}^{-1} \\
&\Leftrightarrow \alpha \mathfrak{D}_{E/F} O_K \subset \mathfrak{D}_{K/E}^{-1} \\
&\Leftrightarrow \alpha \in \mathfrak{D}_{K/E}^{-1} \cdot \mathfrak{D}_{E/F}^{-1}
\end{aligned}$$

This completes the proof.

According to (2-3-2), we know that O_E has an integral basis over O_F. If $\mathfrak{A}$ is any O_E ideal, say, $\mathfrak{A} = \alpha O_E$, and $\{\omega_1, \ldots, \omega_n\}$ is an integral basis, then the set $\{\alpha\omega_1, \ldots, \alpha\omega_n\}$ is a basis for E/F and $\mathfrak{A} = R(\alpha\omega_1) + \cdots + R(\alpha\omega_n)$. Thus, in view of the remarks preceding (3-7-15), $\Delta_{E/F} = \Delta(O_E) = \Delta(\omega_1, \ldots, \omega_n)$ is the discriminant of the extension E/F, and $\Delta(\mathfrak{A}) = \Delta(\alpha\omega_1, \ldots, \alpha\omega_n)$ is the discriminant of the ideal $\mathfrak{A}$. In accordance with our convention, we write $\mathbf{\Delta}_{E/F} = \Delta_{E/F}O_F$. From (3-7-8) it follows that $\Delta(\mathfrak{A}) = [N_{E\to F}(\alpha)]^2\,\Delta_{E/F}$.

Consider again any E ideal $\mathfrak{A} = \alpha O_E$, and put

$$N_{E\to F}(\mathfrak{A}) = (N_{E\to F}\alpha)O_F$$

It is clear that $N_{E\to F}(\mathfrak{P}_E) = \mathfrak{P}_E^{f(E/F)}$, that $N_{E\to F}(\mathfrak{A})$ is the F ideal generated by $\{N_{E\to F}(\beta) | \beta \in \mathfrak{A}\}$, and that $N_{E\to F}$ is a monomorphism of the group of E ideals into the group of F ideals. Note also that, if $F \subset E \subset K$, then $N_{K\to F} = N_{E\to F} \circ N_{K\to E}$. The formula for $\Delta(\mathfrak{A})$ now reads

$$\Delta(\mathfrak{A})O_F = \mathbf{\Delta}_{E/F}[N_{E\to F}(\mathfrak{A})]^2$$

We shall soon give a detailed description of $\mathfrak{D}_{E/F}$ and $\mathbf{\Delta}_{E/F}$; but first it is useful to point out how they are related.

3-7-19. Proposition. $N_{E\to F}(\mathfrak{D}_{E/F}) = \Delta_{E/F}O_F = \mathbf{\Delta}_{E/F}$.

Proof. Consider an integral basis $\{\omega_1, \ldots, \omega_n\}$ for O_E over O_F, and let $\{\omega_1^*, \ldots, \omega_n^*\}$ be the complementary basis. Thus

$$O_E = O_F\omega_1 \oplus \cdots \oplus O_F\omega_n$$

and, according to (3-7-13),

$$O_E^* = O_F\omega_1^* \oplus \cdots \oplus O_F\omega_n^*$$

Therefore, by (3-7-4),

$$\Delta(O_E)\Delta(O_E^*) = \Delta(\omega_1, \ldots, \omega_n)\Delta(\omega_1^*, \ldots, \omega_n^*) = 1$$

Consequently,

$$O_F = \Delta(O_E)\Delta(O_E^*)O_F = [\Delta(O_E)][\Delta_{E/F}][N_{E\to F}(O_E^*)]^2$$

so that $(\Delta_{E/F}O_F)^2 = [N_{E\to F}(\mathfrak{D}_{E/F})]^2$ and $\Delta_{E/F}O_F = N_{E\to F}(\mathfrak{D}_{E/F})$.

The transitivity property of the discriminant is now easily determined.

3-7-20. Corollary. Suppose that $F \subset E \subset K$ and that K/F is separable; then

$$\mathbf{\Delta}_{K/F} = \mathbf{\Delta}_{E/F}^{[K:E]}N_{E\to F}(\mathbf{\Delta}_{K/E})$$

Proof. Because $\mathfrak{D}_{E/F}$ is a principal ideal, it is easy to see that $N_{K\to E}(\mathfrak{D}_{E/F}O_K) = \mathfrak{D}_{E/F}^{[K:E]}$. Therefore,

$$\begin{aligned}\mathbf{\Delta}_{K/F} = N_{K\to F}(\mathfrak{D}_{K/F}) &= N_{E\to F}\{N_{K\to E}(\mathfrak{D}_{K/E}\mathfrak{D}_{E/F})\}\\ &= N_{E\to F}[\mathbf{\Delta}_{K/E}\cdot N_{K\to E}(\mathfrak{D}_{E/F}O_K)]\\ &= N_{E\to F}(\mathbf{\Delta}_{K/E})\mathbf{\Delta}_{E/F}^{[K:E]}\end{aligned}$$

Of course, the proof could also be based on (3-7-10).

The key step in finding $\mathfrak{D}_{E/F}$ and $\mathbf{\Delta}_{E/F}$ is to connect them with the different and discriminant of a single element. More precisely, the following proposition applies.

3-7-21. Proposition. For any $\alpha \in O_E$ put $f_\alpha(x) = f(\alpha, E/F)$. If $E = F(\alpha)$ with $\alpha \in O_E$, then $f'_\alpha(\alpha) \in \mathfrak{D}_{E/F}$ and

i. $\{1, \alpha, \ldots, \alpha^{n-1}\}$ is an integral basis for E/F if and only if $\mathfrak{D}_{E/F} = f'_\alpha(\alpha)O_E$.

ii. $\{1, \alpha, \ldots, \alpha^{n-1}\}$ is an integral basis for E/F if and only if $\mathbf{\Delta}_{E/F} = \Delta(1, \alpha, \ldots, \alpha^{n-1})O_F$.

Proof. We have $f_\alpha(x) = f(\alpha, F)$, and from the properties of $N_{E\to F}$ it follows immediately that

$$\mathfrak{D}_{E/F} = f'_\alpha(\alpha)O_E \Leftrightarrow \mathbf{\Delta}_{E/F} = \Delta(1, \alpha, \ldots, \alpha^{n-1})O_F$$

Furthermore, (3-7-14) says that

$$\begin{aligned} O_F[\alpha] &= O_F \cdot 1 \oplus O_F\alpha \oplus \cdots \oplus O_F\alpha^{n-1} \\ &\subset O_E \subset O_E^* \\ &\subset (O_F[\alpha])^* = \frac{O_F[\alpha]}{f'_\alpha(\alpha)} \subset \frac{O_E}{f'_\alpha(\alpha)} \end{aligned}$$

In particular $f'_\alpha(\alpha) \in \mathfrak{D}_{E/F}$. Moreover,

$$\begin{aligned} \{1, \alpha, \ldots, \alpha^{n-1}\} \text{ is an integral basis} &\Leftrightarrow O_E = O_F[\alpha] \\ &\Leftrightarrow O_E^* = (O_F[\alpha])^* \\ &\qquad = \frac{O_F[\alpha]}{f'_\alpha(\alpha)} = \frac{O_E}{f'_\alpha(\alpha)} \\ &\Leftrightarrow (O_E^*)^{-1} = \left(\frac{O_E}{f'_\alpha(\alpha)}\right)^{-1} \\ &\Leftrightarrow \mathfrak{D}_{E/F} = f'_\alpha(\alpha)O_E \end{aligned}$$

3-7-22. Corollary. We have

(i) $\nu_F(\Delta_{E/F}) = \min\limits_{\alpha_1, \ldots, \alpha_n \in O_E} \{\nu_F[\Delta(\alpha_1, \ldots, \alpha_n)]\}$

Moreover, if E/F has an integral basis of form $1, \alpha, \ldots, \alpha^{n-1}$, then

(ii) $\nu_E(\mathfrak{D}_{E/F}) = \min\limits_{\beta \in O_E} \{\nu_F[f'_\beta(\beta)]\}$

(iii) $\nu_F(\Delta_{E/F}) = \min\limits_{\beta \in O_E} \{\nu_F[\Delta(1, \beta, \ldots, \beta^{n-1})]\}$

Proof. i. Fix an integral basis $\{\beta_1, \ldots, \beta_n\}$ for E/F. If $\alpha_1, \ldots, \alpha_n$ are linearly dependent, then $\Delta(\alpha_1, \ldots, \alpha_n) = 0$. On the other hand, if $\{\alpha_1, \ldots, \alpha_n\}$ is a basis for E/F, then (3-7-15) applies.

ii. If $F(\beta) \neq E$, then $f'_\beta(\beta) = 0$; if $F(\beta) = E$, then, by (3-7-14), $f'_\beta(\beta)O_E^* \subset O_E$, so that $f'_\beta(\beta) \in (O_E^*)^{-1} = \mathfrak{D}_{E/F}$.

The proof of (iii) is now trivial.

One way to guarantee the existence of an integral basis of form $1, \alpha, \ldots, \alpha^{n-1}$ is to require that $\bar{E}/\bar{F}$ be separable [see (3-3-3)]; of course, this additional hypothesis is satisfied in the situations of real interest.

3-7-23. Theorem. Suppose that E/F is a finite separable extension with $\bar{E}/\bar{F}$ separable; then

$$\mathfrak{D}_{E/F} = \mathfrak{P}_E^m = \pi_E^m O_E$$

where the ***differential exponent*** $m = m(E/F)$ has the properties:

(i) E/F is tamely ramified $\Leftrightarrow m = e - 1$.

(ii) E/F is wildly ramified $\Leftrightarrow e \leq m \leq \nu_E(e) + e - 1$.

Proof. We need to compute $m = \nu_E[f'(\alpha)]$ when $\{1, \alpha, \ldots, \alpha^{n-1}\}$ is an integral basis and $f(x) = f(\alpha, F)$.

Suppose that E/F is unramified. Then, according to (3-2-6), α may be chosen so that $\bar{\alpha}$ is a simple root of $\bar{f}(x)$, which says that $f'(\alpha) \in U_E$. Consequently, $m = 0 = e - 1$ in this case.

Suppose next that E/F is totally ramified—so that $e = n$. According to (3-3-1), α may be taken to be any prime element of E. In particular, the minimum polynomial

$$f(x) = f(\alpha, F) = x^e + a_1x^{e-1} + \cdots + a_e$$

is Eisenstein. Consider

$$f'(\alpha) = ea_0\alpha^{e-1} + (e-1)a_1\alpha^{e-2} + \cdots + a_{e-1} \qquad (a_0 = 1)$$

For $r = 0, 1, \ldots, e-1$ we have

$$\nu_E[(e-r)a_r\alpha^{e-r-1}] = e\nu_F(e-r) + e\nu_F(a_r) + e - r - 1$$

and so the individual terms of $f'(\alpha)$ have distinct ordinals. Therefore,

$$\nu_E[f'(\alpha)] = \min_{r=0,1,\ldots,e-1} \{\nu_E[(e-r)a_r\alpha^{e-r-1}]\}$$

If, now, $p = \operatorname{char} \bar{F}$ does not divide e (that is, if E/F is both totally and tamely ramified), then $\nu_F(e) = 0$ and the minimal ordinal is $e - 1$. Consequently, $m = e - 1$ in this case.

If, on the other hand, p divides e (that is, if E/F is both totally and wildly ramified), then $\nu_F(e) \geq 1$ and we can conclude only that

$$e \leq \nu_E[f'(\alpha)] \leq \nu_E(e) + e - 1$$

Now, let us consider the general case. We have

$$\mathfrak{D}_{E/F} = \mathfrak{D}_{E/T}\mathfrak{D}_{T/F} = \mathfrak{D}_{E/T}O_T = \mathfrak{D}_{E/T}$$

so that $m(E/F) = m(E/T)$. Since $e(E/F) = e(E/T)$, the results for the totally ramified case apply, and the proof is complete.

3-7-24. Corollary. Let E/F be a separable extension such that $\bar{E}/\bar{F}$ is separable; then

$$\frac{E}{F} \text{ is unramified} \Leftrightarrow \mathfrak{D}_{E/F} = O_E \Leftrightarrow \Delta_{E/F} = O_F$$

3-7-25. Proposition. Let E/F be a Galois extension such that $\bar{E}/\bar{F}$ is separable; then

$$m\left(\frac{E}{F}\right) = \sum_{r=0}^{\infty} [\#(\mathcal{V}_r) - 1]$$

Proof. Choose $\alpha \in O_E$ so that $\{1, \alpha, \ldots, \alpha^{n-1}\}$ is an integral basis for E/F. With $\nu = \nu_E$, $f(x) = f(\alpha, F)$, $\mathcal{G} = \mathcal{G}(E/F)$, we have

$$m\left(\frac{E}{F}\right) = \nu[f'(\alpha)] = \sum_{\sigma \neq 1 \in \mathcal{G}} \nu(\alpha - \sigma\alpha)$$

For $\sigma \neq 1 \in \mathcal{G}$, let us define an integer $\nu(\sigma) \geq -1$ by the property $\sigma \in \mathcal{V}_{\nu(\sigma)}$, $\sigma \notin \mathcal{V}_{\nu(\sigma)+1}$. In other words,

$$\nu(\sigma) + 1 = \min_{\beta \in O_E} \{\nu(\sigma\beta - \beta)\}$$

In the course of the proof of (3-6-2) it was observed that

$$\nu(\sigma\alpha - \alpha) = \min_{\beta \in O_E} \{\nu(\sigma\beta - \beta)\}$$

Therefore, since $\nu(\sigma) + 1$ is the number of ramification groups (starting with $\mathcal{V}_0$) to which σ belongs, it follows that

$$\sum_{\sigma \neq 1} \nu(\sigma\alpha - \alpha) = \sum_{\sigma \neq 1} [\nu(\sigma) + 1] = \sum_{r=0}^{\infty} [\#(\mathcal{V}_r) - 1]$$

If, for $r \geq 0$, we put $n_r = \#(\mathcal{V}_r)$, then our result may be expressed in the form of "***Hilbert's formula,***"

$$m\left(\frac{E}{F}\right) = n_0 - n_1 + 2(n_1 - n_2) + \cdots + r(n_{r-1} - n_r) + \cdots$$

3-7-26. Example. Let $F = \mathbf{Q}_p$ and $E = \mathbf{Q}_p(\sqrt{p})$. In view of (3-7-25) and (3-6-9), it is trivial to compute $\mathfrak{D}_{E/F}$. However, $\mathfrak{D}_{E/F}$ may also be computed directly. Since $O_E = O_F \cdot 1 + O_F \sqrt{p}$, it is easy to see that, for $p \neq 2$,

$$S\left(\frac{1}{\sqrt{p}} O_E\right) \subset O_F \quad \text{and} \quad S\left(\frac{1}{(\sqrt{p})^2} O_E\right) \not\subset O_F$$

Therefore, $O_E^* = \mathcal{P}_E^{-1}$, and $\mathfrak{D}_{E/F} = \mathcal{P}_E$. On the other hand, if $p = 2$, then

$$S\left(\frac{1}{(\sqrt{2})^3} O_E\right) \subset O_F \quad \text{and} \quad S\left(\frac{1}{(\sqrt{2})^4} O_E\right) \not\subset O_F$$

so that $\mathfrak{D}_{E/F} = \mathcal{P}_E^3$.

EXERCISES

The hypotheses and notation are those which have been in force throughout this chapter.

3-1. Find $\sqrt{-2}$ in $\mathbf{Q}_3$.

3-2. Let $f(x)$ denote a monic polynomial of degree n in $F[x]$ with the factorization

$$f(x) = \prod_1^r (x - \alpha_i)^{m_i} \qquad m_i \geq 1,\ \alpha_i \neq \alpha_j \text{ for } i \neq j,\ \alpha_i \in \Omega$$

and put $\alpha = \alpha_1$. Let $\bar{\nu}$ be the valuation on $F(x)$ [and also on $\Omega(x)$] as described in Exercise 1-3.

a. If $\bar{\nu}(f) \geq M$, then $\nu(\alpha_i) \geq M/n$ for $i = 1, \ldots, r$.

b. If $g(x) \in F[x]$ is monic, of degree n, and sufficiently close to $f(x)$ (in the topology determined by $\bar{\nu}$), then, for each $i = 1, \ldots, r$, $g(x)$ has m_i roots lying as close as desired to α_i.

c. Moreover, if $f(x)$ is separable and irreducible and $\beta \in \Omega$ is a root of $g(x)$ sufficiently close to α [β exists, by (*b*)], then $g(x)$ is irreducible and $F(\beta) = F(\alpha)$.

d. The monic irreducible polynomials of degree n in $F[x]$ form an open set.

3-3. *a.* Give an example to show that Ω need not be complete.

b. Show that the completion of Ω is algebraically closed.

3-4. Use the Newton polygon to show that an Eisenstein polynomial is irreducible and gives rise to a totally ramified extension.

3-5. If E/F is totally ramified and T/F is unramified, then TE/T is totally ramified.

3-6. Give an example to show that the composite of two totally ramified extensions need not be totally ramified.

3-7. If the residue class field $\bar{F}$ is finite, then a subset B of F is compact if and only if it is closed and bounded [in the sense that $\{\varphi(b) | b \in B\}$ is bounded above].

3-8. Discuss the extension $\mathbf{Q}_p(\zeta)/\mathbf{Q}_p$, where ζ is a primitive nth root of unity and n is arbitrary.

3-9. Suppose that $\bar{F}$ is finite and that T/F is unramified; then:

a. If $m \geq 1$ and $u_m \in 1 + \mathcal{P}_F^m$, then there exists $\mu_m \in 1 + \mathcal{P}_T^m$ such that $N\mu_m \equiv u_m \pmod{\mathcal{P}_F^{m+1}}$.

b. If $u \in 1 + \mathcal{P}_F$, then there exists $\mu \in U_T$ such that $N\mu = u$.

c. $N(U_T) = U_F$.

3-10. *a.* If $\sigma \in \mathcal{V}_i$, $\tau \in \mathcal{V}_j$ with $i, j \geq 1$, then $\sigma\tau\sigma^{-1}\tau^{-1} \in \mathcal{V}_{i+j}$.

b. $\mathcal{V}_i/\mathcal{V}_{2i}$ is abelian, and so is $\mathcal{V}_i/\mathcal{V}_{i+1}$.

c. If $p \neq 0$ and $j = \min\{pi, \nu_E(p) + i\} \geq i + 1 \geq 2$, then $\sigma \in \mathcal{V}_i \Rightarrow \sigma^p \in \mathcal{V}_j$.

d. At a discontinuity in the chain of ramification groups, $\mathcal{V}_i/\mathcal{V}_{i+1}$ is abelian of type $(p, \ldots, p)$.

3-11. *a.* If $m(E/F) = 1$, then $e = 2$. When char $\bar{F} = 2$, then $m(E/F) = 3 \Rightarrow e = 2$.

b. Suppose that $e \geq 2$ is given and that e is divisible by char F = char $\bar{F}$; then there exist totally ramified extensions E/F with $e(E/F) = e$ and $m(E/F)$ arbitrarily large.

3-12. Consider the field of formal power series $F = k\langle x\rangle$ as described in (1-9-5). If k' is a finite separable extension of k, then $E = k'\langle x\rangle$ is unramified over F. Moreover, every unramified extension of F arises in this way.

3-13. (*Swan*) Suppose that F is the completion of an algebraic number field at a nonarchimedean prime and let $f(x) \in O_F[x]$ be a monic polynomial of degree n

with roots $\alpha_1, \ldots, \alpha_n$ and such that $\bar{f}(x) \in \bar{F}[x]$ has no repeated roots. Let r be the number of irreducible factors of $\bar{f}(x)$ over $\bar{F}$ and put $\delta(f) = \prod_{i<j} (\alpha_i - \alpha_j)$.

a. If E denotes the splitting field of $f(x)$, then E/F is a cyclic unramified extension.

b. If σ is a generator of $\mathcal{G}(E/F)$, then $\sigma(\delta(f)) = (-1)^{n-r}\delta(f)$.

c. D_f is a square in $F \Leftrightarrow r \equiv n \pmod 2$.

d. Suppose that $F = \mathbf{Q}_2$, so that $\bar{F} = \{0, 1\}$; then $r \equiv n \pmod 2 \Leftrightarrow D_f \equiv 1 \pmod 8$.

e. A polynomial of form $x^{8k} + x^m + 1$ with odd $m > 0$ is reducible over the two-element field when $8k > m$.

4

Ordinary Arithmetic Fields

In this chapter we turn to the problem of unique factorization in an algebraic number field F. The first thing one must do is to select a subring O of F which will serve as an analogue of $\mathbf{Z}$ in $\mathbf{Q}$. Naturally, there are many properties that one wants O to satisfy—for example, its quotient field should be F and $O \cap \mathbf{Q}$ should be $\mathbf{Z}$—and it is more than just semantics which leads to the choice for O of the integral closure of $\mathbf{Z}$ in F. Unfortunately, one soon discovers that O need not be a unique factorization domain. Let us describe the standard counterexample.

Consider the field $F = \mathbf{Q}(\sqrt{-5})$. The first step is to show that

$$O = \mathbf{Z} \oplus \mathbf{Z}\sqrt{-5}$$

Any $\alpha \in F^*$ may be expressed in the form $\alpha = (a + b\sqrt{-5})/c$, where a, b, c are ordinary integers with no common factor and $c > 0$. Of course, $\alpha \in O$ if and only if its conjugate $\bar{\alpha} = (a - b\sqrt{-5})/c \in O$. Therefore, from the minimum polynomial of α we see that

$$\alpha \in O \Leftrightarrow \frac{2a}{c} \in \mathbf{Z} \text{ and } \frac{a^2 + 5b^2}{c^2} \in \mathbf{Z}$$

It is clear that $\mathbf{Z} + \mathbf{Z}\sqrt{-5} \subset O$. On the other hand, we observe that $\alpha \in O$ implies that $4a^2/c^2 \in \mathbf{Z}$ and $20b^2/c^2 \in \mathbf{Z}$. If $p \neq 2$ is a prime dividing c, it follows that $p|a$ and $p|b$, a contradiction. The same kind of argument shows that 4 does not divide c. Hence, $c = 1$ or 2. If $c = 2$, then $(a^2 + 5b^2)/4 \in \mathbf{Z}$. Since $a^2 + 5b^2 \equiv 0 \pmod 4 \Rightarrow a^2 \equiv b^2 \equiv 0 \pmod 4 \Rightarrow 2|a$ and $2|b$, we have a contradiction. The conclusion is that $c = 1$, and we have thus found O.

In order to deal with factorization in O, one makes use of the norm from F to $\mathbf{Q}$. For $\alpha = a + b\sqrt{-5} \in O$ we have $N(\alpha) = a^2 + 5b^2$; so

N maps O into $\mathbf{Z}$ and is multiplicative. If $\alpha \in O$ is a unit, so that $\alpha | 1$, it is immediate that $N(\alpha) = 1$; consequently, ± 1 are the only units of O. Since the equation $a^2 + 5b^2 = 3$ has no solution in integers, it follows that the elements $3, 2 + \sqrt{-5}, 2 - \sqrt{-5} \in O$ are irreducible (an element of O is said to be *irreducible* when it is divisible only by the units and its associates). By viewing the norm as an analogue of the absolute value in $\mathbf{Z}$, one may show that every element of O can be factored into irreducible elements; however, the factorization is not necessarily unique—as may be seen from

$$9 = 3 \cdot 3 = (2 + \sqrt{-5})(2 - \sqrt{-5})$$

Perhaps the first approach to the problem of somehow recapturing unique factorization in an arbitrary algebraic number field was by Kummer; he succeeded for the class of fields of roots of unity. His interest in these fields stemmed from an attempt to deal with Fermat's last theorem by factoring $z^n = x^n + y^n$ in the field of nth roots of unity.

A complete solution of the problem was developed by Dedekind (see [4]), who introduced the notions of algebraic integer and ideal. His ideal-theoretic approach was subsequently refined and axiomatized. Another completely equivalent solution of the problem was given by Kronecker. His objectives were wider in scope than just number-theoretic. Because his techniques centered about the adjunction of indeterminates, they have not been particularly fashionable.

We shall present a valuation-theoretic approach to ideal theory; its origins go back to Hensel, who was a student of Kronecker. The formulation is sufficiently general so that our considerations apply not only to algebraic number fields but also to algebraic function fields in one variable—that is, to finite extensions of fields of rational functions $k(x)$.

4-1. Axioms and Basic Properties

We begin with a definition. An ***ordinary arithmetic field*** (OAF) is a pair $\{F, \mathcal{S}\}$, where F is a field, $\mathcal{S}$ is a nonempty collection of discrete prime divisors of F, and such that the following axioms are satisfied:

I. For each $a \in F$, we have $\nu_P(a) \geq 0$ for almost all $P \in \mathcal{S}$ (where $\nu_P \in P$ is the normalized valuation).

II. Given any $P_1, P_2 \in \mathcal{S}$ with $P_1 \neq P_2$, there exists an element $a \in F$ such that

$$\begin{aligned} \nu_{P_1}(a - 1) &\geq 1 \\ \nu_{P_2}(a) &\geq 1 \\ \nu_P(a) &\geq 0 \text{ for all other } P \in \mathcal{S} \end{aligned}$$

If we put

$$O = O\{\mathcal{S}\} = O\{F, \mathcal{S}\} = \{a \in F | \nu_P(a) \geq 0 \quad \forall P \in \mathcal{S}\}$$

then O is clearly a ring; it is known as the ***ring of integers*** of $\{F, \mathcal{S}\}$.

4-1-1. Proposition. Suppose that D is a principal ideal domain with quotient field F. Let $\mathcal{S}$ denote the set of prime divisors of F which are determined by the prime elements of D; then $\{F, \mathcal{S}\}$ is an OAF and $O = O\{\mathcal{S}\} = D$.

Proof. We recall that a principal ideal domain D is a unique factorization domain and that each prime element $p \in D$ determines a discrete prime divisor $P = P_p \in \mathcal{S}$. In fact, $\nu_P \in P$ is given as follows: $\nu_P(0) = \infty$, and, for any $a \neq 0 \in F$, $\nu_P(a)$ is the exponent to which p appears in the factorization of the element a.

Since $a \neq 0 \in F$ has only a finite number of primes p appearing in its denominator, it is clear that (I) holds. As for (II), consider distinct primes p_1, p_2 of D. There exist $r, s \in D$ such that $rp_1 + sp_2 = 1$, so that $a = sp_2 \in F$ is an element that satisfies (II). The fact that $O = D$ is immediate from the description of the elements of $\mathcal{S}$.

4-1-2. Corollary. If for the rational field $\mathbf{Q}$ we take $\mathcal{S}$ as the set of all nonarchimedean prime divisors of $\mathbf{Q}$, then $\{\mathbf{Q}, \mathcal{S}\}$ is an OAF whose ring of integers is $\mathbf{Z}$.

4-1-3. Corollary. If for the field of rational functions $F = k(x)$ we take $\mathcal{S}$ as the set of all prime divisors which are trivial on k and arise from irreducible polynomials in $k[x]$ (so that only the prime divisor arising from $1/x$ is excluded), then $\{F, \mathcal{S}\}$ is an OAF whose ring of integers is $k[x]$.

4-1-4. Proposition. In the definition of OAF, axiom II may be replaced by either of the following:

II*a*. Given any $P_1, P_2 \in \mathcal{S}$ with $P_1 \neq P_2$ and any integers m_1, m_2, there exists an element $a \in F$ such that

$$\begin{aligned} \nu_{P_1}(a - 1) &\geq m_1 \\ \nu_{P_2}(a) &\geq m_2 \\ \nu_P(a) &\geq 0 \text{ for all other } P \in \mathcal{S} \end{aligned}$$

IIb. If $S = \{P_1, \ldots, P_r\}$ is any finite subset of $\mathcal{S}$, then for any elements $a_1, \ldots, a_r \in F$ and any integers $m_1, \ldots, m_r$ there exists an element $a \in F$ such that

$$\begin{aligned} \nu_{P_i}(a - a_i) &\geq m_i & i &= 1, \ldots, r \\ \nu_P(a) &\geq 0 & P &\notin S, P \in \mathcal{S} \end{aligned}$$

Proof. (II) $\Rightarrow$ (IIa). There is no loss of generality in using a single $m \geq 1$ instead of both m_1 and m_2. Choose $b \in F$ such that $\nu_{P_1}(b - 1) \geq 1$, $\nu_{P_2}(b) \geq 1$, and $\nu_P(b) \geq 0$ for $P \neq P_1, P_2$. Consider the binomial expansion of

$$1 = [b + (1 - b)]^{2m-1}$$

It follows that we can express 1 in the form

$$1 = b^m g(b) + (1 - b)^m h(b)$$

with $g(b)$, $h(b) \in \mathbf{Z}[b]$. Since $b \in O$, both $g(b)$ and $h(b)$ belong to O. It is now clear that the element $a = b^m g(b)$ satisfies the requirements of (IIa).

(IIa) $\Rightarrow$ (IIb). The first step is to show that for each $i = 1, \ldots, r$ there exists an element $b_i \in F$ such that

$$\begin{aligned} \nu_{P_i}(b_i - 1) &\geq m_i & i &= 1, \ldots, r \\ \nu_{P_j}(b_i) &\geq m_j & j &\neq i \\ \nu_P(b_i) &\geq 0 & P &\notin S, P \in \mathcal{S} \end{aligned}$$

To do this, we may assume without loss of generality that $m_i \geq 1$ for $i = 1, \ldots, r$. Fix i momentarily; then for each $j \neq i$ we can find [according to (IIa)] an element $c_{ij} \in F$ such that

$$\begin{aligned} \nu_{P_i}(c_{ij} - 1) &\geq m_i \\ \nu_{P_j}(c_{ij}) &\geq m_j \\ \nu_P(c_{ij}) &\geq 0 \qquad P \neq P_i, P_j \end{aligned}$$

Put

$$b_i = \prod_{j \neq i} c_{ij}$$

Since each $c_{ij} \in 1 + \mathcal{P}_i^{m_i}$ (where as usual $\mathcal{P}_i = \{a \in F | \nu_{P_i}(a) \geq 1\}$), so is b_i—that is, $\nu_{P_i}(b_i - 1) \geq m_i$. The remaining requirements for b_i are clearly satisfied.

We are now in a position to prove (IIb). Expand S so that it includes the finite set of all $P \in \mathcal{S}$ for which some $\nu_P(a_i) < 0$. Let the expanded set be $\bar{S} = \{P_1, \ldots, P_r, P_{r+1}, \ldots, P_n\}$, and put $a_{r+1} = \cdots = a_n = 0$, $m_{r+1} = \cdots = m_n = 0$. We may assume that $m_i \geq 0$ for $i = 1, \ldots, r$. In view of the preliminary result just proved, there exists for each $i = 1, \ldots, n$ an element $b_i \in F$ with

$$\begin{aligned} \nu_{P_i}(b_i - 1) &\geq m_i - \nu_{P_i}(a_i) & \\ \nu_{P_j}(b_i) &\geq m_j - \nu_{P_j}(a_i) & j \neq i, j \in \{1, \ldots, n\} \\ \nu_P(b_i) &\geq 0 & P \notin \bar{S}, P \in \mathcal{S} \end{aligned}$$

(Note that some of these conditions are vacuous, since their right side is $-\infty$.) Taking

$$a = \sum_1^n b_i a_i = \sum_1^r b_i a_i$$

it is easy to verify that a is an element with the desired properties.

4-1-5. Remarks. 1. Axiom I is equivalent to the requirement that for any $a \in F^*$ we have $\nu_P(a) = 0$ for almost all $P \in \mathcal{S}$. To see this, one simply applies (I) to both a and a^{-1}. Axiom IIb is often referred to as the ***strong approximation theorem.*** Its conclusion remains valid when the a_i are taken from the completions F_{P_i}. Note that, if all m_i are ≥ 0 and the a_i belong to O, then a belongs to O; of course, the m_i may be replaced by a single integer m.

2. Suppose that $\mathcal{S}$ is a finite set of discrete prime divisors of a field F; then $\{F, \mathcal{S}\}$ is automatically an OAF. In fact, in this case, the strong approximation theorem and the "ordinary approximation theorem" (1-2-3) both make the same assertion.

3. Consider the OAF $\{\mathbf{Q}, \mathcal{S}\}$ as described in (4-1-2). Translating to the language of congruences, one observes that our axioms become rather familiar statements. In particular, the content of the strong approximation theorem is precisely that of the ***Chinese remainder theorem.*** More precisely, if $a_1, \ldots, a_s$ are integers and $n_1, \ldots, n_s$ are positive integers that are relatively prime in pairs, then the congruences

$$x \equiv a_i \pmod{n_i} \qquad i = 1, \ldots, s$$

have a common integral solution.

4. Suppose that $\{F, \mathcal{S}\}$ is an OAF; then, for any nonempty subset S of $\mathcal{S}$, $\{F, S\}$ is also an OAF. We put

$$O\{S\} = \{a \in F | \nu_P(a) \geq 0 \quad \forall P \in S\} = \bigcap_{P \in S} O_P$$

where, as usual, $O_P = \{a \in F | \nu_P(a) \geq 0\}$. In keeping with the terminology introduced at the beginning of this section, $O\{S\}$ is called the ***ring of S integers*** of F. Clearly $S_1 \subset S_2 \Rightarrow O\{S_1\} \supset O\{S_2\}$.

4-1-6. Corollary. Suppose that $\{F, \mathcal{S}\}$ is an OAF. Consider any finite subset $S = \{P_1, \ldots, P_r\}$ of $\mathcal{S}$, and let $m_1, \ldots, m_r$ be arbitrary integers; then there exists an element $a \in F$ such that

$$\begin{aligned} \nu_{P_i}(a) &= m_i & i &= 1, \ldots, r \\ \nu_P(a) &\geq 0 & P &\notin S, P \in \mathcal{S} \end{aligned}$$

Proof. For each $i = 1, \ldots, r$ choose $a_i \in F$ with $\nu_{P_i}(a_i) = m_i$. Then choose $a \in F$ such that

$$\begin{aligned} \nu_{P_i}(a - a_i) &\geq m_i + 1 & i &= 1, \ldots, r \\ \nu_P(a) &\geq 0 & P &\notin S \end{aligned}$$

Suppose that E is an extension field of the OAF $\{F, \mathcal{S}\}$. We shall denote by $\mathcal{S}^E$ the set of all extensions to E of the prime divisors belonging to $\mathcal{S}$. The same notation will apply for any subset S of $\mathcal{S}$. In particular, if S consists of a single prime divisor P, then $P^E = S^E = \{Q \in \mathcal{S}^E | Q \supset P\}$. If T is a subset of $\mathcal{S}^E$, we shall write T_F for the subset of $\mathcal{S}$ consisting of all the prime divisors of F which are restrictions of prime divisors belonging to T. It is then convenient to refer to $(T_F)^E$ as the set of prime divisors of E gotten by *saturating* T (with respect to F).

4-1-7. Theorem. Let $\{F, \mathcal{S}\}$ be an OAF; then $O\{\mathcal{S}\}$ is integrally closed in its quotient field F. If E/F is a finite extension, then $\{E, \mathcal{S}^E\}$ is an OAF and $O\{\mathcal{S}^E\}$ is the integral closure of $O\{\mathcal{S}\}$ in E. Furthermore, any element $\alpha \in E$ can be written in the form $\alpha = \beta/c$, where $\beta \in O\{\mathcal{S}^E\}$ and $c \in O\{\mathcal{S}\}$.

Proof. According to item 7 of (2-2-9) and Exercise 2-1, we know that each O_P is integrally closed in F and, therefore, so is

$$O\{\mathcal{S}\} = \bigcap_{P \in \mathcal{S}} O_P$$

To see that F is the quotient field of $O\{\mathcal{S}\}$, consider any $a \neq 0 \in F$, and let $S = \{P_1, \ldots, P_r\} \subset \mathcal{S}$ be the finite set of prime divisors for which $\nu_P(a) \neq 0$. By (4-1-6) there exists $b \in F$ such that

$$\begin{aligned} \nu_{P_i}(b) &= |\nu_{P_i}(a)| & i &= 1, \ldots, r \\ \nu_P(b) &\geq 0 & P &\notin S \end{aligned}$$

Thus, $b \in O\{\mathcal{S}\}$, and $c = ab \in O\{\mathcal{S}\}$. If $S = \emptyset$, then $a \in O\{\mathcal{S}\}$, while if $S \neq \emptyset$, then $b \neq 0$ and $a = c/b$.

Consider now the finite extension E/F. Every $P \in \mathcal{S}$ has a finite number of extensions to E, and each one is discrete—hence, $\mathcal{S}^E$ consists of discrete prime divisors.

For any $\alpha \neq 0 \in E$ we let $f(x) = f(\alpha, F)$ and have, therefore, an equation of the form

$$\alpha^n + a_1\alpha^{n-1} + \cdots + a_n = 0 \qquad a_1, \ldots, a_n \in F$$

Applying Axiom I in F, we see that the set $\{a_1, \ldots, a_n\}$ is contained in O_P for almost all $P \in \mathcal{S}$. Consequently, α is integral over O_P for almost all $P \in \mathcal{S}$. We recall from (2-5-4) that, for any $P \in \mathcal{S}$,

$$O\{P^E\} = \bigcap_{Q \supset P} O_Q$$

is the integral closure of O_P in E. It follows immediately that Axiom I holds for $\{E, \mathcal{S}^E\}$.

Moreover,

$$\begin{aligned} \alpha \in E \text{ integral over } O\{\mathcal{S}\} &\Leftrightarrow f(x) \in O\{\mathcal{S}\}[x] \\ &\Leftrightarrow f(x) \in O_P[x] \text{ for all } P \in \mathcal{S} \\ &\Leftrightarrow \alpha \in \bigcap_{Q \supset P} O_Q \text{ for all } P \in \mathcal{S} \\ &\Leftrightarrow \alpha \in \bigcap_{Q} O_Q \text{ as } Q \text{ runs over } \mathcal{S}^E \\ &\Leftrightarrow \alpha \in O\{\mathcal{S}^E\} \end{aligned}$$

Therefore, $O\{\mathcal{S}^E\}$ is the integral closure of $O\{\mathcal{S}\}$ in E. Furthermore, it is an elementary fact about integral dependence [see item 12 of (2-2-9)] that E is the quotient field of $O\{\mathcal{S}^E\}$ and, even more, that any $\alpha \neq 0 \in E$ can be written in the form $\alpha = \beta/c$, where $\beta \in O\{\mathcal{S}^E\}$ and $c \in O\{\mathcal{S}\}$. Of course, by using (4-1-6), one may give a direct valuation-theoretic proof of the existence of $c \in O\{\mathcal{S}\}$ such that $c\alpha = \beta \in O\{\mathcal{S}^E\}$.

It remains only to show that Axiom II holds. This may be accomplished by verifying the strong approximation theorem. Suppose we are given prime divisors $Q_1, \ldots, Q_r \in \mathcal{S}^E$, elements $\alpha_1, \ldots, \alpha_r \in E$, and an integer $m > 0$. Fix some basis $\omega_1, \ldots, \omega_n$ of E over F, and consider the finite set $T = \{Q_1, \ldots, Q_r, Q_{r+1}, \ldots, Q_s\} \subset \mathcal{S}^E$ gotten by adjoining to $\{Q_1, \ldots, Q_r\}$ all $Q \in \mathcal{S}^E$ for which some $\nu_Q(\omega_j) < 0$ and then saturating the resulting set. We put $\alpha_i = 0$ for $i = r + 1 \ldots, s$.

By the ordinary approximation theorem, there exists $\beta \in E$ such that

$$\nu_{Q_i}(\beta - \alpha_i) > m \qquad i = 1, \ldots, s$$

and β may be expressed as $\beta = b_1\omega_1 + \cdots + b_n\omega_n$, where $b_i \in F$. Since T_F is a finite set, there exists (by the strong approximation theorem in $\{F, \mathcal{S}\}$) for each $j = 1, \ldots, n$ an element $a_j \in F$ such that

$$\begin{aligned} \nu_P(a_j - b_j) &> M \qquad && P \in T_F \\ \nu_P(a_j) &\geq 0 && P \in \mathcal{S},\ P \notin T_F \end{aligned}$$

where

$$M > \max_{i,j} \{m + |\nu_{Q_i}(\omega_j)|\}$$

Now, let us put $\alpha = a_1\omega_1 + \cdots + a_n\omega_n$. By using the fact that if $Q \supset P$ then $\nu_Q(a) \geq \nu_P(a)$ for $a \in F$, a straightforward computation shows that

$$\begin{aligned} \nu_{Q_i}(\alpha - \alpha_i) &\geq m \qquad && i = 1, \ldots, s \\ \nu_Q(\alpha) &\geq 0 && Q \in \mathcal{S}^E,\ Q \notin T \end{aligned}$$

This completes the proof.

4-1-8. Corollary. Suppose that F is an algebraic number field and that $\mathcal{S}$ is the set of all nonarchimedean prime divisors of F. Then $\{F, \mathcal{S}\}$ is an OAF, and $O\{\mathcal{S}\}$ is the integral closure of $\mathbf{Z}$ in F.

We shall denote the set of all nonarchimedean (that is, discrete) prime divisors of an algebraic number field F by $\mathfrak{M}_0(F)$. An element of F is said to be an ***algebraic number,*** while an element of $O\{\mathfrak{M}_0(F)\}$ is called an ***algebraic integer.***

4-1-9. Corollary. Suppose that F is an algebraic function field in one variable over k [(by definition, this means that F is a finite extension of a field of rational functions $k(x)$] and that $\mathcal{S}$ is the set of all prime divisors of F which are trivial on k and for which x is an integer; then $\{F, \mathcal{S}\}$ is an OAF.

4-2. Ideals and Divisors

Until further notice, we shall be concerned with a fixed ordinary arithmetic field $\{F, \mathcal{S}\}$.

A nonempty subset $\mathfrak{A}$ of F is said to be a ***fractional*** $\mathcal{S}$ ***ideal*** or an $\mathcal{S}$ ***ideal*** or simply an ***ideal*** when:

i. $\mathfrak{A} \neq (0)$.

ii. $\mathfrak{A}$ is an $O = O\{\mathcal{S}\}$ module.

iii. There exists $c \in F^*$ such that $c\mathfrak{A} \subset O$.

Thus, the notions of $\mathcal{S}$ ideal and O ideal [as described in (1-5-1)] are identical in this situation (note that $O \neq F$ since $\mathcal{S}$ is nonempty), and we say that $\mathfrak{A}$ is ***integral*** when $\mathfrak{A} \subset O$ [that is, when $\mathfrak{A}$ is an ordinary $\neq$ (0) ideal of the ring O]. Since F is the quotient field of O, condition iii could just as well require that the nonzero element c belong to O. For any $a \in F^*$, aO is clearly an $\mathfrak{S}$ ideal—it is said to be a ***principal*** $\mathfrak{S}$ ***ideal.*** It is helpful perhaps to think of an $\mathfrak{S}$ ideal as an O module whose elements have "bounded" denominators. Note that F is not an ideal, since $O \neq F$.

Let $\mathfrak{J} = \mathfrak{J}\{\mathfrak{S}\}$ denote the set of all $\mathfrak{S}$ ideals of F. If $\mathfrak{A}$ and $\mathfrak{B}$ both belong to $\mathfrak{J}$, then so do

$$\mathfrak{A} + \mathfrak{B} = \{a + b | a \in \mathfrak{A},\ b \in \mathfrak{B}\}$$
$$\mathfrak{A} \cdot \mathfrak{B} = \left\{\sum_1^n a_i b_i | a_i \in \mathfrak{A},\ b_i \in \mathfrak{B},\ n = 1, 2, 3, \ldots\right\}$$
$$\mathfrak{A} \cap \mathfrak{B} = \{d | d \in \mathfrak{A},\ d \in \mathfrak{B}\}$$

Note that there is some trivial checking to be done—for example, that $\mathfrak{A} \cap \mathfrak{B} \neq (0)$. Clearly, these three operations on $\mathfrak{J}$ are associative. Moreover, it is natural to define a notion of divisibility for ideals by:

$$\mathfrak{A} | \mathfrak{B} \Leftrightarrow \text{there exists an integral ideal } \mathfrak{C} \text{ such that } \mathfrak{A}\mathfrak{C} = \mathfrak{B}$$

For $a \in F^*$ we put $(a) = aO$ and say that a is divisible by $\mathfrak{A}$ when $\mathfrak{A} | (a)$. The ideal $\mathfrak{A} + \mathfrak{B}$ is known as the ***greatest common divisor*** of the ideals $\mathfrak{A}$ and $\mathfrak{B}$; the integral ideals $\mathfrak{A}$ and $\mathfrak{B}$ are said to be ***relatively prime*** when $\mathfrak{A} + \mathfrak{B} = O$. One calls $(a, b) = aO + bO$ the ***greatest common divisor*** of the elements a and b; the elements a and b of O are said to be ***relatively prime*** when $(a, b) = O$—this is usually denoted by $(a, b) = 1$.

It is clear that $\mathfrak{J}$ is a commutative semigroup under multiplication. Moreover, since O is an ideal and $\mathfrak{A}O = O\mathfrak{A} = \mathfrak{A}$ for all $\mathfrak{A} \in \mathfrak{J}$, we see that $\mathfrak{J}$ has an identity for multiplication. The first objective of this chapter is to show that $\mathfrak{J}$ is, in fact, a group.

For the moment, we note that, since $(a^{-1}O)(aO) = O$ for every $a \in F^*$, it follows that the set $\mathfrak{J}^* = \mathfrak{J}^*\{\mathfrak{S}\}$ of all principal $\mathfrak{S}$ ideals is a group under multiplication. The mapping $a \to aO$ is then a homomorphism of F^* onto $\mathfrak{J}^*$; its kernel is the multiplicative group

$$U = U\{\mathfrak{S}\} = \{a \in F^* | \nu_P(a) = 0 \quad \forall P \in \mathfrak{S}\}$$

which is known as the ***group of*** $\mathfrak{S}$ ***units*** of F. Obviously, the axioms for an OAF provide no information about U. However, in Chapter 5 we shall determine the structure of $U \approx F^*/\mathfrak{J}^*$ when F is an algebraic number field and $\mathfrak{S}$ consists of almost all elements of $\mathfrak{M}_0(F)$.

Now, let us transfer the valuations ν_P to $\mathfrak{J}$.

4-2-1. Lemma. Consider any $\mathfrak{A} \in \mathfrak{J}$, and for each $P \in \mathfrak{S}$ put

$$\nu_P(\mathfrak{A}) = \min_{a \in \mathfrak{A}} \{\nu_P(a)\}$$

Then the functions ν_P on $\mathfrak{J}$ have the following properties:

(1) $\nu_P(aO) = \nu_P(a) \qquad a \in F^*$
(2) $\nu_P(\mathfrak{A}) \in \mathbf{Z}$
(3) $\nu_P(\mathfrak{A}) = 0$ for almost all $P \in \mathfrak{S}$
(4) $\nu_P(\mathfrak{A} + \mathfrak{B}) = \min \{\nu_P(\mathfrak{A}), \nu_P(\mathfrak{B})\}$
(5) $\nu_P(\mathfrak{A}\mathfrak{B}) = \nu_P(\mathfrak{A}) + \nu_P(\mathfrak{B})$
(6) $\nu_P(\mathfrak{A} \cap \mathfrak{B}) \geq \max \{\nu_P(\mathfrak{A}), \nu_P(\mathfrak{B})\}$
(7) $\mathfrak{A}$ is integral $\Leftrightarrow \nu_P(\mathfrak{A}) \geq 0$ for all $P \in \mathfrak{S}$
(8) Given $\mathfrak{A} \in \mathfrak{J}$ and any finite subset $S = \{P_1, \ldots, P_r\}$ of $\mathfrak{S}$; then there exists $a \in \mathfrak{A}$ such that

$$\nu_{P_i}(a) = \nu_{P_i}(\mathfrak{A}) \qquad i = 1, \ldots, r$$

(9) Let $\mathfrak{A} \in \mathfrak{J}$ and an arbitrary element $a \neq 0 \in \mathfrak{A}$ be given; then there exists $b \in \mathfrak{A}$ such that $\nu_P(aO + bO) = \nu_P(\mathfrak{A})$ for all $P \in \mathfrak{S}$

Proof. Since $O = \{b \in F \mid \nu_P(b) \geq 0 \quad \forall P \in \mathfrak{S}\}$, it is immediate that, for any $a \in F^*$, $\nu_P(a) = \nu_P(aO)$. Thus (1) holds, and clearly so does (7).

For the ideal $\mathfrak{A}$, there exists (by the definition) an element $c \in F^*$ with $c\mathfrak{A} \subset O$. Taking any $a \neq 0 \in \mathfrak{A}$, we have $aO \subset \mathfrak{A} \subset c^{-1}O$; and, therefore, for any $P \in \mathfrak{S}$,

$$\infty > \nu_P(a) \geq \nu_P(\mathfrak{A}) \geq \nu_P(c^{-1}) = -\nu_P(c) > -\infty$$

Since $\nu_P(a) = \nu_P(c) = 0$ for almost all $P \in \mathfrak{S}$, we see that both (2) and (3) are valid.

For (4):
$$\begin{aligned}\nu_P(\mathfrak{A} + \mathfrak{B}) &= \min_{a\in\mathfrak{A},b\in\mathfrak{B}} \{\nu_P(a + b)\} \\ &= \min_{a,b} \{\nu_P(a + 0), \nu_P(0 + b), \nu_P(a + b)\} \\ &= \min_{a,b} \{\nu_P(a), \nu_P(b)\} \\ &= \min \{\nu_P(\mathfrak{A}), \nu_P(\mathfrak{B})\}\end{aligned}$$

For (5):
$$\begin{aligned}\nu_P(\mathfrak{A}\mathfrak{B}) &= \min_{a_i\in\mathfrak{A},b_i\in\mathfrak{B}} \left\{\nu_P\left(\sum_i a_ib_i\right)\right\} \\ &= \min_{a\in\mathfrak{A},b\in\mathfrak{B}} \{\nu_P(ab)\} \\ &= \min_a \{\nu_P(a)\} + \min_b \{\nu_P(b)\} \\ &= \nu_P(\mathfrak{A}) + \nu_P(\mathfrak{B})\end{aligned}$$

For (6): Since $\mathfrak{A} \supset \mathfrak{A} \cap \mathfrak{B}$ and $\mathfrak{B} \supset \mathfrak{A} \cap \mathfrak{B}$, the assertion of (6) is trivial. Note that we do not prove equality; this will fall out eventually.

For (8): for each $i = 1, \ldots, r$ there exists $a_i \in \mathfrak{A}$ such that $\nu_{P_i}(a_i) = \nu_{P_i}(\mathfrak{A})$. Choose, then, $b_i \in F$ with

$$\begin{aligned} \nu_{P_i}(b_i) &= 0 & \\ \nu_{P_j}(b_i) &= m & j \neq i \\ \nu_P(b_i) &\geq 0 & P \notin S \end{aligned}$$

where $m > 0$ is large enough so that $\nu_{P_i}(a_j) + m > \nu_{P_i}(a_i)$ for all $j \neq i$. Since $m > 0$, we have $b_i \in O$, so that $a = \Sigma a_i b_i \in \mathfrak{A}$. It is straightforward to verify that $\nu_{P_i}(a) = \nu_{P_i}(a_i) = \nu_{P_i}(\mathfrak{A})$.

For (9): for $a, b \in F^*$ we know by (4) and (1) that

$$\nu_P(aO + bO) = \min \{\nu_P(a), \nu_P(b)\}$$

for every $P \in \mathcal{S}$. Now, given any $a \in \mathfrak{A}$, we have $\nu_P(a) \geq \nu_P(\mathfrak{A})$ for all $P \in \mathcal{S}$. Therefore, it suffices to find $b \in \mathfrak{A}$ such that $\min \{\nu_P(a), \nu_P(b)\} = \nu_P(\mathfrak{A})$ for all $P \in \mathcal{S}$. To accomplish this, let $S = \{P_1, \ldots, P_r\}$ be a finite set such that, for $P \notin S$, $\nu_P(a) = \nu_P(\mathfrak{A}) = 0$. According to (8), there exists $b \in \mathfrak{A}$ with $\nu_{P_i}(b) = \nu_{P_i}(\mathfrak{A})$ $(i = 1, \ldots, r)$. This b "does it."

Consider an $\mathcal{S}$ ideal $\mathfrak{A}$. From what has gone before, we see that $\mathfrak{A}$ determines a function from $\mathcal{S}$ into $\mathbf{Z}$—namely, $\mathfrak{A}: P \to \nu_P(\mathfrak{A})$. This leads us to a natural definition. An ***$\mathcal{S}$-ordinal-number vector*** is a function $f: \mathcal{S} \to \mathbf{Z}$ such that $f(P) = 0$ for almost all $P \in \mathcal{S}$. With the usual definition of addition, the set of all such functions forms an abelian group—the ***group of $\mathcal{S}$-ordinal-number vectors.*** It is denoted by $\mathbf{V} = \mathbf{V}\{\mathcal{S}\}$.

Equivalent to this, but somewhat more convenient notationally, is the following approach: Consider the free abelian group with basis consisting of $\{P | P \in \mathcal{S}\}$. This multiplicative group is called the ***group of $\mathcal{S}$ divisors*** of F and is denoted by $\mathbf{D} = \mathbf{D}\{\mathcal{S}\}$. An element $A \in \mathbf{D}$ is called an $\mathcal{S}$ ***divisor*** of F and has a unique expression

$$A = \prod_{P \in \mathcal{S}} P^{m_P}$$

where each $m_P \in \mathbf{Z}$ and $m_P = 0$ for almost all P. If we agree to write $\nu_P(A) = m_P$ for each $P \in \mathcal{S}$ (no confusion with the former meanings of ν_P should arise), then

$$A = \prod_{P \in \mathcal{S}} P^{\nu_P(A)}$$

In particular, for $A, B \in \mathbf{D}$ we have

$$AB = \prod_P P^{\nu_P(A) + \nu_P(B)} \qquad \text{and} \qquad A^{-1} = \prod_P P^{-\nu_P(A)}$$

(since $I = \Pi P^{0=\nu_P(I)}$ is the identity of **D**). Clearly **V** and **D** are isomorphic, so that anything we do for **D** may be transferred to **V**.

The divisor A is said to be $\mathfrak{S}$-***integral*** when $\nu_P(A) \geq 0$ for all $P \in \mathfrak{S}$. In terms of integrality a notion of divisibility can be defined in **D**; namely, for $A, B \in \mathbf{D}$,

$$A|B \Leftrightarrow BA^{-1} \text{ is integral}$$

Consequently, A divides $B \Leftrightarrow$ there exists an integral $C \in \mathbf{D}$ such that $AC = B \Leftrightarrow \nu_P(A) \leq \nu_P(B)$ for all $P \in \mathfrak{S}$.

Join and meet in **D** are defined by

$$A + B = \prod_P P^{\min\{\nu_P(A),\, \nu_P(B)\}}$$

$$A \cap B = \prod_P P^{\max\{\nu_P(A),\, \nu_P(B)\}}$$

In other words, $A + B$ is the greatest common divisor of A and B—that is, the unique element of **D** which divides A and B and which is divisible by any element that divides both A and B, and $A \cap B$ is the least common multiple of A and B—that is, the unique multiple of A and B which divides every common multiple of A and B. As usual, $(A + B)(A \cap B) = AB$.

4-3. The Fundamental Theorem of OAFs

In this section we develop the connections between $\mathfrak{J}$ and **D**.

Consider any $a \in F^*$, and put

$$\delta(a) = \prod_P P^{\nu_P(a)}$$

Since $\nu_P(a) = 0$ for almost all $P \in \mathfrak{S}$, $\delta(a)$ is an $\mathfrak{S}$ divisor; it is called a ***principal*** $\mathfrak{S}$ ***divisor.*** The mapping $\delta: F^* \to \mathbf{D}$ is clearly a homomorphism with kernel U, and the image group $\delta(F^*) = \mathbf{D}^* = \mathbf{D}^*\{\mathfrak{S}\}$ is called the ***group of principal*** $\mathfrak{S}$ ***divisors.***

Consider next any $\mathfrak{A} \in \mathfrak{J}$, and put

$$\delta(\mathfrak{A}) = \prod_P P^{\nu_P(\mathfrak{A})}$$

By (4-2-1) we know that $\delta(\mathfrak{A}) \in \mathbf{D}$ and that $\delta: \mathfrak{J} \to \mathbf{D}$ is a homomorphism. Moreover, this δ is compatible with the preceding one in the sense that $\delta(aO) = \delta(a)$ for every $a \in F^*$. In particular, $\mathfrak{J}^*$ is isomorphic to $\mathbf{D}^*$ under the map $\delta: \mathfrak{J}^* \to \mathbf{D}^*$, and, in fact, $\mathfrak{J}^* \approx \mathbf{D}^* \approx F^*/U$.

Transferring parts 4, 5, 7, and 9 of (4-2-1) to δ gives the following proposition.

4-3-1. Proposition. The mapping δ of the semigroup with identity $\mathfrak{J}$ into the group **D** has the following properties:

1. δ is a homomorphism; that is, for $\mathfrak{A}, \mathfrak{B} \in \mathfrak{J}$,

$$\delta(\mathfrak{A}\mathfrak{B}) = \delta(\mathfrak{A})\delta(\mathfrak{B})$$

2. δ preserves joins; that is, for $\mathfrak{A}, \mathfrak{B} \in \mathfrak{J}$,

$$\delta(\mathfrak{A} + \mathfrak{B}) = \delta(\mathfrak{A}) + \delta(\mathfrak{B})$$

3. $\mathfrak{A}$ is integral $\Leftrightarrow \delta(\mathfrak{A})$ is integral.
4. Given any $a \neq 0 \in \mathfrak{A}$, there exists $b \in \mathfrak{A}$ such that

$$\delta(aO + bO) = \delta(\mathfrak{A})$$

The next step is to define a mapping $\kappa:\mathbf{D} \to \mathfrak{J}$. Given

$$A = \prod_P P^{\nu_P(A)} \in \mathbf{D}$$

we put

$$\kappa(A) = \{a \in F | \nu_P(a) \geq \nu_P(A) \quad \forall P \in \mathfrak{S}\}$$

It is trivial that $\kappa(A)$ is an O module, and simple applications of (4-1-6) show that $\kappa(A) \in \mathfrak{J}$.

Again, from (4-1-6), we see that $\nu_P\{\kappa(A)\} = \nu_P(A)$ for every $P \in \mathfrak{S}$. This means that the mapping $\kappa:\mathbf{D} \to \mathfrak{J}$ (we do not claim that κ is a homomorphism) is such that $\delta \circ \kappa$ is the identity on **D**. In particular, κ is 1-1, and δ is onto.

What about the composite map $\kappa \circ \delta:\mathfrak{J} \to \mathfrak{J}$? Since, for $a \in F^*$, $aO = \{b \in F | \nu_P(b) \geq \nu_P(a) \quad \forall P \in \mathfrak{S}\}$, it follows that $\kappa\delta(aO) = aO$. This means that $\kappa \circ \delta$ is the identity on $\mathfrak{J}^*$. Now, for any $\mathfrak{A} \in \mathfrak{J}$, there exist $a, b \in \mathfrak{A}$ with $\delta(aO + bO) = \delta(\mathfrak{A})$. Putting $\overline{\mathfrak{A}} = \kappa\delta(\mathfrak{A})$, we have

$$aO + bO \subset \mathfrak{A} \subset \overline{\mathfrak{A}}$$

Consider any $d \in \overline{\mathfrak{A}}$; that is, consider any $d \in F$ such that $\nu_P(d) \geq \nu_P(\mathfrak{A})$ for all $P \in \mathfrak{S}$. Suppose that we have shown that there exist $x, y \in O$ such that $ax + by = d$ (the proof of this is deferred for the moment). If so, then

$$aO + bO = \mathfrak{A} = \overline{\mathfrak{A}}$$

Therefore, $\kappa \circ \delta$ is the identity on $\mathfrak{J}$,

$$\mathfrak{A} = \{c \in F | \nu_P(c) \geq \nu_P(\mathfrak{A}) \quad \forall P \in \mathfrak{S}\}$$

and any $\mathfrak{S}$ ideal may be generated by two of its elements of which the first may be chosen arbitrarily ($\neq 0$).

Thus both κ and δ are 1-1 onto and inverses of each other. Since δ preserves multiplication, so does κ; that is, κ is an isomorphism of the group $\mathbf{D}$ onto the semigroup $\mathfrak{J}$. Therefore, $\mathfrak{J}$ is a group, and the inverse $\mathfrak{A}^{-1}$ of an ideal $\mathfrak{A}$ is characterized by

$$\nu_P(\mathfrak{A}^{-1}) = -\nu_P(\mathfrak{A}) \qquad \forall P \in \mathcal{S}$$

In particular, $O = O^{-1}$ is the identity element of $\mathfrak{J}$ (note that the symbol O^{-1} now has two possible meanings, but there is no danger of confusion), and $\mathfrak{A}\mathfrak{A}^{-1} = O$. Of course, $\mathfrak{J}^*$ and $\mathbf{D}^*$ are isomorphic under δ and κ.

Since $\delta:\mathfrak{J} \rightarrowtail\!\!\twoheadrightarrow \mathbf{D}$ preserves joins, so does $\kappa:\mathbf{D} \rightarrowtail\!\!\twoheadrightarrow \mathfrak{J}$; that is, $\kappa(A + B) = \kappa(A) + \kappa(B)$ for all $A, B \in \mathbf{D}$. Furthermore, κ preserves meets since

$$\begin{aligned}\kappa(A \cap B) &= \{d \in F | \nu_P(d) \geq \max(\nu_P(A), \nu_P(B)) \quad \forall P\}\\ &= \{d \in F|\nu_P(d) \geq \nu_P(A) \quad \forall P\} \cap \{d \in F|\nu_P(d) \geq \nu_P(B) \quad \forall P\}\\ &= \kappa(A) \cap \kappa(B)\end{aligned}$$

Therefore, for $\mathfrak{A}, \mathfrak{B} \in \mathfrak{J}$ we have $\delta(\mathfrak{A} \cap \mathfrak{B}) = \delta(\mathfrak{A}) \cap \delta(\mathfrak{B})$; that is, δ preserves meets.

It is clear that both δ and κ preserve integrality; hence they both preserve divisibility. We have also, for $\mathfrak{A}, \mathfrak{B} \in \mathfrak{J}$,

$$\mathfrak{A} \subset \mathfrak{B} \Leftrightarrow \mathfrak{B}|\mathfrak{A}$$

In fact,

$$\begin{aligned}\mathfrak{A} \subset \mathfrak{B} &\Rightarrow \nu_P(\mathfrak{A}) \geq \nu_P(\mathfrak{B}) \text{ for all } P \in \mathcal{S}\\ &\Rightarrow \delta(\mathfrak{B})|\delta(\mathfrak{A})\\ &\Rightarrow \kappa\delta(\mathfrak{B})|\kappa\delta(\mathfrak{A})\\ &\Rightarrow \mathfrak{B}|\mathfrak{A}\\ &\Rightarrow \text{there exists } \mathfrak{C} \in \mathfrak{J} \text{ with } \mathfrak{C} \subset O \text{ and } \mathfrak{B}\mathfrak{C} = \mathfrak{A}\\ &\Rightarrow \mathfrak{A} = \mathfrak{B}\mathfrak{C} \subset \mathfrak{B}O = \mathfrak{B}\end{aligned}$$

From this it follows easily that every proper prime ideal of the ring O is maximal; in other words, the set of proper prime ideals of O and the set of maximal ideals of O are identical.

For each $P \in \mathcal{S}$, let us now put

$$\mathfrak{P} = \mathfrak{P}_P = \kappa(P) = \{a \in F|\nu_P(a) \geq 1,\ \nu_{P'}(a) \geq 0 \quad \forall P' \neq P\}$$

To put it another way,

$$\mathfrak{P} = O \cap \mathcal{P}$$

where $\mathcal{P} = \mathcal{P}_P = \{a \in F|\nu_P(a) \geq 1\}$. In view of the lattice isomorphism of $\mathfrak{J}$ and $\mathbf{D}$, $\{\mathfrak{P}|P \in \mathcal{S}\}$ is the set of all maximal (i.e., proper prime) ideals of O. Combining all of the preceding remarks, we have the following theorem.

4-3-2. Fundamental theorem of OAFs. Let $\{F, \mathfrak{S}\}$ be an OAF; then:

1. $\mathfrak{J} = \mathfrak{J}\{\mathfrak{S}\}$ is an abelian group under multiplication; its identity is $O = O\{\mathfrak{S}\}$.

2. The mappings $\delta:\mathfrak{J} \to \mathbf{D}$ and $\kappa:\mathbf{D} \to \mathfrak{J}$ which are given by

$$\delta(\mathfrak{A}) = \prod_P P^{\nu_P(\mathfrak{A})} \qquad \text{and} \qquad \kappa(A) = \{a \in F | \nu_P(a) \geq \nu_P(A) \quad \forall P \in \mathfrak{S}\}$$

are isomorphisms which are inverses of each other. They induce isomorphisms $\delta:\mathfrak{J}^* \to \mathbf{D}^*$ and $\kappa:\mathbf{D}^* \to \mathfrak{J}^*$.

3. Both κ and δ preserve joins and meets; they determine a 1-1 correspondence

$$\delta(\mathfrak{P}) = P \leftrightarrow \mathfrak{P} = \kappa(P)$$

between $\mathfrak{S}$ and the set of all proper prime ideals of O.

4. Every $\mathfrak{S}$-ideal $\mathfrak{A}$ has a unique decomposition into prime ideals

$$\mathfrak{A} = \prod_P \mathfrak{P}^{\nu_P(\mathfrak{A})}$$

Its inverse is

$$\mathfrak{A}^{-1} = \prod_P \mathfrak{P}^{-\nu_P(\mathfrak{A})}$$

and

$$\mathfrak{A}\mathfrak{A}^{-1} = O = \prod_P \mathfrak{P}^0$$

5. For $\mathfrak{A}, \mathfrak{B} \in \mathfrak{J}$ we have

$$\mathfrak{A}|\mathfrak{B} \Leftrightarrow \mathfrak{B} \subset \mathfrak{A}$$

6. Given any $a \neq 0 \in \mathfrak{A}$, there exists $b \in \mathfrak{A}$ such that

$$\mathfrak{A} = aO + bO$$

To complete the proof of the fundamental theorem, it remains to prove the following lemma.

4-3-3. Lemma. Suppose that $a, b \in F$ are given and that $d \in F$ is such that

$$\nu_P(d) \geq \min \{\nu_P(a), \nu_P(b)\}$$

for all $P \in \mathfrak{S}$. Then there exist $x, y \in O$ with $ax + by = d$.

Proof. Let us recall first why we want this result. Given $\mathfrak{A} \in \mathfrak{J}$, there exist $a, b \in \mathfrak{A}$ such that $\delta(aO + bO) = \delta(\mathfrak{A})$; in other words, $\nu_P(\mathfrak{A}) = \min \{\nu_P(a), \nu_P(b)\}$ for all $P \in \mathfrak{S}$. We have then $aO + bO \subset \mathfrak{A} \subset \bar{\mathfrak{A}} = \kappa\delta(\mathfrak{A})$. Since any $d \in \bar{\mathfrak{A}}$ satisfies the conditions of the lemma, the validity of the lemma implies that $aO + bO = \mathfrak{A} = \bar{\mathfrak{A}}$.

The proof of the lemma (which is the most difficult step in the proof of the fundamental theorem) is not hard. If either a or b is 0, the lemma is clearly true; so suppose that $a, b \in F^*$. Consider

$$y = \frac{d}{b} - \frac{a}{b}x$$

We wish to find an $x \in O$ for which $y \in O$. For each P, we have

$$\nu_P(y) = \nu_P\left\{\frac{a}{b}\left[\frac{d}{a} - x\right]\right\} = \nu_P\left(\frac{a}{b}\right) + \nu_P\left(x - \frac{d}{a}\right)$$

Let $S = \{P_1, \ldots, P_r\} = \{P \in \mathcal{S} | \nu_P(a/b) < 0\}$. Thus, according to the hypothesis of the lemma, $\nu_P(d/a) \geq \min\{0, \nu_P(b/a)\}$ for every $P \in \mathcal{S}$—and for $P_i \in S$, we have $\nu_{P_i}(d/a) \geq 0$. By the strong approximation theorem, we may choose $x \in F$ such that

$$\nu_{P_i}\left(x - \frac{d}{a}\right) \geq \nu_{P_i}\left(\frac{b}{a}\right) > 0 \qquad i = 1, \ldots, r$$
$$\nu_P(x) \geq 0 \qquad P \notin S$$

It is easy to verify that $x \in O$ and $y \in O$. In fact, for $P_i \in S$,

$$\nu_{P_i}(x) = \nu_{P_i}\left(x - \frac{d}{a} + \frac{d}{a}\right) \geq \min\left\{\nu_{P_i}\left(x - \frac{d}{a}\right), \nu_{P_i}\left(\frac{d}{a}\right)\right\} \geq 0$$
$$\nu_{P_i}(y) = \nu_{P_i}\left(\frac{a}{b}\right) + \nu_{P_i}\left(x - \frac{d}{a}\right) \geq \nu_{P_i}\left(\frac{a}{b}\right) + \nu_{P_i}\left(\frac{b}{a}\right) = 0$$

while, for $P \notin S$,

$$\nu_P(x) \geq 0$$
$$\nu_P(y) = \nu_P\left(\frac{a}{b}\right) + \nu_P\left(x - \frac{d}{a}\right) \geq \nu_P\left(\frac{a}{b}\right) + \min\left\{\nu_P(x), \nu_P\left(\frac{d}{a}\right)\right\}$$
$$\geq \nu_P\left(\frac{a}{b}\right) + \min\left\{0, \nu_P\left(\frac{b}{a}\right)\right\} = 0$$

This completes the proof.

4-3-4. *Exercise.* i. Let elements $a_1, \ldots, a_n, b \in F$ and a divisor $A = \Pi P^{\nu_P(A)}$ be given, and consider the equation

$$y = a_1x_1 + \cdots + a_nx_n + b$$

Then, there exist $x_1, \ldots, x_n \in O$ for which y belongs to $\kappa(A) \Leftrightarrow \nu_P(b) \geq \min\{\nu_P(A), \nu_P(a_1), \ldots, \nu_P(a_n)\}$ for all $P \in \mathcal{S}$.

ii. For $a_1, \ldots, a_n, d \in F$ the Diophantine equation

$$a_1x_1 + \cdots + a_nx_n = d$$

has a solution $x_1, \ldots, x_n \in O \Leftrightarrow \nu_P(d) \geq \min\{\nu_P(a_1), \ldots, \nu_P(a_n)\}$ for all $P \in \mathcal{S} \Leftrightarrow dO \subset a_1O + \cdots + a_nO$.

iii. For $a_1, \ldots, a_n, d \in F$ the equation $a_1x_1 + \cdots + a_nx_n = d$ has a global integral solution (that is, a solution with $x_1, \ldots, x_n \in O$) $\Leftrightarrow$ it has a local integral solution (that is, a solution with all x_i in O_P) for every $P \in \mathcal{S}$. In fact, the local solutions may even come from the completions.

iv. The m equations in n unknowns

$$\sum_{j=1}^{n} a_{ij}x_j = d_i \qquad a_{ij},\ d_i \in F,\ i = 1, \ldots, m$$

have a global integral solution $\Leftrightarrow$ they have a local integral solution for every $P \in \mathcal{S}$. Again, the local solutions may come from the completions.

4-3-5. Example. Consider the field $F = \mathbf{Q}(\sqrt{-5})$ with $\mathcal{S} = \mathfrak{M}_0(F)$, and recall that we have the nonunique factorization into irreducible elements

$$9 = 3 \cdot 3 = (2 + \sqrt{-5})(2 - \sqrt{-5})$$

Let us analyze this further in terms of unique factorization into prime ideals.

For any $\alpha \in O_F = \mathbf{Z} \cdot 1 \oplus \mathbf{Z} \cdot \sqrt{-5}$ we observe [from (2-5-9)] that if $N_{F\to\mathbf{Q}}(\alpha)$ is a unit at the prime $p \in \mathbf{Z}$ then α is a unit at all primes P of F which are extensions of p. Since $N(3) = N(2 \pm \sqrt{-5}) = 9$, we know that $\nu_P(3) = \nu_P(2 \pm \sqrt{-5}) = 0$ for all $P \in \mathcal{S}$ which are not extensions of 3. Moreover, since $x^2 + 5 \equiv (x + 2)(x + 1) \pmod 3$, it follows from extension theory and Hensel's lemma that the prime 3 has two extensions P_1, P_2 to F and that $e\left(\frac{P_1}{3}\right) = e\left(\frac{P_2}{3}\right) = f\left(\frac{P_1}{3}\right) = f\left(\frac{P_2}{3}\right) = 1$. In particular, $\nu_{P_1}(3) = \nu_{P_2}(3) = 1$, and we see that

$$(3) = \mathfrak{P}_1\mathfrak{P}_2$$

Furthermore, we note that $\nu_{P_i}(2 + \sqrt{-5}) + \nu_{P_i}(2 - \sqrt{-5}) = \nu_{P_i}(9) = 2$ for $i = 1$ and 2, so that $0 \leq \nu_{P_i}(2 \pm \sqrt{-5}) \leq 2$. If we have $\nu_{P_i}(2 + \sqrt{-5}) = \nu_{P_i}(2 - \sqrt{-5}) = 1$ for either $i = 1$ or 2 then $0 = \nu_{P_i}(4) \geq \min\{\nu_{P_i}(2 + \sqrt{-5}),\ \nu_{P_i}(2 - \sqrt{-5})\} = 1$, a contradiction. If $\nu_{P_1}(2 + \sqrt{-5}) = 2$ and $\nu_{P_1} = (2 - \sqrt{-5}) = 0$, then since $2 = \nu_3[N(2 + \sqrt{-5})] = \nu_{P_1}(2 + \sqrt{-5}) + \nu_{P_2}(2 + \sqrt{-5})$, it follows that $\nu_{P_2}(2 + \sqrt{-5}) = 0$—and then $\nu_{P_2}(2 - \sqrt{-5}) = 2$. Thus we may index so that $\nu_{P_1}(2 + \sqrt{-5}) = 2$, and the conclusion is that

$$(9) = \mathfrak{P}_1^2\mathfrak{P}_2^2$$
$$\mathfrak{P}_1^2 = (2 + \sqrt{-5}) \qquad \mathfrak{P}_2^2 = (2 - \sqrt{-5})$$
$$\mathfrak{P}_1 = (3, 2 + \sqrt{-5}) \qquad \mathfrak{P}_2 = (3, 2 - \sqrt{-5})$$

4-3-6. Exercise. In $\mathbf{Q}(\sqrt{-5})$ consider the factorizations into irreducible elements

$$\begin{aligned} 6 &= 2 \cdot 3 = (1 + \sqrt{-5})(1 - \sqrt{-5}) \\ 14 &= 7 \cdot 2 = (3 + \sqrt{-5})(3 - \sqrt{-5}) \\ 21 &= 7 \cdot 3 = (1 + 2\sqrt{-5})(1 - 2\sqrt{-5}) \\ 49 &= 7 \cdot 7 = (2 + 3\sqrt{-5})(2 - 3\sqrt{-5}) \end{aligned}$$

and describe the corresponding unique factorizations in terms of prime ideals. [All this will become especially easy when we have proved (4-9-1).]

The remainder of this section is devoted to some simple questions of counting. For this it is convenient to have our next result, which is also of interest in itself.

4-3-7. Proposition. Suppose that $\{F, \mathfrak{S}\}$ is an OAF. Then for every $P \in \mathfrak{S}$ and every integer $r \geq 1$ we have

$$O_P = O + \mathcal{P}^r$$

and an isomorphism of rings

$$\frac{O}{\mathfrak{P}^r} \approx \frac{O_P}{\mathcal{P}^r}$$

Proof. It is clear that $O + \mathcal{P}^r \subset O_P$. On the other hand, given $a \in O_P$ there exists $b \in F$ such that $\nu_P(b - a) \geq r$ and $\nu_{P'}(b) \geq 0$ for all $P' \neq P$; therefore, $a = b + (a - b)$ with $b \in O$ and $a - b \in \mathcal{P}^r$.

Since any $\mathfrak{A} \in \mathfrak{J}$ is of form $\mathfrak{A} = \{a \in F | \nu_P(a) \geq \nu_P(\mathfrak{A}) \quad \forall P \in \mathfrak{S}\}$ it follows that $\mathfrak{P}^r = \mathcal{P}^r \cap O$. Therefore,

$$\frac{O}{\mathfrak{P}^r} = \frac{O}{\mathcal{P}^r \cap O} \approx \frac{O + \mathcal{P}^r}{\mathcal{P}^r} = \frac{O_P}{\mathcal{P}^r}$$

In view of this, we may identify $O/\mathfrak{P}$ with $O_P/\mathcal{P}$ and call the map $O \to O/\mathfrak{P}$ the ***residue class map*** at $\mathfrak{P}$.

4-3-8. Remark. Suppose that for every $P \in \mathfrak{S}$ the residue class field (which we denote by $\bar{F}_P$, since it is also the residue class field of the completion of F at P) is finite. This hypothesis is not necessary for all parts of the discussion, but we prefer to avoid making distinctions.

For any integral ideal $\mathfrak{A}$ and corresponding integral divisor A, we put

$$\mathfrak{N}\mathfrak{A} = \mathfrak{N}A = (O:\mathfrak{A}) = \#\left(\frac{O}{\mathfrak{A}}\right)$$

By determining the structure of the ring $O/\mathfrak{A}$ we shall see that $\mathfrak{N}\mathfrak{A}$ is finite and that it is easily computed.

Note that, for $P \in \mathfrak{S}$ and $r \geq 1$, $\mathfrak{N}(P^r)$ has the meaning just given and also the meaning given in (1-9-7); but these agree, since by (4-3-7

$$\mathfrak{N}(P^r) = \mathfrak{N}(\mathfrak{P}^r) = \mathfrak{N}(\mathcal{P}^r) = \mathfrak{N}(P^r)$$

In particular, we see that

$$\mathfrak{N}(\mathfrak{P}^r) = \mathfrak{N}(P^r) = (\mathfrak{N}P)^r = (\mathfrak{N}\mathfrak{P})^r < \infty$$

Suppose next that $\mathfrak{A}$ and $\mathfrak{B}$ are integral ideals which are relatively prime, so that $\min\{\nu_P(\mathfrak{A}), \nu_P(\mathfrak{B})\} = 0$ for all $P \in \mathfrak{S}$. Consider the natural map

$$O \to \frac{O}{\mathfrak{A}} \oplus \frac{O}{\mathfrak{B}}$$

By using the strong approximation theorem it is easy to see that this map is onto. Moreover, its kernel is clearly $\mathfrak{A} \cap \mathfrak{B} = (\mathfrak{A} \cap \mathfrak{B})(\mathfrak{A} + \mathfrak{B}) = \mathfrak{A}\mathfrak{B}$. Therefore, $\mathfrak{N}(\mathfrak{A}\mathfrak{B}) = \mathfrak{N}(\mathfrak{A})\mathfrak{N}(\mathfrak{B})$. It follows that for $\mathfrak{A} = \Pi\mathfrak{P}^{\nu_P(\mathfrak{A})}$ we have

$$\mathfrak{N}\mathfrak{A} = \Pi(\mathfrak{N}\mathfrak{P})^{\nu_P(\mathfrak{A})} < \infty$$

so that

$$\mathfrak{N}(\mathfrak{A}\mathfrak{B}) = (\mathfrak{N}\mathfrak{A})(\mathfrak{N}\mathfrak{B})$$

even if $\mathfrak{A}$ and $\mathfrak{B}$ are not relatively prime.

One may extend the function $\mathfrak{N}$ to all $\mathfrak{C} \in \mathfrak{J}\{\mathfrak{S}\}$ by writing $\mathfrak{C} = \mathfrak{A}/\mathfrak{B}$ with $\mathfrak{A}$ and $\mathfrak{B}$ both integral and putting

$$\mathfrak{N}\mathfrak{C} = \frac{\mathfrak{N}\mathfrak{A}}{\mathfrak{N}\mathfrak{B}}$$

Of course, this extended $\mathfrak{N}$ is still multiplicative.

4-3-9. Exercise. Suppose that we are in the situation of (4-3-8). For an integral ideal $\mathfrak{A}$, let $\Phi(\mathfrak{A})$ denote the number of elements in the group of units of the ring $O/\mathfrak{A}$. Thus Φ is a generalization of the Euler ϕ function.

i. For $b, c \in O$ the congruence $bx \equiv c \pmod{\mathfrak{A}}$ has a solution in O if and only if the ideal $(b, \mathfrak{A}) = bO + \mathfrak{A}$ divides c. Moreover, any two solutions are congruent modulo the ideal $\mathfrak{A}/(b, \mathfrak{A})$.

ii. For $b \in O$, the residue class of b in $O/\mathfrak{A}$ is a unit $\Leftrightarrow (b, \mathfrak{A}) = 1$. The units of $O/\mathfrak{A}$ are also called the ***prime residue classes*** (mod $\mathfrak{A}$).

iii. For $b \in O$ such that $(b, \mathfrak{A}) = 1$ we have

$$b^{\Phi(\mathfrak{A})} \equiv 1 \pmod{\mathfrak{A}}$$

This generalizes Fermat's theorem.

iv. If $\mathfrak{A}$, $\mathfrak{B}$ are integral ideals with $(\mathfrak{A}, \mathfrak{B}) = 1$, then

$$\Phi(\mathfrak{A}\mathfrak{B}) = \Phi(\mathfrak{A})\Phi(\mathfrak{B})$$

v. Show that

$$\Phi(\mathfrak{A}) = (\mathfrak{N}\mathfrak{A}) \prod_{\mathfrak{P}|\mathfrak{A}} \left(1 - \frac{1}{\mathfrak{N}\mathfrak{P}}\right)$$

and

$$\sum_{\mathfrak{B}|\mathfrak{A}} \Phi(\mathfrak{B}) = \mathfrak{N}\mathfrak{A}$$

vi. For a prime ideal $\mathfrak{P}$ and $r \geq 1$, evaluate $\mathfrak{N}(\mathfrak{P})^r$ and $\Phi(\mathfrak{P}^r)$ directly by finding a full system of representatives from O for the classes of $O/\mathfrak{P}^r$.

4-4. Dedekind Rings

Let $\{F, \mathcal{S}\}$ be an OAF, and let $O = O\{\mathcal{S}\}$ be its ring of integers. From what has gone before, it is immediate that:

i. O is noetherian; that is, O has the ascending chain condition on ideals.

ii. Every proper prime ideal of O is maximal.

iii. O is integrally closed in its quotient field $F \neq O$.

A domain that satisfies these three conditions is commonly known as a ***Dedekind domain,*** or ***Dedekind ring.*** In particular, the ring of integers of any algebraic number field is a Dedekind ring.

The main purpose of this section is to show that the notions of Dedekind ring and OAF are completely equivalent. In particular, we shall prove once again that "classical ideal theory" holds for the ring of integers of an algebraic number field. Since the discussion here is in the nature of a digression, the treatment will be somewhat sketchy and more or less independent of our previous results.

Let R be an integral domain with quotient field $F \neq R$. In view of (1-5-1), we know what is meant by the terms R *ideal, fractional ideal, ideal,* and *integral ideal.* For any $a \in F^*$, aR is clearly an ideal—it is said to be *principal.* Of course, F is not an ideal, and the definition excludes (0) from the class of R ideals. The set of R ideals is closed under the operations of sum, product, and intersection.

If $\mathfrak{A}$ is an ideal, then there exists an integral ideal $\mathfrak{B}$ such that $\mathfrak{A}\mathfrak{B}$ is integral—as a matter of fact, $\mathfrak{B}$ may be taken to be principal.

Suppose that $\mathfrak{A}$ is an R ideal. Putting

$$\mathfrak{A}^{-1} = \{x \in F | x\mathfrak{A} \subset R\}$$

we have $\mathfrak{A}\mathfrak{A}^{-1} \subset R$. If $\mathfrak{A}\mathfrak{A}^{-1} = R$, the ideal $\mathfrak{A}$ is said to be ***invertible.*** If $\mathfrak{A} \subset R$, then $\mathfrak{A}^{-1} \supset R$. Of course, $R^{-1} = R$. One checks easily that $\mathfrak{A}^{-1}$ is an ideal; furthermore, if $a \in F^*$, then $(Ra)^{-1} = Ra^{-1}$ and Ra is an invertible ideal. The set of all ideals is a commutative semigroup (with identity R) under multiplication.

If $\mathfrak{A}$ is an ideal and there exists an ideal $\mathfrak{A}'$ such that $\mathfrak{A}\mathfrak{A}' = R$, then $\mathfrak{A}$ is invertible and $\mathfrak{A}' = \mathfrak{A}^{-1}$. To see this, observe that $\mathfrak{A}' \subset \mathfrak{A}^{-1}$, and hence $R = \mathfrak{A}\mathfrak{A}' \subset \mathfrak{A}\mathfrak{A}^{-1} \subset R$, which says that $\mathfrak{A}$ is invertible; furthermore, $\mathfrak{A}^{-1} = \mathfrak{A}^{-1}R = \mathfrak{A}^{-1}\mathfrak{A}\mathfrak{A}' = R\mathfrak{A}' = \mathfrak{A}'$. If $\mathfrak{A}$ and $\mathfrak{B}$ are invertible, then so is $\mathfrak{A}\mathfrak{B}$, and $(\mathfrak{A}\mathfrak{B})^{-1} = \mathfrak{A}^{-1}\mathfrak{B}^{-1}$. If $\mathfrak{A}$ is invertible, then so is $\mathfrak{A}^{-1}$, and $(\mathfrak{A}^{-1})^{-1} = \mathfrak{A}$. Thus, the set of invertible ideals is a multiplicative group. If a product of ideals is invertible, then so is each factor.

4-4-1. Lemma. If $\mathfrak{A}$ is a nonprime integral ideal, then there exist integral ideals $\mathfrak{B}$ and $\mathfrak{C}$ such that

$$\mathfrak{A} < \mathfrak{B} \qquad \mathfrak{A} < \mathfrak{C} \qquad \mathfrak{B}\mathfrak{C} \subset \mathfrak{A}$$

Proof. Since $\mathfrak{A}$ is not prime, there exist $b, c \in R$ such that $b \notin \mathfrak{A}$, $c \notin \mathfrak{A}$, $bc \in \mathfrak{A}$. Put $\mathfrak{B} = \mathfrak{A} + Rb$ and $\mathfrak{C} = \mathfrak{A} + Rc$.

4-4-2. Lemma. Suppose that R is noetherian. If $\mathfrak{A}$ is a proper integral ideal, then there exist proper prime ideals $\mathfrak{P}_1, \ldots, \mathfrak{P}_r$ of R such that

$$\mathfrak{A} \subset \mathfrak{P}_i \quad \text{for } i = 1, \ldots, r \qquad \text{and} \qquad \mathfrak{P}_1 \cdots \mathfrak{P}_r \subset \mathfrak{A}$$

Proof. Suppose the assertion is false. Choose a maximal element $\mathfrak{A}$ from the set of ideals for which the assertion is false. Thus $\mathfrak{A}$ is not prime. Take $\mathfrak{B}$ and $\mathfrak{C}$ as in (4-4-1). Since the assertion of the lemma is true for both $\mathfrak{B}$ and $\mathfrak{C}$, it is also true for $\mathfrak{A}$, a contradiction.

4-4-3. Lemma. Suppose that R is noetherian and that every proper prime ideal of R is maximal. If $\mathfrak{P}$ is a proper prime ideal of R, then

$$R < \mathfrak{P}^{-1}$$

Proof. Choose any $a \neq 0 \in \mathfrak{P}$; so $(0) \neq Ra \subset \mathfrak{P} < R$. By (4-4-2) there exist proper prime ideals $\mathfrak{P}_1, \ldots, \mathfrak{P}_r$ such that $\mathfrak{P}_1 \cdots \mathfrak{P}_r \subset Ra \subset \mathfrak{P}$—and we may assume that the integer r is minimal. Since

$\mathfrak{P}$ is prime, it contains $\mathfrak{P}_1$, say; and because our prime ideals are maximal, we have $\mathfrak{P} = \mathfrak{P}_1$. From the minimality of r we know that $\mathfrak{P}_2 \cdots \mathfrak{P}_r \not\subset Ra$ (of course, this is true when $r = 1$); so there exists $b \in \mathfrak{P}_2 \cdots \mathfrak{P}_r$ with $b \notin Ra$. Hence, $b\mathfrak{P} \subset \mathfrak{P}\mathfrak{P}_2 \cdots \mathfrak{P}_r \subset Ra$, and $(b/a)\mathfrak{P} \subset R$. Thus, $b/a \in \mathfrak{P}^{-1}$ and $b/a \notin R$.

4-4-4. Lemma. Suppose that R is Dedekind; then every proper prime ideal is invertible.

Proof. Let $\mathfrak{P}$ be a proper prime ideal. We have then $R < \mathfrak{P}^{-1}$ and $\mathfrak{P} = R\mathfrak{P} \subset \mathfrak{P}^{-1}\mathfrak{P} \subset R$. Suppose that $\mathfrak{P}\mathfrak{P}^{-1} = \mathfrak{P}$. If so, then $\mathfrak{P}(\mathfrak{P}^{-1})^n = \mathfrak{P}$ for $n = 1, 2, 3, \ldots$. Therefore, for any $d \neq 0 \in \mathfrak{P}$ and any $c \in \mathfrak{P}^{-1}$ such that $c \notin R$, we have $dc^n \in \mathfrak{P} \subset R$ for all $n \geq 1$. It follows that $dR[c] \subset R$, so that $R[c] \subset Rd^{-1}$ and $R[c]$ is contained in a finitely generated R module. Now, let us recall some elementary facts. If V is a finitely generated unitary module over a domain R and R is noetherian, then V has the ascending chain condition on R submodules; moreover, the ascending chain condition on V is equivalent to the assertion that every submodule of V is finitely generated. Combining these facts with item 1 of (2-2-9), we conclude that c is integral over R; but R is integrally closed, and so $c \in R$, a contradiction. Therefore, $\mathfrak{P}\mathfrak{P}^{-1} = R$.

4-4-5. Theorem. If R is a Dedekind ring, then every fractional ideal can be expressed uniquely (upto order) in the form

$$\mathfrak{P}_1^{m_1} \cdots \mathfrak{P}_r^{m_r} \qquad m_i \in \mathbf{Z}$$

where the $\mathfrak{P}$'s are proper prime ideals of R.

Proof. Consider first an integral ideal $\mathfrak{A}$; we show that there exists a factorization for $\mathfrak{A}$. If $\mathfrak{A} = R$, it is a vacuous product (note that $\mathfrak{P}^0 = R$). Consider then $\mathfrak{A} \neq (0), R$. According to (4-4-2), $\mathfrak{A}$ contains some product of prime ideals; let $n(\mathfrak{A}) \geq 1$ be the minimal number (counting multiplicities) of proper prime ideals needed to accomplish this. The case $n(\mathfrak{A}) = 1$ is trivial; so suppose inductively that $n(\mathfrak{A}) > 1$. Since $\mathfrak{A}$ is contained in some maximal ideal (which is then prime), we have a relation

$$\mathfrak{P}_1 \cdots \mathfrak{P}_{n(\mathfrak{A})} \subset \mathfrak{A} \subset \mathfrak{P}$$

As before, $\mathfrak{P} = \mathfrak{P}_1$, say, and consequently

$$\mathfrak{P}_2 \cdots \mathfrak{P}_{n(\mathfrak{A})} \subset \mathfrak{P}^{-1}\mathfrak{A} \subset \mathfrak{P}^{-1}\mathfrak{P} = R$$

Thus, $\mathfrak{P}^{-1}\mathfrak{A}$ is an integral ideal, and $n(\mathfrak{P}^{-1}\mathfrak{A}) < n(\mathfrak{A})$. Therefore, $\mathfrak{P}^{-1}\mathfrak{A}$ is a product of prime ideals, and so is $\mathfrak{A} = \mathfrak{P}(\mathfrak{P}^{-1}\mathfrak{A})$.

To prove uniqueness, it is clearly more than sufficient to show that, if we are given distinct prime ideals $\mathfrak{P}_1, \ldots, \mathfrak{P}_r$ and integers $m_1, \ldots, m_r, n_1, \ldots, n_r \geq 0$, then

$$\mathfrak{P}_1^{m_1} \cdots \mathfrak{P}_r^{m_r} \subset \mathfrak{P}_1^{n_1} \cdots \mathfrak{P}_r^{n_r} \Rightarrow m_i \geq n_i \text{ for } i = 1, \ldots, r$$

To prove this implication, suppose that $n_1 > 0$; then $\mathfrak{P}_1$ contains the right side, and it follows (from properties of maximal ideals) that $m_1 > 0$. Multiplying by $\mathfrak{P}_1^{-1}$ enables us to repeat the process. This procedure may be continued until all the n_i are exhausted.

Consider finally an arbitrary fractional ideal $\mathfrak{A}$. There exists then an integral ideal $\mathfrak{B}$ such that the ideal $\mathfrak{C} = \mathfrak{A}\mathfrak{B}$ is integral. From the factorizations of $\mathfrak{B}$ and $\mathfrak{C}$ we get a factorization of $\mathfrak{A}$. Moreover, if $\mathfrak{P}_1^{m_1} \cdots \mathfrak{P}_r^{m_r} \subset \mathfrak{P}_1^{n_1} \cdots \mathfrak{P}_r^{n_r}$ (where negative exponents are allowed) then $\mathfrak{P}_1^{m_1+t} \cdots \mathfrak{P}_r^{m_r+t} \subset \mathfrak{P}_1^{n_1+t} \cdots \mathfrak{P}_r^{n_r+t}$ for all integers $t > 0$. Taking t so large that $m_i + t \geq 0$ and $n_i + t \geq 0$ for $i = 1, \ldots, r$, we conclude that $m_i \geq n_i$. This proves uniqueness and completes the proof.

4-4-6. Theorem. Let R be a Dedekind ring with quotient field F; then there exists a collection $\mathfrak{S}$ of discrete prime divisors of F such that $\{F, \mathfrak{S}\}$ is an OAF whose ring of integers $O\{\mathfrak{S}\}$ is R.

Proof. For each proper prime ideal $\mathfrak{P}$ of R we define a function $\nu_P = \nu_{\mathfrak{P}}$ on the group of all ideals of F by

$$\nu_P(\mathfrak{A}) = \text{exponent to which } \mathfrak{P} \text{ appears in the factorization of } \mathfrak{A}$$

Thus, $\mathfrak{A}$ has the unique expression

$$\mathfrak{A} = \prod_{\mathfrak{P}} \mathfrak{P}^{\nu_P(\mathfrak{A})}$$

and in particular

$$R = \prod_{\mathfrak{P}} \mathfrak{P}^0$$

One checks easily that the functions ν_P satisfy the following properties:

(1) $\nu_P(\mathfrak{A}) = 0$ for almost all P

(2) $\mathfrak{A} \subset \mathfrak{B} \Leftrightarrow \nu_P(\mathfrak{A}) \geq \nu_P(\mathfrak{B})$ for all P
Thus, $\mathfrak{A} = \mathfrak{B} \Leftrightarrow \nu_P(\mathfrak{A}) = \nu_P(\mathfrak{B})$ for all P

(3) $\nu_P(\mathfrak{A}\mathfrak{B}) = \nu_P(\mathfrak{A}) + \nu_P(\mathfrak{B})$ for all P

(4) $\nu_P(\mathfrak{A} + \mathfrak{B}) = \min \{\nu_P(\mathfrak{A}), \nu_P(\mathfrak{B})\}$ for all P
Thus, $\mathfrak{A} + \mathfrak{B} = \Pi\mathfrak{P}^{\min\{\nu_P(\mathfrak{A}), \nu_P(\mathfrak{B})\}}$

(5) If $\mathfrak{P}_i, \mathfrak{P}_j$ are prime ideals, then $\nu_{P_i}(\mathfrak{P}_j) = \delta_{ij}$

For $a \neq 0 \in F$, let us put $\nu_P(a) = \nu_P(Ra)$; naturally, we also take $\nu_P(0) = \infty$. Clearly, each ν_P is a discrete (normalized) valuation of F, and from (5) the corresponding prime divisors P are distinct. The set $\mathfrak{S} = \{P|P \leftrightarrow \mathfrak{P}\}$ is nonempty, and Axiom I for an OAF is satisfied. As for Axiom II, consider $P_1 \neq P_2 \in \mathfrak{S}$. Since $\mathfrak{P}_1 \neq \mathfrak{P}_2$, it is immediate that $\mathfrak{P}_1 + \mathfrak{P}_2 = R$. Thus, there exist $b \in \mathfrak{P}_1$ and $a \in \mathfrak{P}_2$ with $b + a = 1$. This is the required a. Hence, $\{F, \mathfrak{S}\}$ is an OAF, and obviously $O = R$.

4-4-7. Proposition. Let R be an integral domain with quotient field $F \neq R$; then the following conditions are equivalent:

i. R is Dedekind.

ii. Every proper integral ideal is a product of maximal ideals.

iii. Every fractional ideal is invertible; that is, the fractional ideals form a group.

Proof. (i) $\Rightarrow$ (ii). This has been done.

(ii) $\Rightarrow$ (iii). It suffices to show that any maximal ideal $\mathfrak{P}$ is invertible. Consider $a \neq 0 \in \mathfrak{P}$. We have then $Ra = \Pi\mathfrak{P}_i \subset \mathfrak{P}$ where the $\mathfrak{P}_i$ are maximal. Thus $\mathfrak{P} = \mathfrak{P}_1$, say. Since Ra is invertible and any factor of an invertible ideal is also invertible, $\mathfrak{P}$ is invertible.

(iii) $\Rightarrow$ (i). *R is noetherian.* Let $\mathfrak{A}$ be a proper ideal of R. Since $\mathfrak{A}\mathfrak{A}^{-1} = R$, we may write $1 = \Sigma a_i b_i$ with $a_i \in \mathfrak{A}$, $b_i \in \mathfrak{A}^{-1}$. Thus, $\mathfrak{A} = \Sigma R a_i$, so that any ideal is a finitely generated R module. Hence [as observed in the proof of (4-4-4)], R is noetherian.

Every proper prime ideal $\mathfrak{P}$ is maximal. Suppose we have an ideal $\mathfrak{A}$ with $\mathfrak{P} < \mathfrak{A} \subset R$. Then $\mathfrak{A}^{-1}\mathfrak{P} \subset \mathfrak{A}^{-1}\mathfrak{A} = R$. Because $\mathfrak{P} = (\mathfrak{A})(\mathfrak{A}^{-1}\mathfrak{P})$ is prime and $\mathfrak{A} \not\subset \mathfrak{P}$, we have $\mathfrak{A}^{-1}\mathfrak{P} \subset \mathfrak{P}$. Hence, $\mathfrak{A}^{-1} = \mathfrak{A}^{-1}\mathfrak{P}\mathfrak{P}^{-1} \subset \mathfrak{P}\mathfrak{P}^{-1} = R$ and $R \subset \mathfrak{A}$. Therefore, $R = \mathfrak{A}$, and $\mathfrak{P}$ is maximal.

R is integrally closed. Let $a \in F^*$ be integral over R. We know from item 8 of (2-2-9) that there exists $c \neq 0 \in R$ such that $ca^n \in R$ for $n = 1, 2, 3, \ldots$. Let $\mathfrak{A} = R[a]$. Since $c\mathfrak{A} \subset R$, $\mathfrak{A}$ is a fractional ideal with $\mathfrak{A}^2 = \mathfrak{A}$. By invertibility, $\mathfrak{A} = R$, and so $a \in R$.

In order to round out our sketch of the classical approach, it remains to show that the ring of integers in any algebraic number field is a Dedekind ring. To do this, it suffices to observe that **Z** (or any principal ideal domain) is a Dedekind ring and to prove the following result.

4-4-8. Theorem. Let R be a Dedekind ring with quotient field F. Suppose that E is a finite separable extension of F and that S is the integral closure of R in E; then S is Dedekind with quotient field E.

Proof. From parts 5, 6, and 12 of (2-2-9) we see that S is integrally closed in its quotient field E and that $S \neq E$.

To show that S is noetherian, it is sufficient [in view of the elementary facts quoted in the proof of (4-4-4)] to show that:

If $[E:F] = n$, then S is contained in an R module with n generators

It follows from part 12 of (2-2-9) that one may choose a basis $\alpha_1, \ldots, \alpha_n$ for E over F whose elements all belong to S. Let Ω be an algebraic closure of E, let O_Ω denote the integral closure of R in Ω, and let $\Gamma(F, E \rightarrow \Omega) = \{\sigma_1, \ldots, \sigma_n\}$. Since distinct homomorphisms of a group into the multiplicative group of a field are independent (for us this means that if $\beta_1, \ldots, \beta_n \in \Omega$ are such that

$$\sum_1^n \beta_i(\sigma_i\alpha) = 0$$

for all $\alpha \in E$ then $\beta_1 = \cdots = \beta_n = 0$), it follows easily that the $n \times n$ matrix $(\sigma_i\alpha_j)$ has an inverse.

Consider $\beta \in S$; we may write $\beta = x_1\alpha_1 + \cdots + x_n\alpha_n$ with $x_i \in F$. Therefore,

$$\sigma_i\beta = x_1\sigma_i(\alpha_1) + \cdots + x_n\sigma_i(\alpha_n) \qquad i = 1, \ldots, n$$

and by part 10 of (2-2-9) all $\sigma_i\beta$, $\sigma_i\alpha_j \in O_\Omega$.

In solving for the x_j according to Cramer's rule, we get solutions (for $j = 1, \ldots, n$) of form

$$x_j = \frac{\gamma_j}{\det(\sigma_i\alpha_j)} = \frac{\gamma_j \det(\sigma_i\alpha_j)}{\Delta(\alpha_1, \ldots, \alpha_n)}$$

Since γ_j, $\det(\sigma_i\alpha_j) \in O_\Omega$, we have by part 9 of (2-2-9)

$$x_j\,\Delta(\alpha_1, \ldots, \alpha_n) \in F \cap O_\Omega = R$$

Putting $\Delta(\alpha_1, \ldots, \alpha_n) = \Delta$, we have

$$S \subset \sum_1^n R\frac{\alpha_i}{\Delta}$$

and this part of the proof is complete.

It remains to prove that every proper prime ideal $\mathfrak{P}$ of S is maximal. The ideal $\tilde{\mathfrak{P}} = \mathfrak{P} \cap R$ of R is clearly prime. It is proper. In fact, $\tilde{\mathfrak{P}} = R \Rightarrow \mathfrak{P} \supset R \Rightarrow 1 \in \mathfrak{P} \Rightarrow \mathfrak{P} = S$, a contradiction; and if $\alpha \neq 0 \in \mathfrak{P}$, then, substituting α in $f(\alpha, F)$, we have a relation of form $\alpha^m + a_1\alpha^{m-1} + \cdots + a_m = 0$ with $a_m \neq 0 \in R$ and all $a_i \in R$—since $a_m \in S\alpha \subset \mathfrak{P}$, we see that $\tilde{\mathfrak{P}} \neq (0)$.

Let $\pi: S \twoheadrightarrow S/\mathfrak{P}$ be the natural homomorphism. When restricted to R, the kernel of π is $\tilde{\mathfrak{P}}$—that is, $R/\tilde{\mathfrak{P}} \approx \pi R$. Since $\tilde{\mathfrak{P}}$ is maximal, πR is a field. One checks easily that $S/\mathfrak{P}$ is integral over πR, and therefore by part 6 of (2-2-9) $S/\mathfrak{P}$ is a field. Hence $\mathfrak{P}$ is maximal. This completes the proof.

4-4-9. Exercise. Interpret the factorization of elements mentioned in (4-3-5) and (4-3-6) in terms of prime ideals, without using valuation theory.

4-5. Over-rings of O

4-5-1. Proposition. Let $\{F, \mathcal{S}\}$ be an OAF; then $\mathcal{S}$ is the set of all nonarchimedean prime divisors Q of F for which O_Q contains $O = O\{\mathcal{S}\}$.

Proof. Given such a Q, we choose any $\varphi_Q \in Q$ and put

$$\mathfrak{Q} = \{a \in F | \varphi_Q(a) < 1\}$$

Clearly, $O \cap \mathfrak{Q}$ is a prime ideal of O. Since $O \cap \mathfrak{Q} = O \Rightarrow O \subset \mathfrak{Q} \Rightarrow \varphi_Q(1) < 1$ and $O \cap \mathfrak{Q} = (0) \Rightarrow \varphi_Q(a) = 1$ for every $a \neq 0 \in O \Rightarrow \varphi_Q(a) = 1$ for every $a \in F^* \Rightarrow Q$ is trivial, we see that $O \cap \mathfrak{Q}$ is, in fact, a proper prime ideal of O. We have, therefore, $O \cap \mathfrak{Q} = \mathfrak{P} = O \cap \mathcal{P}$, where $\mathfrak{P}$ is the prime ideal corresponding to $P \in \mathcal{S}$.

Consider now any $a \in F$ with $\varphi_P(a) < 1$; that is, consider any $a \in \mathcal{P}$. Put $a = b/c$ with $b, c \in O$. Since $\varphi_P(b) < \varphi_P(c)$, we may (by multiplying by a suitable element of F^*) assume that $\varphi_P(c) = 1$ and $\varphi_P(b) < 1$. From

$$\varphi_P(b) < 1 \Rightarrow b \in O \cap \mathcal{P} = O \cap \mathfrak{Q} \Rightarrow \varphi_Q(b) < 1$$
$$\varphi_P(c) = 1 \Rightarrow c \in O,\ c \notin O \cap \mathfrak{Q} \Rightarrow \varphi_Q(c) = 1$$

we conclude that $\varphi_P(a) < 1 \Rightarrow \varphi_Q(a) < 1$. Hence, by (1-1-4), $P = Q$.

If S is any nonempty subset of $\mathcal{S}$, then $O\{S\} \supset O = O\{\mathcal{S}\}$. Suppose, on the other hand, that $O' \neq F$ is any over-ring of O. Let $\mathcal{S}'$ denote the set of all nonarchimedean prime divisors P of F such that $O_P \supset O'$. From (4-5-1) we know that $\mathcal{S}' \subset \mathcal{S}$. Let $\mathcal{S}''$ denote the complement of $\mathcal{S}'$ in $\mathcal{S}$—that is,

$$\mathcal{S} = \mathcal{S}' \cup \mathcal{S}'' \qquad \text{disjoint}$$

4-5-2. Lemma. $P \in \mathcal{S}'' \Rightarrow \mathfrak{P}^{-m} \subset O'$ for all $m > 0$. Symbolically, we then write $\mathfrak{P}^{-\infty} \subset O'$.

Proof. Since $P \in \mathcal{S}'' \subset \mathcal{S}$, we have $O_P \supset O$ and $O_P \not\supset O'$. There exist then $a \in O'$ and $n \in \mathbf{Z}$ with $\nu_P(a) = -n < 0$. Clearly $O + aO \subset O'$. Since for any $Q \in \mathcal{S}$ we have

$$\nu_Q(O + aO) = \min\{\nu_Q(1), \nu_Q(a)\} = \min\{0, \nu_Q(a)\}$$

it follows that

$$\mathfrak{P}^{-n} \subset O + aO \subset O'$$

For any $j > 0$, the same procedure applies to a^j and yields $\mathfrak{P}^{-nj} \subset O'$. If $0 < m_1 < m_2$, then $\mathfrak{P}^{-m_1} \subset \mathfrak{P}^{-m_2}$; therefore, $\mathfrak{P}^{-m} \subset O'$ for all $m > 0$.

4-5-3. Proposition. Let $\{F, \mathcal{S}\}$ be an OAF. Then there is a 1-1 order inverting correspondence between the over-rings $\neq F$ of $O = O\{\mathcal{S}\}$ and the nonempty subsets of $\mathcal{S}$. The correspondence is described as follows:

(i) If $\mathcal{S}' \subset \mathcal{S}$, then $\mathcal{S}' \leftrightarrow O' = O\{\mathcal{S}'\}$.

(ii) If $O' \supset O$, then $O' \leftrightarrow \mathcal{S}'$, where $\mathcal{S}'$ is the set of all nonarchimedean prime divisors P of F such that $O_P \supset O'$.

Proof. Given the over-ring $O' \neq F$ of O, we show that $O' = O\{\mathcal{S}'\}$. From the definition of $\mathcal{S}'$, it is clear that $O' \subset O\{\mathcal{S}'\}$. Of course, we do not know at this stage that $\mathcal{S}' \neq \phi$; so, for $\mathcal{S}' = \phi$, we put $O\{\mathcal{S}'\} = F$.

Consider then any $a \in O\{\mathcal{S}'\}$. Let $S = \{P_1, \ldots, P_r\}$ be the finite set of prime divisors from $\mathcal{S}$ such that $\nu_{P_i}(a) = -n_i < 0$. Clearly, $S \subset \mathcal{S}''$. If $S = \phi$, then $a \in O \subset O'$. If $S \neq \phi$, then, from (4-5-2) and the fact that $\nu_P(O + aO) = \min\{0, \nu_P(a)\}$ for all $P \in \mathcal{S}$, it follows that

$$a \in O + aO = \mathfrak{P}_1^{-n_1} + \cdots + \mathfrak{P}_r^{-n_r} \subset O'$$

Thus, $O\{\mathcal{S}'\} \subset O'$, and in particular $\mathcal{S}'$ is nonempty. Symbolically, we may write

$$O' = \sum_{P \in \mathcal{S}''} \mathfrak{P}^{-\infty}$$

The proof is easily completed by making use of (4-5-1).

As a consequence of this result, we have (in the language of Section 4-4) the following proposition.

4-5-4. Proposition. If an over-ring of a Dedekind ring is strictly contained in the quotient field, then it too is a Dedekind ring.

4-5-5. Remark. Suppose that $O = O\{F, \mathcal{S}\} = O\{\mathcal{S}\}$ is a Dedekind ring and that $O' = O\{F, \mathcal{S}'\} = O\{\mathcal{S}'\}$ is an over-ring with $O' \neq O, F$. We wish to describe the relation between the O' arithmetic and the O arithmetic.

Let us define a homomorphism $\lambda: \mathbf{D}\{\mathcal{S}\} \rightarrow \mathbf{D}\{\mathcal{S}'\}$ by putting, for $P \in \mathcal{S}$,

$$\lambda(P) = P \qquad \text{if } P \in \mathcal{S}'$$
$$\lambda(P) = 1 \qquad \text{if } P \notin \mathcal{S}'$$

and extending from the generators. In more detail, λ maps

$$A = \prod_{P \in \mathcal{S}} P^{\nu_P(A)} \in \mathbf{D}\{\mathcal{S}\}$$

to

$$A' = \lambda(A) = \prod_{P \in \mathcal{S}'} P^{\nu_P(A)} = \prod_{P \in \mathcal{S}'} P^{\nu_P(A')} \in \mathbf{D}\{\mathcal{S}'\}$$

It is clear that λ is a homomorphism onto and that A is in the kernel of λ if and only if $\nu_P(A) = 0$ for all $P \in \mathcal{S}'$. Of course, λ preserves joins, meets, divisibility, and integrality.

In virtue of the fundamental theorem of OAFs, λ may be carried over uniquely to a homomorphism $\lambda : \mathfrak{I}\{\mathcal{S}\} \twoheadrightarrow \mathfrak{I}\{\mathcal{S}'\}$. This means that all possible commutativities hold in the diagram

$$\begin{array}{ccc} \mathbf{D}\{\mathcal{S}\} & \underset{\delta}{\overset{\kappa}{\rightleftarrows}} & \mathfrak{I}\{\mathcal{S}\} \\ \lambda\downarrow & & \lambda\downarrow \\ \mathbf{D}\{\mathcal{S}'\} & \underset{\delta}{\overset{\kappa}{\rightleftarrows}} & \mathfrak{I}\{\mathcal{S}'\} \end{array}$$

To be precise, if $\mathfrak{P}$ is the prime ideal of O corresponding to $P \in \mathcal{S}$ and $\mathfrak{P}'$ is the prime ideal of O' corresponding to $P \in \mathcal{S}'$, then λ maps

$$\mathfrak{A} = \prod_{P \in \mathcal{S}} \mathfrak{P}^{\nu_P(\mathfrak{A})} \in \mathfrak{I}\{\mathcal{S}\}$$

to

$$\mathfrak{A}' = \lambda(\mathfrak{A}) = \prod_{P \in \mathcal{S}'} \mathfrak{P}'^{\nu_P(\mathfrak{A})} = \prod_{P \in \mathcal{S}'} \mathfrak{P}'^{\nu_P(\mathfrak{A}')} \in \mathfrak{I}\{\mathcal{S}'\}$$

In particular, if $P \in \mathcal{S}'$, then $\lambda(\mathfrak{P}) = \mathfrak{P}'$. Moreover, for $\mathfrak{A}, \mathfrak{B} \in \mathfrak{I}\{\mathcal{S}\}$ we have $(\mathfrak{A} + \mathfrak{B})' = \mathfrak{A}' + \mathfrak{B}'$, $(\mathfrak{A} \cap \mathfrak{B})' = \mathfrak{A}' \cap \mathfrak{B}'$, $\mathfrak{A}|\mathfrak{B} \Rightarrow \mathfrak{A}'|\mathfrak{B}'$, and $\mathfrak{A}$ integral $\Rightarrow \mathfrak{A}'$ integral.

The important thing is that this artificial map of ideals λ has a natural algebraic interpretation—namely,

$$\lambda(\mathfrak{A}) = \mathfrak{A}' = \mathfrak{A}O'$$

To see this, note that $\mathfrak{A}$ may be expressed in the form $\mathfrak{A} = aO + bO$ for some choice of $a, b \in \mathfrak{A}$. Since

$$\lambda(aO) = (aO)' = \{c \in F | \nu_P(c) \geq \nu_P(a) \quad \forall P \in \mathcal{S}'\} = aO'$$

it follows that

$$\begin{aligned} \lambda(\mathfrak{A}) &= \lambda(aO + bO) = \lambda(aO) + \lambda(bO) \\ &= aO' + bO' = (aO + bO)O' = \mathfrak{A}O' \end{aligned}$$

In particular, it follows that $\mathfrak{P}O' = \mathfrak{P}'$ for $P \in \mathcal{S}'$, $\mathfrak{P}O' = O'$ for $P \notin \mathcal{S}'$, and $\mathfrak{A}O' = O'$ if and only if $\nu_P(\mathfrak{A}) = 0$ for all $P \in \mathcal{S}'$.

In order to define a mapping that goes in the opposite direction, let $\mathfrak{S}''$ be the complement of $\mathfrak{S}'$ in $\mathfrak{S}$, and put $O'' = O\{\mathfrak{S}''\}$. It may be noted that $O' \cap O'' = O$. Since for $\mathfrak{C} \in \mathfrak{J}\{\mathfrak{S}'\}$ we have

$$\mathfrak{C} \cap O'' = \{a \in F | \nu_P(a) \geq \nu_P(\mathfrak{C}) \quad \forall P \in \mathfrak{S}' \text{ and } \nu_P(a) \geq 0 \quad \forall P \in \mathfrak{S}''\}$$

it is clear that $\nu_P(\mathfrak{C} \cap O'') = \nu_P(\mathfrak{C})$ for all $P \in \mathfrak{S}'$ and that

$$(\mathfrak{C} \cap O'')O' = \mathfrak{C}$$

This means that the map

$$\mathfrak{C} = \prod_{P \in \mathfrak{S}'} \mathfrak{P}'^{\nu_P(\mathfrak{C})} \rightarrow \mathfrak{C} \cap O'' = \prod_{P \in \mathfrak{S}'} \mathfrak{P}^{\nu_P(\mathfrak{C})}$$

of $\mathfrak{J}\{\mathfrak{S}'\} \rightarrow \mathfrak{J}\{\mathfrak{S}\}$ is an isomorphism into whose effect is undone by λ.

If $\mathfrak{C}$ is integral, then $\mathfrak{C} \cap O'' = \mathfrak{C} \cap O' \cap O'' = \mathfrak{C} \cap O$ and

$$(\mathfrak{C} \cap O)O' = \mathfrak{C}$$

4-6. Class Number

Given an ordinary arithmetic field $\{F, \mathfrak{S}\}$ we have by the fundamental theorem a group isomorphism

$$\frac{\mathbf{D}}{\mathbf{D}^*} \approx \frac{\mathfrak{J}}{\mathfrak{J}^*}$$

where all of the objects are taken with respect to $\mathfrak{S}$. These groups are called the ***group of $\mathfrak{S}$ divisor classes*** and the ***group of $\mathfrak{S}$ ideal classes,*** respectively. Their common order is denoted by $h = h\{\mathfrak{S}\}$ and is called the $\mathfrak{S}$ ***class number.*** In Chapter 5, we shall see that, if F is an algebraic number field and $\mathfrak{S}$ consists of almost all the non-archimedean primes of F, then h is finite.

The size of h is a measure of how far $O = O\{\mathfrak{S}\}$ is from being a principal ideal domain. In fact,

$h = 1 \Leftrightarrow \mathfrak{J} = \mathfrak{J}^* \Leftrightarrow$ every fractional ideal is principal $\Leftrightarrow$ every integral ideal is principal $\Leftrightarrow O$ is a principal ideal domain

If the set $\mathfrak{S}$ is finite, then O is a principal ideal domain. In fact, if $\mathfrak{A}$ is then any $\mathfrak{S}$ ideal, there exists $a \in \mathfrak{A}$ such that $\nu_P(a) = \nu_P(\mathfrak{A})$ for all $P \in \mathfrak{S}$, so that $\mathfrak{A} = aO$.

4-6-1. Proposition. O is a unique factorization domain $\Leftrightarrow O$ is a principal ideal domain.

Proof. $\Leftarrow$: It is well known that any principal ideal domain is a unique factorization domain.

$\Rightarrow$: Take any prime element $p \in O$; so pO is a proper prime ideal of O, and therefore corresponds to a prime divisor $P_p \in \mathcal{S}$. We have

$$pO = \{a \in F | \nu_{P_p}(a) \geq 1 \text{ and } \nu_P(a) \geq 0 \quad \forall P \neq P_p \in \mathcal{S}\}$$

Putting $\mathcal{S}' = \{P_p | p \text{ prime in } O\}$, we know that $\mathcal{S}' \subset \mathcal{S}$. From the fact that $O\{\mathcal{S}'\} = O\{\mathcal{S}\}$, it follows that $\mathcal{S}' = \mathcal{S}$. In other words, $\{pO\}$ is the set of all proper prime ideals of O, and they are all principal. Since we have unique factorization of ideals, O is indeed a principal ideal domain.

4-6-2. Corollary. O is a unique factorization domain $\Leftrightarrow h = 1$.

4-6-3. Proposition. Every ideal class contains an integral ideal; in fact, if an integral ideal $\mathfrak{M}$ is given, then every ideal class contains an integral ideal which is prime to $\mathfrak{M}$.

Proof. Let $\mathfrak{B}$ be any ideal belonging to the class in question, and consider the finite set

$$S = \{P | \nu_P(\mathfrak{M}) \neq 0\} \cup \{P | \nu_P(\mathfrak{B}) \neq 0\}$$

Choose $a \in F$ such that

$$\begin{aligned} \nu_P(a) &= -\nu_P(\mathfrak{B}) \qquad && P \in S \\ \nu_P(a) &\geq 0 && P \notin S \end{aligned}$$

Then $\mathfrak{A} = a\mathfrak{B}$ is an integral ideal with $(\mathfrak{A}, \mathfrak{M}) = 1$.

4-7. Mappings of Ideals

Suppose that $\{F, \mathcal{S}\}$ is an OAF and that E/F is a finite extension of degree n. We write $\mathcal{S} = \mathcal{S}(F)$ and let $\mathcal{S}(E)$ denote the set of all extensions to E of the primes in $\mathcal{S}(F)$. Then $\{E, \mathcal{S}(E)\}$ is an OAF, and the ring of integers $O_E = O\{\mathcal{S}(E)\}$ is the integral closure of $O_F = O\{\mathcal{S}(F)\}$ in E. All finite extensions of F will be understood to come from a fixed algebraic closure Ω of F. This is the basic framework within which we shall operate throughout the remainder of this chapter; the objective is to study the relations between the arithmetics of E and F. The first step (to which this section is devoted) is to define some mappings between $\mathfrak{I}\{\mathcal{S}(F)\}$ and $\mathfrak{I}\{\mathcal{S}(E)\}$.

Consider the inclusion map $i_{F \to E}: F^* \to E^*$. Since divisors (or ideals) are in a sense generalizations of field elements, it is reasonable for us to try to extend this map in a consistent way to a mapping

$i_{F\to E}:\mathbf{D}\{\mathcal{S}(F)\} \to \mathbf{D}\{\mathcal{S}(E)\}$. In other words, we want the following diagram (in which the δ maps are as in Section 4-3) to commute:

$$\begin{array}{ccc} E^* & \xrightarrow{\delta_E} & \mathbf{D}\{\mathcal{S}(E)\} \\ {\scriptstyle i_{F\to E}}\uparrow & & \uparrow{\scriptstyle i_{F\to E}} \\ F^* & \xrightarrow{\delta_F} & \mathbf{D}\{\mathcal{S}(F)\} \end{array}$$

For each $P \in \mathcal{S}(F)$, let us put

$$i_{F\to E}(P) = \prod_{Q\supset P} Q^{e(Q/P)}$$

Extending $i_{F\to E}$ from the generators of $\mathbf{D}\{\mathcal{S}(F)\}$ to the entire group, it follows that for

$$A = \prod_{P\in\mathcal{S}(F)} P^{\nu_P(A)} \in \mathbf{D}\{\mathcal{S}(F)\}$$

we have

$$i_{F\to E}(A) = \prod_{P\in\mathcal{S}(F)} \prod_{Q\supset P} Q^{e(Q/P)\,\nu_P(A)}$$

Because $i_{F\to E}$ is essentially a listing of the e's, it is immediate that it is a monomorphism which preserves joins, meets, divisibility, and integrality. As for the commutativity of the diagram, it is easy to verify—in fact, for $a \in F^*$,

$$\begin{aligned} i_{F\to E}\delta_F(a) &= i_{F\to E}\Big(\prod_{P\in\mathcal{S}(F)} P^{\nu_P(a)}\Big) = \prod_{P\in\mathcal{S}(F)} \prod_{Q\supset P} Q^{e(Q/P)\,\nu_P(a)} \\ &= \prod_{Q\in\mathcal{S}(E)} Q^{\nu_Q(a)} = \delta_E(a) = \delta_E i_{F\to E}(a) \end{aligned}$$

In view of the isomorphisms between the divisor group and the ideal group, the mapping $i_{F\to E}$ may be transferred to the ideal groups. In summary, we have the following result.

4-7-1. Proposition. Let $\mathfrak{P}$ denote the prime ideal of O_F corresponding to $P \in \mathcal{S}(F)$, and let $\mathfrak{Q}$ denote the prime ideal of O_E corresponding to $Q \in \mathcal{S}(E)$. The mapping $i_{F\to E}:\mathfrak{I}\{\mathcal{S}(F)\} \to \mathfrak{I}\{\mathcal{S}(E)\}$ such that for

$$\mathfrak{A} = \prod_{P\in\mathcal{S}(F)} \mathfrak{P}^{\nu_P(\mathfrak{A})} \in \mathfrak{I}\{\mathcal{S}(F)\}$$

we have

$$i_{F\to E}(\mathfrak{A}) = \prod_{P\in\mathcal{S}(F)} \prod_{Q\supset P} \mathfrak{Q}^{e(Q/P)\,\nu_P(\mathfrak{A})}$$

is a monomorphism with the properties:

(i) $i_{F\to E}(\mathfrak{A} + \mathfrak{B}) = i_{F\to E}(\mathfrak{A}) + i_{F\to E}(\mathfrak{B})$

(ii) $i_{F\to E}(\mathfrak{A} \cap \mathfrak{B}) = i_{F\to E}(\mathfrak{A}) \cap i_{F\to E}(\mathfrak{B})$

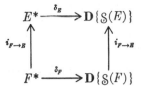

(iii) $\mathfrak{A}|\mathfrak{B} \Leftrightarrow i_{F\to E}(\mathfrak{A})|i_{F\to E}(\mathfrak{B})$

(iv) $\mathfrak{A}$ is integral $\Leftrightarrow i_{F\to E}(\mathfrak{A})$ is integral

(v) $i_{F\to E}(\mathfrak{A}) = \mathfrak{A}O_E = \prod_{Q\in \mathcal{S}(E)} \mathfrak{Q}^{\nu_Q(\mathfrak{A}O_E)}$ and in particular,

$$\mathfrak{P}O_E = \prod_{Q\supset P} \mathfrak{Q}^{e(Q/P)}$$

Proof. The important assertion here—and the only one still requiring proof—is the last one. Suppose first that $\mathfrak{A}$ is principal, $\mathfrak{A} = aO_F$, $a \in F^*$. From the commutative diagram:

$$\begin{array}{ccccc} E^* & \xrightarrow{\delta_E} & \mathbf{D}\{\mathcal{S}(E)\} & \xrightarrow{\kappa_E} & \mathfrak{J}\{\mathcal{S}(E)\} \\ \uparrow {\scriptstyle i_{F\to E}} & & \uparrow {\scriptstyle i_{F\to E}} & & \uparrow {\scriptstyle i_{F\to E}} \\ F^* & \xrightarrow{\delta_F} & \mathbf{D}\{\mathcal{S}(F)\} & \xrightarrow{\kappa_F} & \mathfrak{J}\{\mathcal{S}(F)\} \end{array}$$

and the discussion of Section 4-3, it follows that

$$\begin{aligned} i_{F\to E}(\mathfrak{A}) &= i_{F\to E}(aO_F) = i_{F\to E}\kappa_F\delta_F(aO_F) = i_{F\to E}\kappa_F\delta_F(a) \\ &= \kappa_E\delta_E i_{F\to E}(a) = \kappa_E\delta_E(a) = aO_E = \mathfrak{A}O_E \end{aligned}$$

Now, an arbitrary ideal $\mathfrak{A}$ may be expressed, as usual, in the form $\mathfrak{A} = aO_F + bO_F$ with $a, b \in F^*$; and then

$$\begin{aligned} i_{F\to E}(\mathfrak{A}) &= i_{F\to E}(aO_F + bO_F) = i_{F\to E}(aO_F) + i_{F\to E}(bO_F) \\ &= aO_E + bO_E = (aO_F + bO_F)O_E = \mathfrak{A}O_E \end{aligned}$$

One may note that the kernel of the argument used above is that two homomorphisms of ideal groups are identical when they preserve joins and agree on the principal ideals.

4-7-2. Proposition. If $\mathfrak{A} \in \mathfrak{J}\{\mathcal{S}(F)\}$, then $\mathfrak{A}O_E \cap F = \mathfrak{A}$.

Proof. From what has gone before, we know that, for $Q \in \mathcal{S}(E)$, $\nu_Q(\mathfrak{A}O_E) = e(Q/P)\nu_P(\mathfrak{A})$, where P is the restriction of Q to F. Therefore,

$$\begin{aligned} \mathfrak{A}O_E \cap F &= \{a \in F | \nu_Q(a) \geq \nu_Q(\mathfrak{A}O_E) \quad \forall Q \in \mathcal{S}(E)\} \\ &= \{a \in F | \nu_P(a) \geq \nu_P(\mathfrak{A}) \quad \forall P \in \mathcal{S}(F)\} = \mathfrak{A} \end{aligned}$$

Thus, intersection with F provides a left inverse for the mapping $i_{F\to E}$.

4-7-3. Remark. Consider a prime ideal $\mathfrak{P}$ of O_F. We may write

$$\mathfrak{P}O_E = \prod_1^r \mathfrak{Q}_i^{e(Q_i/P)}$$

where $Q_1, \ldots, Q_r$ are the extensions of P to E. Thus, $\mathfrak{P} \subset \mathfrak{Q}_i$, and, from the expression for an ideal in terms of valuations, it is immediate that $\mathfrak{P} = \mathfrak{Q}_i \cap F = \mathfrak{Q}_i \cap O_F$. The $\mathfrak{Q}_i$ are said to ***lie above*** $\mathfrak{P}$. On the other hand, suppose that $\mathfrak{Q}$ is a prime ideal of O_E. It is clear that $\mathfrak{Q} \cap F = \mathfrak{Q} \cap O_F$ is a prime ideal $\mathfrak{P}$ of O_F ($\mathfrak{P}$ is said to ***lie below*** $\mathfrak{Q}$), and since $\mathfrak{P} \subset \mathfrak{P}O_E \subset \mathfrak{Q}$, $\mathfrak{Q}$ appears in the factorization of $\mathfrak{P}O_E$.

The preceding results may now be used to describe the classical scheme for identification of ideals in different fields. Let $\mathfrak{F}$ denote the set of all finite extensions of F. We say that the ideals $\mathfrak{A}_1 \in \mathfrak{I}\{\mathfrak{S}(E_1)\}$ and $\mathfrak{A}_2 \in \mathfrak{I}\{\mathfrak{S}(E_2)\}$ (where $E_1, E_2 \in \mathfrak{F}$) are ***equivalent***—$\mathfrak{A}_1 \simeq \mathfrak{A}_2$—if there exists a field $E \in \mathfrak{F}$ containing both E_1 and E_2 and such that $i_{E_1\to E}(\mathfrak{A}_1) = i_{E_2\to E}(\mathfrak{A}_2)$ (in other words, $\mathfrak{A}_1O_E = \mathfrak{A}_2O_E$). One notes immediately that

$$\mathfrak{A}_1 \simeq \mathfrak{A}_2 \Leftrightarrow \mathfrak{A}_1O_{E_1E_2} = \mathfrak{A}_2O_{E_1E_2}$$

In fact, $\mathfrak{A}_1 \simeq \mathfrak{A}_2 \Rightarrow \mathfrak{A}_1O_E = \mathfrak{A}_2O_E \Rightarrow \mathfrak{A}_1O_E \cap E_1E_2 = \mathfrak{A}_2O_E \cap E_1E_2 \Rightarrow (\mathfrak{A}_1O_{E_1E_2})O_E \cap E_1E_2 = (\mathfrak{A}_2O_{E_1E_2})O_E \cap E_1E_2 \Rightarrow \mathfrak{A}_1O_{E_1E_2} = \mathfrak{A}_2O_{E_1E_2}$. Thus, the question of the equivalence of $\mathfrak{A}_1$ and $\mathfrak{A}_2$ can be settled in the field E_1E_2.

It is trivial to observe that $\simeq$ is indeed an equivalence relation and that, if $E_1 = E_2$, then $\mathfrak{A}_1 \simeq \mathfrak{A}_2 \Leftrightarrow \mathfrak{A}_1 = \mathfrak{A}_2$. The notion of equality of ideals may therefore be extended by identifying equivalent ideals—that is, for $\mathfrak{A}_1 \in \mathfrak{I}\{\mathfrak{S}(E_1)\}$ and $\mathfrak{A}_2 \in \mathfrak{I}\{\mathfrak{S}(E_2)\}$ one says that $\mathfrak{A}_1 = \mathfrak{A}_2$ when $\mathfrak{A}_1 \simeq \mathfrak{A}_2$. According to a fancier terminology, we have formed the direct limit of the groups $\mathfrak{I}\{\mathfrak{S}(E)\}$ (which are indexed by the directed set $\mathfrak{F}$) with respect to the maps $i_{E_1\to E_2}$ for $E_1 \subset E_2$.

Suppose that $\mathfrak{A}_1 \in \mathfrak{I}\{\mathfrak{S}(E_1)\}$, $\mathfrak{A}_2 \in \mathfrak{I}\{\mathfrak{S}(E_2)\}$ $(E_1 \neq E_2)$; it is clear what meanings should be attached to $\mathfrak{A}_1\mathfrak{A}_2$, $\mathfrak{A}_1 + \mathfrak{A}_2$, $\mathfrak{A}_1 \cap \mathfrak{A}_2$, $\mathfrak{A}_1 \subset \mathfrak{A}_2$, or $\mathfrak{A}_1|\mathfrak{A}_2$. One simply moves $\mathfrak{A}_1$ and $\mathfrak{A}_2$ up to the field E_1E_2—or to any larger $E \in \mathfrak{F}$—and deals with them there. Of course, the result is independent of the field $E \supset E_1E_2$ in which one operates. It should be noted that, although ideals can be treated without reference to the field of definition, their properties often do depend on the field of reference. For example, in moving up to a bigger field, a principal ideal remains principal, and two relatively prime ideals remain relatively prime; on the other hand, a prime ideal need not remain prime, and a nonprincipal ideal may well become principal.

A useful connection between this discussion and (4-5-5) is provided by our next result.

4-7-4. Proposition. Suppose that $\mathfrak{S}'(F)$ is a subset of $\mathfrak{S}(F)$ and that $\mathfrak{S}'(E) \subset \mathfrak{S}(E)$ is the set of extensions to E of all $P \in \mathfrak{S}'(F)$. If we put

$O'_F = O\{\mathcal{S}'(F)\}$ and $O'_E = O\{\mathcal{S}'(E)\}$, then

$$O'_E = O_E \cdot O'_F$$

In particular, if $\mathfrak{A} \in \mathfrak{J}\{\mathcal{S}(E)\}$, then

$$\mathfrak{A}O'_E = \mathfrak{A}O'_F$$

Proof. It is clear that $O_E \cdot O'_F$ is a ring with $O_E \subset O_E \cdot O'_F \subset O'_E$. Let $\mathfrak{I}$ be the subset of $\mathcal{S}(E)$ corresponding to $O_E O'_F$. For $Q \in \mathfrak{I}$, let P be its restriction to F; so $P \in \mathcal{S}'(F)$, and $Q \in \mathcal{S}'(E)$. Therefore, $\mathcal{S}'(E) = \mathfrak{I}$, and $O'_E = O_E \cdot O'_F$.

Now, let us extend the norm mapping $N_{E\to F}:E^* \to F^*$ to a map $N_{E\to F}:\mathbf{D}\{\mathcal{S}(E)\} \to \mathbf{D}\{\mathcal{S}(F)\}$. Suppose that for a prime divisor $Q \in \mathcal{S}(E)$ we put

$$N_{E\to F}(Q) = P^{g(Q/P)f(Q/P)}$$

[where $P \in \mathcal{S}(F)$ is the restriction of Q to F] and extend from the generators. The map $N_{E\to F}$ is then a homomorphism such that for

$$A = \prod_{Q \in \mathcal{S}(E)} Q^{\nu_Q(A)} \in \mathbf{D}\{\mathcal{S}(E)\}$$

we have

$$N_{E\to F}(A) = \prod_{P \in \mathcal{S}(F)} \prod_{Q \supset P} P^{g(Q/P)f(Q/P)\nu_Q(A)}$$

$$= \prod_{P \in \mathcal{S}(F)} P^{\sum_{Q\supset P} g(Q/P)f(Q/P)\nu_Q(A)} \in \mathbf{D}\{\mathcal{S}(F)\}$$

Since for $\alpha \in E^*$ and $P \in \mathcal{S}(F)$ we know [by (2-5-8)] that

$$\nu_P[N_{E\to F}(\alpha)] = \sum_{Q\supset P} g(Q/P)f(Q/P)\nu_Q(\alpha)$$

it follows that the diagram

$$\begin{array}{ccc} E^* & \xrightarrow{\delta_E} & \mathbf{D}\{\mathcal{S}(E)\} \\ \downarrow N_{E\to F} & & \downarrow N_{E\to F} \\ F^* & \xrightarrow{\delta_F} & \mathbf{D}\{\mathcal{S}(F)\} \end{array}$$

is commutative. It may also be observed that this norm mapping preserves divisibility and integrality (but not join or meet). When $N_{E\to F}$ is transferred to the corresponding ideal groups, we have the following result.

4-7-5. Proposition. The mapping

$$N_{E\to F} : \mathfrak{I}\{\mathcal{S}(E)\} \to \mathfrak{I}\{\mathcal{S}(F)\}$$

such that for $\mathfrak{A} \in \mathfrak{I}\{\mathcal{S}(E)\}$ we have

$$N_{E\to F}(\mathfrak{A}) = \prod_{P\in\mathcal{S}(F)} \mathfrak{P}^{\sum_{Q\supset P} g(Q/P)f(Q/P)\nu_Q(\mathfrak{A})}$$

is a homomorphism with the following properties:

i. $\mathfrak{A} \subset \mathfrak{B} \Rightarrow N_{E\to F}(\mathfrak{A}) \subset N_{E\to F}(\mathfrak{B})$.

ii. $\mathfrak{A}|\mathfrak{B} \Rightarrow N_{E\to F}(\mathfrak{A})|N_{E\to F}(\mathfrak{B})$.

iii. $\mathfrak{A}$ integral $\Rightarrow N_{E\to F}(\mathfrak{A})$ integral; in fact, when $\mathfrak{A}$ is integral, $N_{E\to F}(\mathfrak{A}) \subset \mathfrak{A}$.

iv. $N_{E\to F}(\alpha O_E) = (N_{E\to F}\alpha)O_F \qquad \alpha \in E^*$.

v. $N_{E\to F}(\mathfrak{A})$ is the greatest common divisor of the set of ideals $\{N_{E\to F}(\alpha)O_F|\alpha \in \mathfrak{A}\}$.

vi. If $\mathfrak{A} \in \mathfrak{I}\{\mathcal{S}(F)\}$, then $N_{E\to F}i_{F\to E}(\mathfrak{A}) = \mathfrak{A}^n$.

vii. For $F \subset E \subset K$, $N_{K\to F} = N_{E\to F} \circ N_{K\to E}$.

Proof. Assertions (i), (ii), and the first part of (iii) are immediate consequences of the formula for $N_{E\to F}$. If $\mathfrak{A}$ is integral, then it is clear from the expression for $i_{F\to E}N_{E\to F}(\mathfrak{A}) = (N_{E\to F}(\mathfrak{A}))O_E$ that $N_{E\to F}(\mathfrak{A}) \subset \mathfrak{A}$; this proves (iii). The commutativity of the diagram

$$\begin{array}{ccccc}
E^* & \xrightarrow{\delta_E} & \mathbf{D}\{\mathcal{S}(E)\} & \xrightarrow{\kappa_E} & \mathfrak{I}\{\mathcal{S}(E)\} \\
\downarrow{\scriptstyle N_{E\to F}} & & \downarrow{\scriptstyle N_{E\to F}} & & \downarrow{\scriptstyle N_{E\to F}} \\
F^* & \xrightarrow{\delta_F} & \mathbf{D}\{\mathcal{S}(F)\} & \xrightarrow{\kappa_F} & \mathfrak{I}\{\mathcal{S}(F)\}
\end{array}$$

says that, for $\alpha \in E^*$, $N_{E\to F}(\alpha O_E) = N_{E\to F}\kappa_E\delta_E(\alpha) = \kappa_F\delta_F N_{E\to F}(\alpha) = (N_{E\to F}\alpha)O_F$. This proves (iv).

To prove (v), we must show that, for every $P \in \mathcal{S}(F)$,

$$\nu_P[N_{E\to F}(\mathfrak{A})] = \min_{\alpha\in\mathfrak{A}} \{\nu_P[N_{E\to F}(\alpha)]\}$$

From the relations

$$\mathfrak{A} = \{\alpha \in E|\nu_Q(\alpha) \geq \nu_Q(\mathfrak{A}) \quad \forall Q \in \mathcal{S}(E)\}$$

$$\nu_P[N_{E\to F}(\mathfrak{A})] = \sum_{Q\supset P} g\left(\frac{Q}{P}\right) f\left(\frac{Q}{P}\right) \nu_Q(\mathfrak{A})$$

$$\nu_P[N_{E\to F}(\alpha)] = \sum_{Q\supset P} g\left(\frac{Q}{P}\right) f\left(\frac{Q}{P}\right) \nu_Q(\alpha)$$

it follows that $\nu_P[N_{E\to F}(\mathfrak{A})] \leq \min_{\alpha\in\mathfrak{A}} \{\nu_P[N_{E\to F}(\alpha)]\}$. On the other hand, given P, there exists $\alpha \in \mathfrak{A}$ such that $\nu_Q(\alpha) = \nu_Q(\mathfrak{A})$ for all $Q \supset P$; so $\nu_P[N_{E\to F}(\mathfrak{A})] = \nu_P[N_{E\to F}(\alpha)]$.

To prove (vi), it suffices to note that

$$N_{E\to F} i_{F\to E}(\mathfrak{P}) = N_{E\to F}\left(\prod_{Q\supset P} \mathfrak{Q}^{e(Q/P)}\right) = \mathfrak{P}^{\sum_{Q\supset P} g(Q/P)e(Q/P)f(Q/P)} = \mathfrak{P}^n$$

The proof of (vii) is straightforward.

The connection of our norm with the classical definition of the norm of an ideal may now be given.

4-7-6. Proposition. For $\mathfrak{A} \in \mathfrak{J}\{\mathfrak{S}(E)\}$ we have

$$N_{E\to F}(\mathfrak{A}) = \prod_\sigma (\sigma\mathfrak{A})^{[E:F]_{\text{ins}}} \qquad \sigma \in \Gamma(F, E \to \Omega)$$

Proof. This formula is the formal analogue of the formula for the norm of an element. Its interpretation is that $\sigma\mathfrak{A} = \{\sigma\alpha | \alpha \in \mathfrak{A}\}$ is the "conjugate" ideal to $\mathfrak{A}$ in the conjugate field σE, and all the ideals are then viewed in the normal extension K/F generated by E/F.

We put

$$M(\mathfrak{A}) = \prod_\sigma (\sigma\mathfrak{A})^{[E:F]_{\text{ins}}}$$

and note that M is a homomorphism of $\mathfrak{J}\{\mathfrak{S}(E)\} \to \mathfrak{J}\{\mathfrak{S}(K)\}$ which takes integral ideals into integral ideals. It is easy to see that M and $N = N_{E\to F}$ agree on principal ideals—in fact, for $\alpha \in E^*$,

$$\begin{aligned} M(\alpha O_E) &= \prod_\sigma [\sigma(\alpha O_E)]^{[E:F]_{\text{ins}}} \\ &= \prod_\sigma (\sigma\alpha)^{[E:F]_{\text{ins}}} \prod_\sigma O_{\sigma E}^{[E:F]_{\text{ins}}} && (\text{since } \sigma O_E = O_{\sigma E}) \\ &= N_{E\to F}(\alpha) O_K && (\text{since } O_{\sigma E} = O_K) \\ &= N_{E\to F}(\alpha) O_F \\ &= N_{E\to F}(\alpha O_E) \end{aligned}$$

We see then that, for any $\alpha \in \mathfrak{A}$, $M(\mathfrak{A}) \supset M(\alpha O_E) = N(\alpha O_E)$; therefore, according to (4-7-5v), $M(\mathfrak{A}) \supset N(\mathfrak{A})$, and $M(\mathfrak{A})|N(\mathfrak{A})$.

To show that M and N are identical, it suffices to prove that $M(\mathfrak{Q}) = N(\mathfrak{Q})$ for all prime ideals $\mathfrak{Q}$ of O_E. Fix $\mathfrak{Q}$, and choose $\alpha \neq 0 \in \mathfrak{Q}$. Consequently, $\alpha O_E \subset \mathfrak{Q}$, and there exists an integral ideal $\mathfrak{A}$ such that

$$\alpha O_E = \mathfrak{Q}\mathfrak{A}$$

Applying both M and N, we have $M(\mathfrak{Q})M(\mathfrak{A}) = N(\mathfrak{Q})N(\mathfrak{A})$. Now the relations $M(\mathfrak{A})|N(\mathfrak{A})$, $M(\mathfrak{Q})|N(\mathfrak{Q})$ allow us to conclude that $M(\mathfrak{Q}) = N(\mathfrak{Q})$. This completes the proof.

4-7-7. Remark. The norm $N_{E\to F}$ is related to the counting norm $\mathfrak{N}$ discussed in (4-3-8). To see this, suppose that all residue class fields are finite, and let $\mathfrak{Q}$ be a prime ideal of O_E with $\mathfrak{P} = \mathfrak{Q} \cap O_F$ the prime ideal of O_F below it. Because the residue class field $O_E/\mathfrak{Q}$ is a vector space of dimension $f(Q/P)$ over the residue class field $O_F/\mathfrak{P}$, it follows that

$$\mathfrak{N}_E(\mathfrak{Q}) = \mathfrak{N}_F(\mathfrak{P})^{f(Q/P)} = \mathfrak{N}_F(\mathfrak{P}^{f(Q/P)})$$

where the subscript attached to $\mathfrak{N}$ indicates the field in which the norm $\mathfrak{N}$ is taken.

Thus, if $g(Q/P) = 1$, then $\mathfrak{N}_E(\mathfrak{Q}) = \mathfrak{N}_F[N_{E\to F}(\mathfrak{Q})]$, and if all g's are 1, then for any $\mathfrak{A} \in \mathfrak{J}\{\mathfrak{S}(E)\}$ we have

$$\mathfrak{N}_E(\mathfrak{A}) = \mathfrak{N}_F[N_{E\to F}(\mathfrak{A})]$$

Suppose that $F = \mathbf{Q}$ and $\mathfrak{S} = \mathfrak{M}_0(\mathbf{Q})$. Thus, $O_{\mathbf{Q}} = \mathbf{Z}$, and the prime ideals $\mathfrak{P}$ of $O_{\mathbf{Q}}$ are of form $\mathfrak{P} = \mathfrak{P}_p = p\mathbf{Z}$, p prime. Of course, if $\mathfrak{B}$ is an integral $\mathbf{Z}$ ideal, then $\mathfrak{N}_{\mathbf{Q}}(\mathfrak{B})\mathbf{Z} = \mathfrak{B}$. Since all g's are 1, it follows from the above that for integral $\mathfrak{A} \in \mathfrak{J}\{\mathfrak{S}(E)\}$ we have

$$\mathfrak{N}_E(\mathfrak{A})\mathbf{Z} = N_{E\to\mathbf{Q}}(\mathfrak{A})$$

Therefore, for $\alpha \neq 0 \in \mathfrak{A}$, $\mathfrak{N}_E(\alpha O_E)\mathbf{Z} = N_{E\to\mathbf{Q}}(\alpha O_E) = N_{E\to\mathbf{Q}}(\alpha)\mathbf{Z}$—or, what is the same,

$$|N_{E\to\mathbf{Q}}(\alpha)| = \mathfrak{N}_E(\alpha O_E)$$

Another important mapping of ideals is provided by the trace function; this will be discussed in the next section.

4-7-8. Exercise. Suppose that the class number of $\{F, \mathfrak{S}\}$ is finite; then there exists a finite extension E of F such that in E every ideal of F becomes principal.

4-8. Different and Discriminant

In this section we assume (in addition to our standard hypotheses) that the extension E/F is ***separable.*** Because O_F is integrally closed in its quotient field F and O_E is the integral closure of O_F in E [see (4-1-7)], all the general results of Section 3-7 apply. In addition we shall see from the properties of OAFs [among them the fact that

$\mathfrak{J}\{\mathcal{S}(E)\}$ and $\mathfrak{J}\{\mathcal{S}(F)\}$ are groups] that many of the results for local fields proved in Section 3-7 have analogues in the present global situation. In order to be able to apply the results of local ramification theory—and to keep our hypotheses uniform—we assume also that all the ***residue class field extensions for E/F are separable.***

4-8-1. Proposition. Suppose that $\mathfrak{A} \in \mathfrak{J}\{\mathcal{S}(E)\}$ and that

$$\mathfrak{A}^* = \{\beta \in E | S_{E \to F}(\beta\mathfrak{A}) \subset O_F\}$$

is the complementary set; then:

i. $S(\mathfrak{A}) \in \mathfrak{J}\{\mathcal{S}(F)\}$.
ii. O_E^* is an O_E ideal containing O_E.
iii. $S(\mathfrak{A}) \subset O_F \Leftrightarrow \mathfrak{A} \subset O_E^*$.
iv. $(O_E^*)^{-1}$ is an integral O_E ideal.
v. $\mathfrak{A}^* = O_E^*\mathfrak{A}^{-1}$; that is, $\mathfrak{A}\mathfrak{A}^* = O_E^*$.
vi. $\mathfrak{A}^* \in \mathfrak{J}\{\mathcal{S}(E)\}$, and $\mathfrak{A}^{-1} \subset \mathfrak{A}^*$.
vii. $\mathfrak{A}$ is an integral O_E ideal $\Leftrightarrow (\mathfrak{A}^*)^{-1} \subset (O_E^*)^{-1}$.
viii. $(\mathfrak{A}^*)^* = \mathfrak{A}$.
ix. $S(O_E^*) = O_F$.
x. $O_E^* = S_{E \to F}^{-1}(O_F)$.

Proof. The statement of this proposition is identical with that of (3-7-16), and so is the proof.

Continuing the parallel with Section 3-7, we make a definition and state two results which require no further words of proof. For $\mathfrak{A} \in \mathfrak{J}\{\mathcal{S}(E)\}$, the E ideal $\mathfrak{D}(\mathfrak{A}) = (\mathfrak{A}^*)^{-1}$ is called the ***different of*** $\mathfrak{A}$. The E ideal

$$\mathfrak{D}_{E/F} = \mathfrak{D}(O_E) = (O_E^*)^{-1} = \{\alpha \in E | \alpha O_E^* \subset O_E\}$$

is called the ***different of*** E/F.

4-8-2. Corollary. i. $\mathfrak{D}_{E/F}$ is an integral E ideal.
ii. $\mathfrak{D}(\mathfrak{A}) = \mathfrak{A}\mathfrak{D}_{E/F}$; in particular, $\mathfrak{D}(\mathfrak{A}) \subset \mathfrak{A}$.
iii. $\mathfrak{A}$ is integral $\Leftrightarrow \mathfrak{D}(\mathfrak{A}) \subset \mathfrak{D}_{E/F}$.
iv. $S(\mathfrak{D}_{E/F}^{-1}) = O_F$ and $\mathfrak{D}_{E/F}^{-1} = S_{E \to F}^{-1}(O_F)$.
v. $\mathfrak{A} \subset \mathfrak{D}_{E/F}^{-1} \Leftrightarrow S(\mathfrak{A})$ is an integral F ideal.

4-8-3. Proposition. Suppose that $F \subset E \subset K$ with K/F separable; then

$$\mathfrak{D}_{K/F} = \mathfrak{D}_{K/E} \cdot \mathfrak{D}_{E/F}$$

Consider any $Q \in \mathcal{S}(E)$, and let $P \in \mathcal{S}(F)$ be its restriction to F. The completions E_Q and F_P are at our disposal, and we know that E_Q/F_P is a separable extension of degree $n(Q/P)$; if necessary, we may take $E_Q = F_P \cdot E$. In particular, this local extension has a different

$\mathfrak{D}_{E_Q/F_P}$. We shall use the symbol ν_Q for the normalized valuation of both E and E_Q and shall write $\bar{O}_Q = \{\alpha \in E_Q | \nu_Q(\alpha) \geq 0\}$; so $\bar{O}_Q$ is the closure of $O_Q = \{\alpha \in E | \nu_Q(\alpha) \geq 0\}$ in E_Q. Similar notation is to be used for F and F_P.

4-8-4. Theorem. The global different is the product of the local differents—that is,

$$\mathfrak{D}_{E/F} = \prod_{Q \in \mathcal{S}(E)} \mathfrak{D}_{E_Q/F_P}$$

Proof. As it stands, the left side $\mathfrak{D}_{E/F}$ is an integral O_E ideal, but the right side is merely a formal expression. One cannot interpret the formal expression as an O_E ideal until it is known that almost all $\nu_Q(\mathfrak{D}_{E_Q/F_P}) = 0$. Thus, the assertion of the theorem is really that for each $Q \in \mathcal{S}(E)$ we have

$$(*) \qquad \nu_Q(\mathfrak{D}_{E/F}) = \nu_Q(\mathfrak{D}_{E_Q/F_P})$$

To prove (*) let us show first that for each $Q \in \mathcal{S}(E)$ we have $\nu_Q(\mathfrak{D}_{E/F}) \leq \nu_Q(\mathfrak{D}_{E_Q/F_P})$. Equivalently, we may show that $\nu_Q(\mathfrak{D}_{E/F}^{-1}) \geq \nu_Q(\mathfrak{D}_{E_Q/F_P}^{-1})$ or that $\mathfrak{D}_{E/F}^{-1} \subset \mathfrak{D}_{E_Q/F_P}^{-1} \subset E_Q$. Consider any $\alpha \in \mathfrak{D}_{E/F}^{-1}$, so $S_{E\to F}(\alpha O_E) \subset O_F \subset O_P$. It must be shown that $S_{E_Q\to F_P}(\alpha\beta) \in \bar{O}_P$ for any $\beta \in \bar{O}_Q$. Let $Q = Q_1, Q_2, \ldots, Q_r$ be the extensions of P to E. As pointed out in item 1 of (4-1-5) there exists $\gamma \in E$ such that

$$\begin{aligned} \nu_Q(\gamma - \beta) &> M & & \\ \nu_{Q_i}(\gamma - 0) &> M & \qquad i &= 2, \ldots, r \\ \nu_{Q'}(\gamma) &\geq 0 & \qquad Q' &\neq Q_1, \ldots, Q_r \end{aligned}$$

where M is a large positive integer which remains to be specified. Now let us consider the relation

$$S_{E_Q\to F_P}(\alpha\beta) = S_{E\to F}(\alpha\gamma) - \sum_{i=2}^{r} S_{E_{Q_i}\to F_P}(\alpha\gamma) + S_{E_Q\to F_P}(\alpha(\beta - \gamma))$$

By taking M sufficiently large, we can guarantee that $\gamma \in O_E$ and that each term on the right-hand side belongs to $\bar{O}_P$. This takes care of the first part of the proof.

It remains to show that $\nu_Q(\mathfrak{D}_{E_Q/F_P}^{-1}) \geq \nu_Q(\mathfrak{D}_{E/F}^{-1})$ for every $Q \in \mathcal{S}(E)$. To do this, fix Q, and select an $\alpha \in E$ for which $\nu_Q(\alpha) = \nu_Q(\mathfrak{D}_{E_Q/F_P}^{-1})$ and $\nu_{Q'}(\alpha) \geq 0$ for all $Q' \neq Q$. It suffices to show that $\alpha \in \mathfrak{D}_{E/F}^{-1}$—in other words, that $S_{E\to F}(\alpha\beta) \in O_F$ for any $\beta \in O_E$. Consider any $P' \neq P \in \mathcal{S}(F)$, and let $Q_1', \ldots, Q_s'$ be its extensions to E; then

$$S_{E\to F}(\alpha\beta) = \sum_{1}^{s} S_{E_{Q_i'}\to F_{P'}}(\alpha\beta) \in \bar{O}_{P'} \cap F = O_{P'}$$

Furthermore, we have

$$S_{E\to F}(\alpha\beta) = S_{E_Q\to F_P}(\alpha\beta) + \sum_2^r S_{E_{Q_i}\to F_P}(\alpha\beta) \in \bar{O}_P \cap F = O_P$$

Therefore, $S_{E\to F}(\alpha\beta) \in O_F$, and the proof is complete.

4-8-5. Remark. Consider any $P \in \mathfrak{S}(F)$. If we put $\mathfrak{S}'(F) = P$, then, in accordance with the notation of (4-7-4), $\{Q \in \mathfrak{S}(E) | Q \supset P\} = \mathfrak{S}'(E)$. This situation, in which one is concerned with a single prime divisor in the base field F and all its extensions to E, is known as the ***semilocal*** case. Let us write

$$O_{P,E} = O'_E = \bigcap_{Q\supset P} O_Q = \{\alpha \in E | \nu_Q(\alpha) \geq 0 \quad \forall Q \supset P\}$$

and $O_P = O'_F = \{a \in F | \nu_P(a) \geq 0\}$. Since O_P is integrally closed in its quotient field F and $O_{P,E}$ is the integral closure of O_P in E, the different exists for this situation. We denote it by $\mathfrak{D}_{O_{P,E}/O_P}$ or simply by $\mathfrak{D}_P$ and call it the ***different of*** E/F ***at*** P. In order not to lose track of the rings of integers involved, the previous global different $\mathfrak{D}_{E/F}$ should perhaps be denoted by $\mathfrak{D}_{O_E/O_F}$. When applied to this new situation, (4-8-4) says that

$$\mathfrak{D}_P = \mathfrak{D}_{O_{P,E}/O_P} = \prod_{Q\supset P} \mathfrak{D}_{E_Q/F_P}$$

We conclude, therefore, that

$$\mathfrak{D}_{E/F} = \mathfrak{D}_{O_E/O_F} = \prod_{P\in\mathfrak{S}(F)} \mathfrak{D}_{O_{P,E}/O_P} = \prod_{P\in\mathfrak{S}(F)} \mathfrak{D}_P$$

—that is, the global different is (symbolically) the product of the differents at all P.

Suppose that $Q_1, \ldots, Q_r \in \mathfrak{S}(E)$ are the extensions of $P \in \mathfrak{S}(F)$. We say that: Q_i is ***ramified*** over F when $e(Q_i/P) > 1$; Q_i is ***unramified*** over F when $e(Q_i/P) = 1$; P is ***ramified*** in E when some $e(Q_i/P) > 1$; P is ***unramified*** in E when all $e(Q_i/P) = 1$; P ***splits completely*** in E when all $e(Q_i/P) = 1$ and all $f(Q_i/P) = 1$—that is, when $E_{Q_i} = F_P$ for $i = 1, \ldots, r$. We also say that the extension E/F is ***unramified*** when every $P \in \mathfrak{S}(F)$ is unramified. Of course, the same definitions may be carried over to the corresponding prime ideals.

4-8-6. Theorem (Dedekind). There are only a finite number of primes $Q \in \mathfrak{S}(E)$ that are ramified over F. In fact, $Q \in \mathfrak{S}(E)$ is ramified over F if and only if $\nu_Q(\mathfrak{D}_{E/F}) \geq 1$. Moreover, the extension E/F is unramified if and only if $\mathfrak{D}_{E/F} = O_E$.

Proof. Immediate from (4-8-4) and (3-7-24).

Note that the application of (3-7-24) is the first place in this chapter where separability of the residue class field extensions enters. Note also that $\nu_Q(\mathfrak{D}_{E/F}) = m(E_Q/F_P)$, where $m(E_Q/F_P)$ is the differential exponent for the extension E_Q/F_P.

In particular, (4-8-6) says that for algebraic number fields a prime ideal of E is ramified over F if and only if it divides the different.

What about the notion of discriminant in our global situation? We would like to define the discriminant for an arbitrary O_E ideal $\mathfrak{A}$; and in keeping with the definition of the discriminant of a module (see Section 3-7), it is necessary that $\mathfrak{A}$ have a basis over O_F consisting of n elements. In other words, there should exist elements $\alpha_1, \ldots, \alpha_n \in \mathfrak{A}$ such that $\mathfrak{A} = O_F\alpha_1 \oplus \cdots \oplus O_F\alpha_n$; of course, $\alpha_1, \ldots, \alpha_n$ are then linearly independent over F. It is clear that O_E contains n elements that are linearly independent. Moreover, according to the proof of (4-4-8), O_E is contained in an O_F module that has n generators. Therefore, if we assume that O_F is a principal ideal domain, then it follows from standard results on torsion-free modules over a principal ideal domain (see Exercise 4-6 or N. Bourbaki's "Algèbre," chap. VII, Hermann & Cie, Paris, 1952) that there exists an integral basis for O_E over O_F. Furthermore, it is easy to show that for any O_E ideal $\mathfrak{A}$ there exists $c \in O_F$ such that $c\mathfrak{A} \subset O_E \subset (1/c)\mathfrak{A}$ and so it follows that $\mathfrak{A}$ has a basis over O_F and that any such basis has n elements. Thus, when O_F is a principal ideal domain and $\mathfrak{A} = O_F\alpha_1 \oplus \cdots \oplus O_F\alpha_n$, then $\Delta(\mathfrak{A}) = \Delta(\alpha_1, \ldots, \alpha_n)$ is called the ***discriminant of the ideal*** $\mathfrak{A}$ and $\Delta_{E/F} = \Delta(O_E)$ is the ***discriminant of the extension*** E/F. As observed in Section 3-7, $\Delta(\mathfrak{A})$ is determined upto the square of a unit of O_F.

In the general case, where there are no restrictions on O_F, suppose that $\mathcal{S}'(F)$ is a subset of $\mathcal{S}(F)$ consisting of a single prime divisor P. Then O_P is a principal ideal domain (in fact, so is $O_{P,E}$), and, in view of the discussion above, discriminants of $O_{P,E}$ ideals exist. Because of this, it is possible to extend the definition of discriminant to the case where O_F is not a principal ideal domain. Let us give the details.

Let $\mathfrak{A}$ be an O_E ideal, and for each $P \in \mathcal{S}(F)$ consider the $O_{P,E}$ ideal $\mathfrak{A}O_{P,E} = \mathfrak{A}O_P$ [see (4-7-4)]. Since O_P is a principal ideal domain, the discriminant of $\mathfrak{A}O_P$, which we denote by $\Delta_P(\mathfrak{A}O_P)$, is well defined. $\Delta_P(\mathfrak{A}O_P)$ may be called the ***discriminant of*** $\mathfrak{A}$ ***at*** P. We put

$$\mathbf{\Delta}(\mathfrak{A}) = \prod_{P \in \mathcal{S}(F)} \mathfrak{P}^{\nu_P[\Delta_P(\mathfrak{A}O_P)]}$$

and call it the ***discriminant of*** $\mathfrak{A}$. As expected, we call $\mathbf{\Delta}_{E/F} = \mathbf{\Delta}(O_E)$ the ***discriminant of the extension*** E/F.

Before going further, it must be shown that $\mathbf{\Delta}(\mathfrak{A})$ is indeed an O_F ideal and also that this definition of discriminant is compatible with the previous one. The latter statement means that, if O_F is a principal ideal domain, then $\mathbf{\Delta}(\mathfrak{A}) = \Delta(\mathfrak{A})O_F$; for this it will be more than sufficient to prove (4-8-7). Once these facts are known, $\mathbf{\Delta}$ can play the role assigned to it in Section 3-7. The notations $\mathbf{\Delta}_{E/F} = \mathbf{\Delta}_{O_E/O_F}$ and $\mathbf{\Delta}_P = \mathbf{\Delta}_{O_{P,E}/O_P}$ will be used in a manner consistent with the corresponding notations for the different [see (4-8-5)].

4-8-7. Proposition. If $\mathfrak{A}$ has a basis over O_F, then

$$\mathbf{\Delta}(\mathfrak{A}) = \Delta(\mathfrak{A})O_F$$

Proof. It suffices to show that

$$\nu_P[\Delta_P(\mathfrak{A}O_P)] = \nu_P[\Delta(\mathfrak{A})]$$

for all $P \in \mathcal{S}(F)$. Suppose that $\{\alpha_1, \ldots, \alpha_n\}$ is a basis of $\mathfrak{A}$, so that $\mathfrak{A} = O_F\alpha_1 \oplus \cdots \oplus O_F\alpha_n$ and $\mathfrak{A}O_P = O_P\alpha_1 \oplus \cdots \oplus O_P\alpha_n$ for any P. Therefore,

$$\Delta(\mathfrak{A}) = \Delta(\alpha_1, \ldots, \alpha_n) = \Delta_P(\mathfrak{A}O_P)$$

(the units stay under control), and we are finished.

In order to prove that $\mathbf{\Delta}(\mathfrak{A})$ is an O_F ideal, we need one preliminary result; and for this it is convenient to say that the elements $\alpha_1, \ldots, \alpha_n \in E$ form an ***integral basis at*** P when they are an integral basis for $O_{P,E}$ over O_P—that is, when $O_{P,E} = O_P\alpha_1 \oplus \cdots \oplus O_P\alpha_n$.

4-8-8. Proposition. If $\alpha_1, \ldots, \alpha_n \in O_E$ are linearly independent over F, then $\{\alpha_1, \ldots, \alpha_n\}$ is an integral basis at almost all $P \in \mathcal{S}(F)$; in fact, $\{\alpha_1, \ldots, \alpha_n\}$ is an integral basis at every P for which $\nu_P[\Delta(\alpha_1, \ldots, \alpha_n)] = 0$. Moreover, $\{\alpha_1, \ldots, \alpha_n\}$ is an integral basis for O_E over O_F if and only if $\{\alpha_1, \ldots, \alpha_n\}$ is an integral basis at every $P \in \mathcal{S}(F)$.

Proof. Put $\Delta = \Delta(\alpha_1, \ldots, \alpha_n)$. The fact that $\nu_P(\Delta) = 0$ for almost all P means that Δ is a unit of O_P for almost all P. For such P, we see from the proof of (4-4-8) that $O_{P,E} \subset O_P \dfrac{\alpha_1}{\Delta} \oplus \cdots \oplus O_P \dfrac{\alpha_n}{\Delta}$ $= O_P\alpha_1 \oplus \cdots \oplus O_P\alpha_n \subset O_PO_E = O_{P,E}$.

To prove the second part, observe first that

$$O_E = O_F\alpha_1 \oplus \cdots \oplus O_F\alpha_n \Rightarrow O_{P,E} = O_EO_P = O_P\alpha_1 \oplus \cdots \oplus O_P\alpha_n$$

As for the converse, we simply write $\beta \in O_E$ in the form $\beta = \Sigma a_i\alpha_i$ with $a_i \in F$. Since $\nu_P(a_i) \geq 0$ for $i = 1, \ldots, n$ and all $P \in \mathcal{S}(F)$, we have $a_i \in O_F$ and $O_E = O_F\alpha_1 \oplus \cdots \oplus O_F\alpha_n$.

4-8-9. Proposition. $\boldsymbol{\Delta}(\mathfrak{A})$ is an O_F ideal.

Proof. We must show that $\nu_P[\Delta_P(\mathfrak{A}O_P)] = 0$ for almost all $P \in \mathcal{S}(F)$. Choose $\alpha_1, \ldots, \alpha_n \in \mathfrak{A}$ which are linearly independent over F and such that $\alpha_1, \ldots, \alpha_n \in O_E$; so $\{\alpha_1, \ldots, \alpha_n\}$ is an integral basis at P for almost all P. On the other hand, because $\mathfrak{A}$ is an O_E ideal, it follows from the explicit form for the mapping $\mathfrak{A} \to \mathfrak{A}O_{P,E} = \mathfrak{A}O_P$ [see (4-5-5)] that $\mathfrak{A}O_P = O_{P,E}$ for almost all P. We conclude, therefore, that $\Delta_P(\mathfrak{A}O_P) = \Delta(\alpha_1, \ldots, \alpha_n)$ for almost all P, and the proof is complete.

Now we are in a position to investigate the properties of the discriminant and to relate it to the different.

4-8-10. Proposition. For any O_E ideal $\mathfrak{A}$ we have

$$\boldsymbol{\Delta}(\mathfrak{A}) = [N_{E\to F}(\mathfrak{A})]^2 \boldsymbol{\Delta}_{E/F}$$

Proof. It suffices to show that

$$\nu_P[\Delta_P(\mathfrak{A}O_P)] = \nu_P[N_{E\to F}(\mathfrak{A})]^2 + \nu_P[\Delta_P(O_{P,E})]$$

for all $P \in \mathcal{S}(F)$. Fix P, and let $\{\omega_1, \ldots, \omega_n\}$ be an integral basis at P. Select $\alpha \in \mathfrak{A}$ such that $\nu_Q(\alpha) = \nu_Q(\mathfrak{A})$ for all $Q \supset P$; so

$$\mathfrak{A}O_P = \mathfrak{A}O_{P,E} = \alpha O_{P,E} = O_P(\alpha\omega_1) \oplus \cdots \oplus O_P(\alpha\omega_n)$$

Now,

$$\begin{aligned} \Delta(\alpha\omega_1, \ldots, \alpha\omega_n) &= \det\{\sigma_i(\alpha\omega_j)\}^2 \qquad \sigma_i \in \Gamma(F, E \to \Omega) \\ &= \left(\prod_i \sigma_i\alpha\right)^2 (\det\{\sigma_i(\omega_j)\})^2 \\ &= (N_{E\to F}\alpha)^2 \Delta(\omega_1, \ldots, \omega_n) \end{aligned}$$

Since $\Delta_P(O_{P,E}) = \Delta(\omega_1, \ldots, \omega_n)$, $\Delta_P(\mathfrak{A}O_P) = \Delta(\alpha\omega_1, \ldots, \alpha\omega_n)$, and $\nu_P[N_{E\to F}(\mathfrak{A})] = \nu_P[N_{E\to F}(\mathfrak{A}O_{P,E})] = \nu_P[N_{E\to F}(\alpha O_{P,E})] = \nu_P[N_{E\to F}(\alpha)]$ the proof is complete.

An important result, which in other formulations is often taken as the definition of $\boldsymbol{\Delta}_{E/F}$, is the following.

4-8-11. Proposition. $N_{E\to F}(\mathfrak{D}_{E/F}) = \boldsymbol{\Delta}_{E/F}$.

Proof. We must show that

$$\nu_P[N_{E\to F}(\mathfrak{D}_{E/F})] = \nu_P[\Delta_P(O_{P,E})]$$

for all $P \in \mathcal{S}(F)$. Let $\{\omega_1, \ldots, \omega_n\}$ be an integral basis for $O_{P,E}$ over O_P. With trivial changes of notation, the proof of (3-7-19) goes through (even though we do not have completeness here), to yield

$$N_{E\to F}(\mathfrak{D}_P) = \Delta_P(O_{P,E})O_P = \boldsymbol{\Delta}_P$$

Since $\nu_P[N_{E\to E}(\mathfrak{D}_{E/F})] = \nu_P[N_{E\to F}(\mathfrak{D}_P)]$, the proof is complete.

4-8-12. Proposition. Suppose that $F \subset E \subset K$ and that K/F is separable; then

$$\mathbf{\Delta}_{K/F} = \mathbf{\Delta}_{E/F}^{[K:E]} N_{E\to F}(\mathbf{\Delta}_{K/E})$$

Proof. From (4-7-5vi) we know that $N_{K\to E}(\mathfrak{D}_{E/F}O_K) = \mathfrak{D}_{E/F}^{[K:E]}$; therefore, the proof of (3-7-20) carries over.

4-8-13. Proposition. We have

(i) $$\mathbf{\Delta}_{E/F} = \prod_{Q\in\mathcal{S}(E)} \mathbf{\Delta}_{E_Q/F_P}$$

(ii) $$\mathbf{\Delta}_P = \prod_{Q\supset P} \mathbf{\Delta}_{E_Q/F_P} \qquad \text{for all } P \in \mathcal{S}(F)$$

(iii) $$\mathbf{\Delta}_{E/F} = \prod_{P\in\mathcal{S}(F)} \mathbf{\Delta}_P$$

Proof. From (4-8-11) and

$$\mathfrak{D}_{E/F} = \prod_{Q\in\mathcal{S}(E)} \mathfrak{Q}^{\nu_Q(\mathfrak{D}_{E_Q/F_P})}$$

we know that

$$\mathbf{\Delta}_{E/F} = \prod_{P\in\mathcal{S}(F)} \mathfrak{P}^{\nu_P(\mathbf{\Delta}_{E/F})} = \prod_{P\in\mathcal{S}(F)} \mathfrak{P}^{\sum_{Q\supset P} f(Q/P)\nu_Q(\mathfrak{D}_{E_Q/F_P})}$$

For each $P \in \mathcal{S}(F)$ we have then

$$\nu_P(\mathbf{\Delta}_{E/F}) = \sum_{Q\supset P} f\left(\frac{Q}{P}\right) \nu_Q(\mathfrak{D}_{E_Q/F_P}) = \sum_{Q\supset P} \nu_P[N_{E_Q\to F_P}(\mathfrak{D}_{E_Q/F_P})]$$
$$= \sum_{Q\supset P} \nu_P(\mathbf{\Delta}_{E_Q/F_P}) = \sum_{Q\supset P} m\left(\frac{E_Q}{F_P}\right)$$

This shows that the global discriminant is indeed (symbolically) the product of all the local discriminants and takes care of all three assertions.

4-8-14. Theorem (Dedekind). There are only a finite number of primes $P \in \mathcal{S}(F)$ that ramify in E. In fact, $P \in \mathcal{S}(F)$ ramifies in E if and only if $\nu_P(\mathbf{\Delta}_{E/F}) \geq 1$. Moreover, the extension E/F is unramified if and only if $\mathbf{\Delta}_{E/F} = O_F$.

Proof. Immediate.

In particular, we see that, for algebraic number fields, a prime ideal of F is ramified in E if and only if it divides the discriminant.

Note that the different gives more information about ramification than the discriminant—however, one often settles for less because the discriminant is easier to compute. Of course, the computation of the discriminant always comes down to the case of a principal ideal domain. Let us pursue this a bit further.

4-8-15. Remark. Suppose that O_F is a principal ideal domain and that we wish to find $\Delta_{E/F}$. For this it is necessary to locate an integral basis for O_E over O_F; we describe a procedure which is often helpful in doing so. Choose $\eta_1, \ldots, \eta_n \in O_E$ such that $\Delta(\eta_1, \ldots, \eta_n) \neq 0$—that is, such that $\eta_1, \ldots, \eta_n$ are a basis for E/F. If $\Delta(\eta_1, \ldots, \eta_n) \in O_F$ is square-free, then $\eta_1, \ldots, \eta_n$ is clearly an integral basis and we are finished. If $\Delta(\eta_1, \ldots, \eta_n) \in O_F$ is not square-free, then the only candidates for $\Delta_{E/F}$ are the finite number of divisors of $\Delta(\eta_1, \ldots, \eta_n)$, which differ from it (multiplicatively and up to squares of units) by a perfect square. Suppose that $\eta_1, \ldots, \eta_n$ is not an integral basis; then there exists $\alpha \neq 0 \in O_E$ such that

$$\alpha = c_1\eta_1 + \cdots + c_n\eta_n$$

with $c_i \in F$ and not all $c_i \in O_F$. Furthermore, we may assume that

$$\min_i \{\nu_P(c_i)\} \leq 0$$

for all $P \in \mathfrak{S}(F)$. To see this, note first that $\min_i \{\nu_P(c_i)\} = 0$ for almost all P. Since every $P \in \mathfrak{S}(F)$ corresponds to a unique prime $p \in O_F$, we can, for each of the primes P in the remaining finite set, divide the c_i by a suitable positive power of p (if necessary) to guarantee that $\min_i \{\nu_P(c_i)\} \leq 0$. Because O_F is a principal ideal domain, we may choose $m \neq 0 \in O_F$ such that

$$\nu_P(m) = -\min \{\nu_P(c_i)\}$$

for all $P \in \mathfrak{S}(F)$. Then $b_i = mc_i \in O_F$ and $\min_i \{\nu_P(b_i)\} = 0$ for all P; so the b_i are relatively prime. Since the $c_i = b_i/m$ are not all in O_F, we know that m is not a unit. Let p be a prime factor of m, and put $\beta = (m/p)\alpha \in O_E$; so $\beta = (b_1\eta_1 + \cdots + b_n\eta_n)/p$, and p does not divide some b_i, say, $p \nmid b_1$. We have then $rb_1 + sp = 1$ for some choice of $r, s \in O_F$. Put

$$\gamma_1 = r\beta + s\eta_1 \in O_E \qquad \text{and} \qquad \gamma_2 = \eta_2, \ldots, \gamma_n = \eta_n$$

so $\gamma_1 = (\eta_1 + a_2\eta_2 + \cdots + a_n\eta_n)/p$ with $a_i \in O_F$, and

$$\Delta(\gamma_1, \ldots, \gamma_n) = \frac{1}{p^2}\,\Delta(\eta_1, \ldots, \eta_n) \in O_F$$

In particular, p is a prime such that p^2 divides $\Delta(\eta_1, \ldots, \eta_n)$. This process may be repeated, starting with $\gamma_1, \ldots, \gamma_n$. Because $\Delta(\gamma_1, \ldots, \gamma_n)$ divides $\Delta(\eta_1, \ldots, \eta_n)$ and they are not associates, the process terminates after a finite number of steps.

In this context, something may be said about a basis for an integral ideal.

4-8-16. Proposition. Suppose that O_F is a principal ideal domain, and let $\{\omega_1, \ldots, \omega_n\}$ be an integral basis for E/F. If $\mathfrak{A}$ is an integral ideal of E, then there exists a basis $\{\alpha_1, \ldots, \alpha_n\}$ of $\mathfrak{A}$ of form

$$\begin{aligned} \alpha_1 &= a_{11}\omega_1 \\ \alpha_2 &= a_{21}\omega_1 + a_{22}\omega_2 \\ &\cdots\cdots\cdots\cdots \\ \alpha_n &= a_{n1}\omega_1 + a_{n2}\omega_2 + \cdots + a_{nn}\omega_n \end{aligned}$$

with $a_{ij} \in O_F$ and $a_{ii} \neq 0$ for $i = 1, \ldots, n$. Such a basis is said to be ***canonical.***

In particular, if $F = \mathbf{Q}$ and $O_F = \mathbf{Z}$, then we may take $a_{ii} > 0$ so that

$$\mathfrak{N}\mathfrak{A} = \prod_1^n a_{ii}$$

—in fact, a full system of representatives for the elements of $O_E/\mathfrak{A}$ is given by

$$\{x_1\omega_1 + \cdots + x_n\omega_n | 0 \leq x_i < a_{ii}\}$$

Furthermore, if $\{\gamma_1, \ldots, \gamma_n\}$ is any basis for $\mathfrak{A}$ and we write

$$\gamma_i = \sum_{j=1}^n c_{ij}\omega_j$$

with $c_{ij} \in \mathbf{Z}$, then

$$\mathfrak{N}\mathfrak{A} = |\det(c_{ij})|$$

Proof. We sketch a procedure for finding a canonical basis. If all $\alpha \in \mathfrak{A}$ are expressed in terms of the integral basis $\{\omega_1, \ldots, \omega_n\}$, then the coefficients of ω_n that occur form a nonzero ideal of O_F. Let $a_{nn} \neq 0 \in O_F$ be a generator of this ideal, and choose an element $\alpha_n \in \mathfrak{A}$ which has a_{nn} as the coefficient of ω_n. Now, there exist elements of $\mathfrak{A}$ with coefficient of ω_n equal 0. If we take all such elements of $\mathfrak{A}$, the coefficients of ω_{n-1} form an ideal of O_F. This enables us to choose α_{n-1}. This process continues, and we do get all $a_{ii} \neq 0$.

From the method of construction of the canonical basis $\{\alpha_1, \ldots, \alpha_n\}$, it is clear that when $O_F = \mathbf{Z}$ we may take $a_{ii} > 0$ and that every element of O_E is congruent (mod $\mathfrak{A}$) to a unique element of form $x_1\omega_1 + \cdots + x_n\omega_n$ with $0 \leq x_i < a_{ii}$; thus $\mathfrak{N}\mathfrak{A} = \Pi a_{ii}$.

To prove the final assertion, we note that $+1$ is the only square of a unit in $\mathbf{Z}$, so that from (3-7-1) and (3-7-15) it follows that

$$\begin{aligned}[\det (c_{ij})]^2\Delta(\omega_1, \ldots, \omega_n) &= \Delta(\gamma_1, \ldots, \gamma_n) = \Delta(\alpha_1, \ldots, \alpha_n)\\ &= [\det (a_{ij})]^2\Delta(\omega_1, \ldots, \omega_n)\\ &= (\mathfrak{N}\mathfrak{A})^2\Delta(\omega_1, \ldots, \omega_n)\end{aligned}$$

4-8-17. Proposition. Let $\alpha_1, \ldots, \alpha_n$ denote arbitrary elements of O_E; then:

i. $\Delta(\alpha_1, \ldots, \alpha_n) \in \mathbf{\Delta}_{E/F}$.

ii. $\nu_P(\mathbf{\Delta}_{E/F}) = \min \{\nu_P[\Delta(\alpha_1, \ldots, \alpha_n)]\}$ for all $P \in \mathcal{S}(F)$; in other words, $\mathbf{\Delta}_{E/F}$ is the greatest common divisor of all $\Delta(\alpha_1, \ldots, \alpha_n)$.

iii. $\{\alpha_1, \ldots, \alpha_n\}$ is an integral basis at P if and only if $\mathbf{\Delta}_P = \Delta(\alpha_1, \ldots, \alpha_n)O_P$.

iv. $\{\alpha_1, \ldots, \alpha_n\}$ is an integral basis at P if and only if $\nu_P(\mathbf{\Delta}_{E/F}) = \nu_P[\Delta(\alpha_1, \ldots, \alpha_n)]$.

v. $\{\alpha_1, \ldots, \alpha_n\}$ is an integral basis for E/F if and only if $\mathbf{\Delta}_{E/F} = \Delta(\alpha_1, \ldots, \alpha_n)O_F$.

Proof. Since $\Delta(\alpha_1, \ldots, \alpha_n) = 0 \Leftrightarrow \alpha_1, \ldots, \alpha_n$ are linearly dependent over F, it follows that all the assertions need to be considered only when $\alpha_1, \ldots, \alpha_n$ are linearly independent.

i. Since $O_{P,E}$ has an integral basis over O_P and $\{\alpha_1, \ldots, \alpha_n\} \subset O_{P,E}$ for all $P \in \mathcal{S}(F)$, it follows from (3-7-15) that $\Delta(\alpha_1, \ldots, \alpha_n) \in \mathbf{\Delta}_P$.

ii. It suffices to show that for each $P \in \mathcal{S}(F)$ there exists an integral basis for $O_{P,E}$ over O_P consisting of elements of O_E. To do this, let $\beta_1, \ldots, \beta_n \in O_{P,E}$ be an integral basis at P. According to (4-1-6), there exists a unit μ of $O_{P,E}$ such that $\alpha_i = \mu\beta_i \in O_E$ for $i = 1, \ldots, n$. The elements $\alpha_1, \ldots, \alpha_n$ serve the purpose.

The remaining assertions are immediate from (3-7-15), (4-8-13), and (4-8-8).

4-8-18. Proposition. For any $\theta \in O_E$ let $f_\theta(x) = f(\theta, E/F)$ and $\mathfrak{D}(\theta) = f'_\theta(\theta)$; then

i. $f'_\theta(\theta) \in \mathfrak{D}_{E/F}$.

ii. $1, \theta, \ldots, \theta^{n-1}$ is an integral basis for E/F if and only if $\mathfrak{D}_{E/F} = f'_\theta(\theta)O_E$.

iii. $\nu_Q(\mathfrak{D}_{E/F}) = \min\limits_{\theta \in O_E} \{\nu_Q[\mathfrak{D}(\theta)]\}$ for all $Q \in \mathcal{S}(E)$; in other words, $\mathfrak{D}_{E/F}$ is the greatest common divisor of the differents $\mathfrak{D}(\theta)$ of all $\theta \in O_E$.

Proof. We know that $f'_\theta(\theta) = 0 \Leftrightarrow F(\theta) \neq E$, and it follows that for the proof we have to consider only the $\theta \in O_E$ such that $E = F(\theta)$.

An examination of the proof of (3-7-21) shows that assertions i and ii are valid.

To prove (iii) we must show that, given Q (with restriction P to F), there exists $\theta \epsilon O_E$ such that

$$\nu_Q[\mathfrak{D}(\theta)] = \nu_Q(\mathfrak{D}_{E/F}) = m\left(\frac{E_Q}{F_P}\right)$$

Let $Q = Q_1, Q_2, \ldots, Q_r$ be the extensions of P to E. It is useful to revert to some of the notation of (2-4-6). Thus, we have composition maps μ_i for $\{E, F, F_P, \Omega\}$ such that $Q_i = Q_{\mu_i}$ and $E_{Q_i} = F_P(\mu_i E)$. Furthermore, we shall use the same normalized valuation ν_{Q_i} on both E and its completion E_{Q_i}, so that, for $\alpha \epsilon E$,

$$\nu_{Q_i}(\alpha) = \nu_{Q_i}(\mu_i\alpha)$$

The extension of ν_{Q_i} to the algebraically closed field Ω will also be denoted by ν_{Q_i}.

We recall from (3-3-3) that there exists a unit β of E_{Q_1} such that $\{1, \beta, \ldots, \beta^{n(Q_1/P)-1}\}$ is an integral basis for E_{Q_1} over F_P. In view of item 1 of (4-1-5) there exists $\theta \epsilon E$ such that

$$\begin{aligned} \nu_{Q_1}(\theta - \beta) &\geq 2 \\ \nu_{Q_i}(\theta) &> 1 \qquad i = 2, \ldots, r \\ \nu_{Q'}(\theta) &\geq 0 \qquad Q' \notin \{Q_1, \ldots, Q_r\} \end{aligned}$$

[Strictly speaking, the first relation says that $\nu_{Q_1}(\mu_1\theta - \beta) \geq 2$.] Note that, for $i \neq 1$, $\nu_{Q_i}(\theta) = \nu_{Q_i}(\mu_i(\theta)) > 1$ implies that $\nu_{Q_1}(\mu_i\theta) > 0$.

We observe that $\theta \epsilon O_E$, $\nu_{Q_1}(\mu_1\theta) = 0$, $\nu_{Q_1}(\mu_i\theta) > 0$ for $i \neq 1$ and [according to (3-3-4)] that

$$\{(\mu_1\theta)^s = \mu_1\theta^s | s = 0, 1, \ldots, n(Q_1/P) - 1\}$$

is an integral basis for E_{Q_1} over F_P. Let $f(x) = f(\theta, E/F)$ and $f_i(x) = f(\mu_i\theta, E_{Q_i}/F_P)$ for $i = 1, \ldots, r$, so that $f(x) = f_1(x) \cdots f_r(x)$. Since $f_1(\mu_1\theta) = 0$, we have

$$f'(\mu_1\theta) = f_1'(\mu_1\theta)f_2(\mu_1\theta) \cdots f_r(\mu_1\theta)$$

From (3-7-21) we know that

$$\nu_{Q_1}[f_1'(\mu_1\theta)] = \nu_{Q_1}(\mathfrak{D}_{E_{Q_1}/F_P}) = m(E_{Q_1}/F_P)$$

It suffices, therefore, to show that $\nu_{Q_1}[f_i(\mu_1\theta)] = 0$ for $i = 2, \ldots, r$.

For each of the $n(Q_i/P)$ conjugates of $\mu_i\theta$ over F_P—denote them by $(\mu_i\theta)^{(\rho)}$ with $\rho = 1, \ldots, n(Q_i/P)$—we have

$$\nu_{Q_1}[(\mu_i\theta)^{(\rho)}] = \nu_{Q_1}(\mu_i\theta) > 0$$

Because
$$f_i(\mu_i\theta, E_{Q_i}/F_P) = \prod_\rho [x - (\mu_i\theta)^{(\rho)}]$$
it follows that
$$\nu_{Q_1}[f_i(\mu_1\theta)] = \sum_\rho \nu_{Q_1}[\mu_1\theta - (\mu_i\theta)^{(\rho)}] = 0$$
This completes the proof.

We conclude this section with some remarks about the discriminant Δ_E of an algebraic number field E. By Δ_E (which is often called the ***absolute discriminant*** of E) we mean, of course, the discriminant of the extension $E/\mathbf{Q}$, where the ring of integers in $\mathbf{Q}$ is $\mathbf{Z}$. Since ± 1 are the only units of $\mathbf{Z}$, the discriminant Δ_E is determined without ambiguity. If $\{\alpha_1, \ldots, \alpha_n\} \subset O_E$ is a basis for $E/\mathbf{Q}$ and $\{\omega_1, \ldots, \omega_n\}$ is an integral basis, then
$$|\Delta(\omega_1, \ldots, \omega_n)| \leq |\Delta(\alpha_1, \ldots, \alpha_n)|$$
This is why an integral basis is often called a ***minimal basis.***

4-8-19. Proposition. Suppose that E is an algebraic number field. The sign of Δ_E is $(-1)^{r_2}$, where r_2 is the number of complex archimedean prime divisors of E. Furthermore, we have
$$\Delta_E \equiv 0 \text{ or } 1 \pmod 4$$

Proof. The integer r_2 has already appeared in (2-5-2). We write $E = \mathbf{Q}(\alpha)$ with $\alpha \in O_E$ and observe that it suffices to determine the sign of
$$\Delta(1, \alpha, \ldots, \alpha^{n-1}) = \prod_{i<j} (\alpha_i - \alpha_j)^2$$
where $\{\sigma_1 = 1, \sigma_2, \ldots, \sigma_n\} = \Gamma(\mathbf{Q}, E \to \Omega)$ and $\alpha_i = \sigma_i\alpha$. If both α_i and α_j are real, then $(\alpha_i - \alpha_j)^2 > 0$. If one of $\{\alpha_i, \alpha_j\}$ is real and the other complex, then the product contains $(\alpha_i - \alpha_j)^2[\pm \overline{(\alpha_i - \alpha_j)}]^2 > 0$. If both α_i and α_j are complex, then $(\alpha_i - \alpha_j)^2[\pm \overline{(\alpha_i - \alpha_j)}]^2 > 0$ except when $\alpha_j = \overline{\alpha_i}$; there are exactly r_2 such cases in our product, and for each one $\alpha_i - \alpha_j$ is pure imaginary and $(\alpha_i - \alpha_j)^2 < 0$.

For the remaining part, which is known as the *theorem of Stickelberger*, we take an integral basis $\{\omega_1, \ldots, \omega_n\}$ and let $\sigma_1, \ldots, \sigma_n$ be as above. Now, in the expansion of $\det \{\sigma_i\omega_j\}$, let A be the sum of all the terms that appear with an even permutation of $\{1, \ldots, n\}$, and let B be the sum of all terms that go with an odd permutation. Thus,
$$\begin{aligned}\Delta_E &= \Delta(\omega_1, \ldots, \omega_n) = [\det \{\sigma_i\omega_j\}]^2 \\ &= (A - B)^2 = (A + B)^2 - 4AB\end{aligned}$$

Denoting the splitting field of $E/\mathbf{Q}$ by K, we see that $A + B$ and AB remain fixed under every $\sigma \in \mathcal{G}(K/\mathbf{Q})$. Hence, $A + B \in \mathbf{Z}$, and $AB \in \mathbf{Z}$; so $\Delta_E \equiv 0$ or $1 \pmod 4$.

The same argument shows that for any $\mathfrak{A} = \mathbf{Z}\alpha_1 \oplus \cdots \oplus \mathbf{Z}\alpha_n \subset O_E$ we have $\Delta(\mathfrak{A}) \equiv 0$ or $1 \pmod 4$. In Chapter 5 we shall see that $|\Delta_E| > 1$.

4-8-20. Examples. Consider an algebraic number field $E = \mathbf{Q}(\alpha)$, where the minimum polynomial of α over $\mathbf{Q}$ is denoted by $f(x) \in \mathbf{Z}[x]$. If $f(x) = x^3 - x - 1$, then $\Delta(1, \alpha, \alpha^2)$ is the discriminant of $f(x)$ and is equal to -23 (see Section 3-7). Since 23 is a prime, it follows that $\Delta_E = -23$ and that $\{1, \alpha, \alpha^2\}$ is an integral basis. In the same way, for $f(x) = x^3 + x + 1$, we get $\Delta_E = -31$.

If $f(x) = x^3 - 2x^2 + 2$, then $\Delta(1, \alpha, \alpha^2) = (4)(-11)$ so that $\Delta_E = -44$ or -11. Since $-44 \equiv 0 \pmod 4$ and $-11 \equiv 1 \pmod 4$, Stickelberger is of no help. We know, however, that $\{1, \alpha, \alpha^2\}$ is an integral basis at every prime p except possibly $p = 2$ or 11 [see (4-8-8)]. Now, $x^3 - 2x^2 + 2$ is an Eisenstein polynomial over the 2-adic completion $\mathbf{Q}_2$. It follows, therefore, that the prime 2 has a unique extension to E and that for it $e = 3$. In particular, the prime 2 is ramified in E, and 2 divides Δ_E; hence, $\Delta_E = -44$ and $\{1, \alpha, \alpha^2\}$ is an integral basis at every prime.

Suppose $f(x) = x^3 + x^2 - 2x + 8$; then $\Delta(1, \alpha, \alpha^2) = (2^2)(-503)$ so that $\Delta_E = (2^2)(-503)$ or -503 (503 is a prime). To examine the critical prime 2, one may look at the Newton polygon of $f(x)$ over $\mathbf{Q}_2$ and see that $f(x)$ splits into three linear factors over $\mathbf{Q}_2$. This means, in particular, that the prime 2 has three extensions to E. Since $[E:\mathbf{Q}] = 3$, it follows that 2 is unramified; so $\Delta_E = -503$. Therefore, $\{1, \alpha, \alpha^2\}$ is not an integral basis; however, we can find an integral basis. The procedure described in (4-8-15) leads to the existence of an element $\beta \in O_E$ of form $\beta = (b_1 + b_2\alpha + b_3\alpha^2)/2$, where b_1, b_2, b_3 are 0 or 1 and not all 0. Consider $\beta = (\alpha + \alpha^2)/2$. From

$$\alpha^3 + \alpha^2 - 2\alpha + 8 = 0$$

we see that $(\alpha + \alpha^2)/2 = 1 - 4/\alpha$ and

$$8 + 2\left(\frac{4}{\alpha}\right) - \left(\frac{4}{\alpha}\right)^2 + \left(\frac{4}{\alpha}\right)^3 = 0$$

Thus $4/\alpha \in O_E$, $\beta \in O_E$, and $\beta \notin \mathbf{Z} \cdot 1 \oplus \mathbf{Z} \cdot \alpha \oplus \mathbf{Z} \cdot \alpha^2$. By computing α^2, $\alpha\beta$, and β^2 one sees that $\mathbf{Z} \cdot 1 \oplus \mathbf{Z}\alpha \oplus \mathbf{Z}\beta$ is a ring strictly containing $\mathbf{Z} \cdot 1 \oplus \mathbf{Z}\alpha \oplus \mathbf{Z}\alpha^2$. From (3-7-15) it follows that $\{1, \alpha, \beta\}$ is an integral basis.

4-9. Factoring Prime Ideals in an Extension Field

Suppose that $P \in \mathcal{S}(F)$ and that $Q_1, \ldots, Q_r$ are the extensions of P to E. Thus, $\mathfrak{Q}_1, \ldots, \mathfrak{Q}_r$ are the prime ideals of O_E which lie over $\mathfrak{P}$, and we recall from (4-7-1) that

$$\mathfrak{P}O_E = \prod_1^r \mathfrak{Q}_i^{e(Q_i/P)}$$

is the decomposition of $\mathfrak{P}$ in E. The discussion in this section centers around a theorem of Kummer which provides a procedure for constructing the prime ideals $\mathfrak{Q}_i$ of O_E and finding all $e(Q_i/P)$ and $f(Q_i/P)$.

It is advisable to make precise some of the notation that will occur. We have the completions F_P and E_{Q_i}, their rings of integers $\bar{O}_P$ and $\bar{O}_{Q_i}$, their prime ideals $\bar{\mathcal{P}}$ and $\bar{\mathcal{Q}}_i$, and the residue class fields $\bar{F}_P = \bar{O}_P/\bar{\mathcal{P}}$ and $\bar{E}_{Q_i} = \bar{O}_Q/\bar{\mathcal{Q}}_i$. In virtue of the standard identifications, we know that $\bar{F}_P = O_P/\mathcal{P} = O_F/\mathfrak{P}$ is contained in $\bar{E}_{Q_i} = O_{Q_i}/\mathcal{Q}_i = O_E/\mathfrak{Q}_i$. Let $\psi : \bar{O}_P \twoheadrightarrow \bar{F}_P$ be the residue class map; for $a \in \bar{O}_P$ we shall also write $\psi(a) = \bar{a}$. Of course, $O_F \subset O_P \subset \bar{O}_P$, and we denote the restrictions of ψ to O_P and O_F by ψ also. By our identifications, this means that $\psi : O_P \twoheadrightarrow \bar{F}_P$ has kernel $\mathcal{P}$ and $\psi : O_F \twoheadrightarrow \bar{F}_P$ has kernel $\mathfrak{P}$. As usual, ψ determines a map $f(x) \in \bar{O}_P[x] \rightarrow \bar{f}(x) \in \bar{F}_P[x]$ with corresponding restrictions to $O_P[x]$ and $O_F[x]$.

In the same way, let $\psi_i : \bar{O}_{Q_i} \twoheadrightarrow \bar{E}_{Q_i}$ be the residue class map, and denote its restrictions to O_{Q_i} and O_E by ψ_i. Strictly speaking, O_{Q_i} and its subring O_E are imbedded in $\bar{O}_{Q_i}$ via the composition map μ_i (see Section 2-4); so for $\alpha \in O_{Q_i}$ we have $\psi_i(\alpha) = \psi_i(\mu_i\alpha)$. The identifications permit us to view each ψ_i as an extension of ψ.

4-9-1. Theorem (Kummer). Suppose that E/F is separable and that $\theta \in O_E$ is such that $E = F(\theta)$, and let $f(x) = f(\theta, F)$. If $\{1, \theta, \ldots, \theta^{n-1}\}$ is an integral basis for O_E/O_F, then the irreducible factorization of $\bar{f}(x)$ over $\bar{F}_P$ is of form

$$\bar{f}(x) = G_1(x)^{e_1} \cdots G_r(x)^{e_r}$$

where $G_i(x) \neq G_j(x)$ for $i \neq j$, $\deg G_i(x) = f(Q_i/P)$, and $e_i = e(Q_i/P)$. Moreover, if $g_i(x) \in O_F[x]$ is a monic polynomial such that $\bar{g}_i(x) = G_i(x)$, then

$$\mathfrak{Q}_i = (\mathfrak{P}, g_i(\theta))$$

Proof. According to (2-4-5) and (2-4-6) the factorization of $f(x)$ over F_P into distinct irreducible factors takes the form

$$f(x) = f_1(x) \cdots f_r(x)$$

where $f_i(x) = f(\mu_i\theta, F_P)$ and $\deg f_i(x) = n(Q_i/P)$. From (2-2-5) we know that $\bar{f}_i(x) \in \bar{F}_P[x]$ is a power of an irreducible polynomial $G_i(x) \in \bar{F}_P[x]$—say, $\bar{f}_i(x) = G_i(x)^{e_i}$. It is easy to see that $\bar{E}_{Q_i} = \bar{F}_P(\psi_i(\theta))$ and that $G_i(x)$ is the minimum polynomial of $\psi_i(\theta) = \psi_i(\mu_i\theta)$ over $\bar{F}_P$; so $\deg G_i(x) = [\bar{E}_{Q_i}:\bar{F}_P] = f(Q_i/P)$. Since $\deg \bar{f}_i(x) = (e_i) \deg G_i(x)$, it follows that $e_i = e(Q_i/P)$.

Choose any monic polynomial $g_i(x) \in O_F[x]$ such that $\bar{g}_i(x) = G_i(x)$; so $\deg g_i(x) = f(Q_i/P)$. Because $\psi_i[g_i(\theta)] = G_i(\psi_i(\theta))$, it is immediate that

$$(\mathfrak{P}, g_i(\theta)) = \mathfrak{P}O_E + g_i(\theta)O_E \subset \mathfrak{Q}_i$$

Conversely, consider any $\alpha \in \mathfrak{Q}_i$, so $\alpha = a_0 + a_1\theta + \cdots + a_{n-1}\theta^{n-1}$ with $a_i \in O_F$, and put $h(x) = a_0 + a_1x + \cdots + a_{n-1}x^{n-1} \in O_F[x]$. Now there exist $q(x), r(x) \in O_F[x]$ with $\deg r(x) < \deg g_i(x) = f(Q_i/P)$ such that $h(x) = q(x)g_i(x) + r(x)$. From $\bar{h}(\psi_i(\theta)) = \psi_i(\alpha) = 0$ and $\bar{g}_i(\psi_i(\theta)) = 0$ it follows that $\bar{r}(x) = 0$, so that $r(x) \in \mathfrak{P}[x]$. Therefore,

$$\alpha = h(\theta) = q(\theta)g_i(\theta) + r(\theta) \in g_i(\theta)O_E + \mathfrak{P}O_E$$

and $\mathfrak{Q}_i$ is as described.

Suppose finally that $G_i(x) = G_j(x)$ for $i \neq j$. Then we may take $g_i(x) = g_j(x)$ and conclude that $\mathfrak{Q}_i = \mathfrak{Q}_j$, a contradiction. This completes the proof.

This result facilitates the study of the prime ideals in a large class of algebraic number fields—namely, from the factorization of $\bar{f}(x)$ over $\bar{F}_P$ we may read off the e's and f's and exhibit the prime ideals explicitly. The major difficulty is that not every algebraic number field has an integral basis over $\mathbf{Q}$ of form $\{1, \theta, \ldots, \theta^{n-1}\}$; however, even in this situation, not everything is lost, as may be seen from our next result.

4-9-2. Exercise. Suppose that E/F is separable and that $\theta \in O_E$ is such that $E = F(\theta)$; then Kummer's theorem holds for almost all P. In fact, Kummer's theorem holds for all P at which $\{1, \theta, \ldots, \theta^{n-1}\}$ is an integral basis, or, equivalently, for all P such that $\nu_P(\Delta_{E/F}) = \nu_P[\Delta(1, \theta, \ldots, \theta^{n-1})]$ [see (4-8-17)].

Thus, by varying the choice of θ, it is often possible to get information about different primes.

4-9-3. Remark. Suppose that k is a field containing $\bar{F}_P$ and that we have a homomorphism

$$\lambda: O_E \twoheadrightarrow k$$

which is an extension of ψ. The kernel of λ is then a maximal prime ideal $\mathfrak{Q}$ of O_E and $\mathfrak{Q} \supset \mathfrak{P}$. Therefore, $\mathfrak{Q}$ is one of the $\mathfrak{Q}_i$, and we may identify k with $O_E/\mathfrak{Q}_i$ and λ with ψ_i.

It is clear that such maps λ (often known as ***places***) may be used to get information about the ideals lying over $\mathfrak{P}$. When an integral basis for O_E/O_F is available—say, $O_E = O_F\omega_1 \oplus \cdots \oplus O_F\omega_n$—then one can try to construct such a λ by specifying the $\lambda(\omega_i)$ in such a way that $\lambda(\omega_i\omega_j) = \lambda(\omega_i)\lambda(\omega_j)$. Of course, it is then necessary to see whether or not $\lambda(O_E)$ is a field. Sometimes it is enough to work with even a subring of O_E.

4-9-4. Example (Dedekind). We give an example of a cubic field which does not have an integral basis of form $\{1, \theta, \theta^2\}$. Let $E = \mathbf{Q}(\alpha)$, where α is a root of the irreducible polynomial $f(x) = x^3 + x^2 - 2x + 8$. We have seen in (4-8-20) that the prime 2 has three extensions Q_1, Q_2, Q_3 to E—and, of course, that $e(Q_i/2) = f(Q_i/2) = 1$. Suppose that $O_E = \mathbf{Z} \cdot 1 \oplus \mathbf{Z} \cdot \theta + \mathbf{Z} \cdot \theta^2$. Then each ψ_i is completely determined as soon as we know $\psi_i(\theta)$. Since $\bar{E}_{Q_i} = \{0, 1\}$, $\psi_i(\theta)$ must be 0 or 1; so there are only two possibilities for ψ_i, a contradiction. The same argument shows that there does not exist an integral basis at 2 of form $1, \theta, \theta^2$.

We recall further, from (4-8-20), that $\Delta_E = -503$ and that if $\beta = 1 - 4/\alpha = (\alpha + \alpha^2)/2$ then $O_E = \mathbf{Z} \cdot 1 \oplus \mathbf{Z} \cdot \alpha + \mathbf{Z} \cdot \beta$; moreover, multiplication in O_E is described via the relations $\alpha^2 = 2\beta - \alpha$, $\alpha\beta = \alpha - 4$, $\beta^2 = \beta - 2\alpha - 2$. Let us try to describe the ideals $\mathfrak{Q}_1$, $\mathfrak{Q}_2$, $\mathfrak{Q}_3$ lying over $2\mathbf{Z}$. It is clear that each ψ_i must map α and β to 0 or 1. The possibility that $\psi_i(\alpha) = 1$ and $\psi_i(\beta) = 0$ is excluded because then $\psi_i(\alpha\beta) = 0$, while $\psi_i(\alpha - 4) = 1$. Therefore, we may index things so that

$$\begin{aligned} \psi_1&: \alpha \to 1, \beta \to 1 \\ \psi_2&: \alpha \to 0, \beta \to 1 \\ \psi_3&: \alpha \to 0, \beta \to 0 \end{aligned}$$

In particular, $\alpha \notin \mathfrak{Q}_1$, $\beta \notin \mathfrak{Q}_1$, $\alpha \in \mathfrak{Q}_2$, $\beta \notin \mathfrak{Q}_2$, $\alpha \in \mathfrak{Q}_3$, $\beta \in \mathfrak{Q}_3$, and so on.

For every $a \in \mathbf{Z}$, $f(x + a) = (x + a)^3 + (x + a)^2 - 2(x + a) + 8$ is the minimum polynomial of $\alpha - a$, and it has constant term $f(a)$; therefore, $N(\alpha - a) = (-1)^3 f(a)$. If $\nu_p[N(\alpha - a)] = 0$, then [by (2-5-8)] $\alpha - a$ is a unit at every prime of E that lies over p.

Since $N(\alpha) = -8$ and $\alpha \notin \mathfrak{Q}_1$, $\alpha \in \mathfrak{Q}_2$, $\alpha \in \mathfrak{Q}_3$, it follows that the factorization of (α) must be of form

$$(\alpha) = \mathfrak{Q}_2^{s_2}\mathfrak{Q}_3^{s_3} \qquad s_2, s_3 \geq 1$$

We also have $N(\beta - 1) = N(-4)/N(\alpha) = 8$ and $\beta - 1 \in \mathfrak{Q}_1$, $\beta - 1 \in \mathfrak{Q}_2$, $\beta - 1 \notin \mathfrak{Q}_3$, so that the factorization of $(\beta - 1)$ looks like

$$(\beta - 1) = \mathfrak{Q}_1^{t_1}\mathfrak{Q}_2^{t_2} \qquad t_1, t_2 \geq 1$$

Since $[\alpha(\beta - 1)]O_E = (-4)O_E = \mathfrak{Q}_1^2\mathfrak{Q}_2^2\mathfrak{Q}_3^2$ and

$$[\alpha(\beta - 1)]O_E = \mathfrak{Q}_1^{t_1}\mathfrak{Q}_2^{s_2+t_2}\mathfrak{Q}_3^{s_3}$$

we conclude that

$$(\alpha) = \mathfrak{Q}_2\mathfrak{Q}_3^2$$
$$(\beta - 1) = \mathfrak{Q}_1^2\mathfrak{Q}_2$$

In particular, $\nu_{Q_1}(\alpha) = 0$, $\nu_{Q_2}(\alpha) = 1$, $\nu_{Q_3}(\alpha) = 2$.

Let us continue in the same vein. Since $N(\alpha - 1) = -8$ and $\alpha - 1 \in \mathfrak{Q}_1$, $\alpha - 1 \notin \mathfrak{Q}_2$, $\alpha - 1 \notin \mathfrak{Q}_3$, it follows that $(\alpha - 1) = \mathfrak{Q}_1^s$; and by taking norms $s = 3$; so

$$(\alpha - 1) = \mathfrak{Q}_1^3$$

We note also that $\nu_{Q_1}(\alpha \pm 2) = 0$, $\nu_{Q_2}(\alpha \pm 2) \geq 1$, $\nu_{Q_3}(\alpha \pm 2) = 1$, $N(\alpha + 2) = -8$, $N(\alpha - 2) = -16$ and, therefore,

$$(\alpha + 2) = \mathfrak{Q}_2^2\mathfrak{Q}_3$$
$$(\alpha - 2) = \mathfrak{Q}_2^3\mathfrak{Q}_3$$

Finally, we note that $N(\alpha + 3) = 4$, $\nu_{Q_1}(\alpha + 3) \geq 0$, $\nu_{Q_2}(\alpha + 3) = 0$, $\nu_{Q_3}(\alpha + 3) = 0$, so that

$$(\alpha + 3) = \mathfrak{Q}_1^2$$

These facts may be combined to show that $\mathfrak{Q}_1$, $\mathfrak{Q}_2$, $\mathfrak{Q}_3$ are principal ideals; in fact,

$$\mathfrak{Q}_1 = \left(\frac{\alpha - 1}{\alpha + 3}\right)$$
$$\mathfrak{Q}_2 = \left(\frac{\alpha - 2}{\alpha + 2}\right)$$
$$\mathfrak{Q}_3 = \left(2\left(\frac{\alpha + 2}{\alpha - 2}\right)\left(\frac{\alpha + 3}{\alpha - 1}\right)\right)$$

If we wish to consider other prime ideals of O_E, it becomes necessary to distinguish them according to the prime p over which they lie. Thus, $\mathfrak{Q}_1$, $\mathfrak{Q}_2$, $\mathfrak{Q}_3$ will now be written as $\mathfrak{Q}_1^{(2)}$, $\mathfrak{Q}_2^{(2)}$, $\mathfrak{Q}_3^{(2)}$. Because $\{1, \alpha, \alpha^2\}$ is an integral basis at every prime $p \neq 2$, we may use (4-9-2). Since $x^3 + x^2 - 2x + 8$ is irreducible (mod 3), the prime 3 has a unique extension $Q_1^{(3)}$ to E for which $e = 1$ and $f = 3$. In other words,

$$(3) = \mathfrak{Q}_1^{(3)}$$

so that 3 remains prime and $\mathfrak{Q}_1^{(3)}$ is principal. Since $x^3 + x^2 - 2x + 8 \equiv (x + 1)(x^2 - 2) \pmod 5$, the prime 5 has two extensions $Q_1^{(5)}$,

$Q_2^{(5)}$ to E and $e(Q_1^{(5)}/5) = e(Q_2^{(5)}/5) = 1$, $f(Q_1^{(5)}/5) = 1$, $f(Q_2^{(5)}/5) = 2$. This implies that

$$(5) = \mathfrak{Q}_1^{(5)}\mathfrak{Q}_2^{(5)}$$

Now, $N(\alpha + 1) = -10$, and $\alpha + 1 \in \mathfrak{Q}_1^{(2)}$, $\alpha + 1 \notin \mathfrak{Q}_2^{(2)}$, $\alpha + 1 \notin \mathfrak{Q}_3^{(2)}$. Then because $\mathfrak{N}(\mathfrak{Q}_1^{(5)}) = 5$ and $\mathfrak{N}(\mathfrak{Q}_2^{(5)}) = 25$, we conclude that

$$(\alpha + 1) = \mathfrak{Q}_1^{(2)}\mathfrak{Q}_1^{(5)}$$

Since $\mathfrak{Q}_1^{(2)}$ is principal, so are $\mathfrak{Q}_1^{(5)}$ and $\mathfrak{Q}_2^{(5)}$.

It may be left to the reader to verify that 7, 11, and 13 remain prime in E.

4-9-5. Exercise. Let $\mathfrak{P}$ be a prime ideal of the algebraic number field F with $f(\mathfrak{P}/p) = f$. Suppose that $g(x) \in \mathbf{Z}[x]$ is of degree f and irreducible (mod p)—that is, $\bar{g}(x)$ is irreducible over $\mathbf{Z}/(p)$. Then there exists $\theta \in O_F$ such that

$$\mathfrak{P} = (p, g(\theta))$$

Hint: First show that there exists $\alpha \in O_F$ such that $\nu_{\mathfrak{P}}[g(\alpha)] = 1$.

4-10. Hilbert Theory

We assume in this section (in addition to the standard hypotheses) that E/F is a Galois extension and that all residue class field extensions are separable. This enables us to describe certain intermediate fields in terms of which more detailed information about the decomposition of ideals may be given.

Consider any $\sigma \in \mathcal{G}(E/F)$. According to (2-5-5), we know that, for each $P \in \mathcal{S}(F)$, σ^* is a permutation of the set $\{Q \in \mathcal{S}(E) | Q \supset P\}$. Therefore, σ^* may be viewed as a permutation of $\mathcal{S}(E)$, and it then determines an automorphism (also denoted by σ^*) of the group $\mathbf{D}\{\mathcal{S}(E)\}$. This automorphism clearly preserves joins, meets, divisibility, and integrality.

In virtue of the fundamental theorem of OAFs there corresponds to σ^* an automorphism of $\mathfrak{I}\{\mathcal{S}(E)\}$ which permutes the prime ideals of O_E; we wish to describe this automorphism directly. Because $\nu_{\sigma^* Q} = \nu_Q \circ \sigma$, it will turn out that it is more natural to deal with $(\sigma^{-1})^* = (\sigma^*)^{-1}$. For $Q \in \mathcal{S}(E)$, it is easy to see that $\sigma O_Q = O_{(\sigma^{-1})^* Q}$, and then it follows that $\sigma O_E = O_E$. If $\mathfrak{A}$ is any E ideal, then $\sigma\mathfrak{A} = \{\sigma\alpha | \alpha \in \mathfrak{A}\}$ is also an E ideal; it is known as a ***conjugate ideal*** of $\mathfrak{A}$. If $\mathfrak{Q}$ is a prime ideal of O_E, the automorphism σ clearly carries it into the maximal prime ideal $\sigma\mathfrak{Q}$ of O_E. Thus σ permutes the prime ideals of O_E. Moreover, if $\mathfrak{Q}$ lies above the prime ideal $\mathfrak{P}$ of O_F (that is, if $\mathfrak{Q} \cap O_F = \mathfrak{Q} \cap F = \mathfrak{P}$), then so does $\sigma\mathfrak{Q}$; in fact, $\mathfrak{Q} \cap O_F = \mathfrak{P}$ implies that $\sigma\mathfrak{Q} \cap O_F = \sigma(\mathfrak{Q} \cap O_F) = \sigma\mathfrak{P} = \mathfrak{P}$. There is no difficulty in verifying that σ is an automorphism of $\mathfrak{J}\{\mathcal{S}(E)\}$ which preserves joins, meets, divisibility, and integrality. Since $\mathfrak{Q} = \{\alpha \in O_E | \nu_Q(\alpha) \geq 1\}$, it follows that $\sigma\mathfrak{Q} = \{\sigma\alpha \in O_E | \nu_Q(\alpha) \geq 1\} = \{\beta \in O_E | \nu_Q(\sigma^{-1}\beta) \geq 1\} = \{\beta \in O_E | \nu_{(\sigma^{-1})^*Q}(\beta) \geq 1\}$. In other words, if $\mathfrak{Q}$ and Q correspond, then so do $\sigma\mathfrak{Q}$ and $(\sigma^{-1})^*Q$; thus, σ is the map of ideals corresponding to the map $(\sigma^{-1})^*$ of divisors.

Now, let us fix $Q \in \mathcal{S}(E)$ with corresponding prime ideal $\mathfrak{Q}$ of O_E. Let $P \in \mathcal{S}(F)$ be the restriction of Q to F, so that the corresponding prime ideal of O_F is $\mathfrak{P} = \mathfrak{Q} \cap O_F$. This notation will remain in force throughout.

In order to study the connections between $\mathfrak{P}$ and $\mathfrak{Q}$, we put

$$\mathcal{Z} = \{\sigma \in \mathcal{G} | \sigma\mathfrak{Q} = \mathfrak{Q}\}$$

If it becomes necessary to make the notation more precise, we shall write one of $\mathcal{Z}_{\mathfrak{Q}} = \mathcal{Z}_Q = \mathcal{Z}_{Q/P} = \mathcal{Z}_{\mathfrak{Q}/\mathfrak{P}}$ for $\mathcal{Z}$. It is clear that $\mathcal{Z}$ is a subgroup of $\mathcal{G}$; it is called the ***decomposition group*** (*Zerlegungsgruppe*) of $\mathfrak{Q}$ over F (or of Q, or of Q/P, . . .). There is no harm in referring to Q or Q/P because

$$\mathcal{Z} = \{\sigma \in \mathcal{G} | (\sigma^{-1})^*Q = Q\} = \{\sigma \in \mathcal{G} | \sigma^*Q = Q\}$$

The fixed field Z of $\mathcal{Z}$ is called the ***decomposition field*** of $\mathfrak{Q}$ over F. Of course, E/Z is a Galois extension with $\mathcal{G}(E/Z) = \mathcal{Z}$.

Furthermore, let us put, for each integer $r \geq 0$,

$$\begin{aligned}\mathcal{V}^{(r)} &= \{\sigma \in \mathcal{G} | \sigma\alpha \equiv \alpha \pmod{\mathfrak{Q}^{r+1}} \quad \forall \alpha \in O_E\} \\ &= \{\sigma \in \mathcal{G} | (\sigma - 1)O_E \subset \mathfrak{Q}^{r+1}\}\end{aligned}$$

Again, when it is necessary to do so, we shall write one of $\mathcal{V}^{(r)}_{\mathfrak{Q}} = \mathcal{V}^{(r)}_Q = \mathcal{V}^{(r)}_{Q/P} = \mathcal{V}^{(r)}_{\mathfrak{Q}/\mathfrak{P}}$ for $\mathcal{V}^{(r)}$. It is easy to see that $\mathcal{V}^{(r)}$ is a group—the ***rth ramification group*** of $\mathfrak{Q}$ over F. For reasons which will soon appear, it is customary to write $\mathcal{V}^{(0)} = \mathfrak{I} = \mathfrak{I}_{\mathfrak{Q}} = \mathfrak{I}_Q = \mathfrak{I}_{Q/P} = \mathfrak{I}_{\mathfrak{Q}/\mathfrak{P}}$ and call it the ***inertia group*** (*Trägheitsgruppe*) of $\mathfrak{Q}$ over F. From the chain of groups

$$\{1\} \subset \cdots \subset \mathcal{V}^{(r)} \subset \mathcal{V}^{(r-1)} \subset \cdots \subset \mathcal{V}^{(1)} \subset \mathcal{V}^{(0)} = \mathfrak{I} \subset \mathcal{Z} \subset \mathcal{G}$$

we get by Galois theory the corresponding tower of fields

$$F \subset Z \subset T = V^{(0)} \subset V^{(1)} \subset \cdots \subset V^{(r-1)} \subset V^{(r)} \subset \cdots \subset E$$

where $\mathfrak{V}^{(r)} = \mathfrak{G}(E/V^{(r)})$ (and in particular $\mathfrak{J} = \mathfrak{G}(E/T)$). $V^{(r)}$ is called the ***rth ramification field*** (and T the ***inertia field***) for $\mathfrak{Q}$ over F.

4-10-1. Proposition. For $\tau \in \mathfrak{G}$ we have

$$\mathfrak{Z}_{\tau\mathfrak{Q}} = \tau\mathfrak{Z}_{\mathfrak{Q}}\tau^{-1} \qquad \text{and} \qquad \mathfrak{V}^{(r)}_{\tau\mathfrak{Q}} = \tau\mathfrak{V}^{(r)}_{\mathfrak{Q}}\tau^{-1}$$

In particular, $\mathfrak{V}^{(r)}$ is a normal subgroup of $\mathfrak{Z}$ and $V^{(r)}/Z$ is a Galois extension for every $r \geq 0$.

Proof. We have

$$\sigma \in \mathfrak{Z}_{\tau\mathfrak{Q}} \Leftrightarrow \sigma\tau\mathfrak{Q} = \tau\mathfrak{Q} \Leftrightarrow \tau^{-1}\sigma\tau\mathfrak{Q} = \mathfrak{Q} \Leftrightarrow \tau^{-1}\sigma\tau \in \mathfrak{Z}_{\mathfrak{Q}} \Leftrightarrow \sigma \in \tau\mathfrak{Z}_{\mathfrak{Q}}\tau^{-1}$$

Moreover,

$$\begin{aligned}\sigma \in \mathfrak{V}^{(r)}_{\tau\mathfrak{Q}} &\Leftrightarrow (\sigma - 1)O_E \subset (\tau\mathfrak{Q})^{r+1} \Leftrightarrow (\sigma - 1)\tau O_E \subset \tau(\mathfrak{Q}^{r+1})\\ &\Leftrightarrow \tau^{-1}(\sigma - 1)\tau O_E \subset \mathfrak{Q}^{r+1} \Leftrightarrow (\tau^{-1}\sigma\tau - 1)O_E \subset \mathfrak{Q}^{r+1}\\ &\Leftrightarrow \tau^{-1}\sigma\tau \in \mathfrak{V}^{(r)}_{\mathfrak{Q}} \Leftrightarrow \sigma \in \tau\mathfrak{V}^{(r)}_{\mathfrak{Q}}\tau^{-1}\end{aligned}$$

4-10-2. Remark. Let H denote any field between E and F, and let $\mathfrak{H} = \mathfrak{G}(E/H)$ be the corresponding subgroup of $\mathfrak{G}$. Let Q_H denote the restriction of Q to H, and let $\mathfrak{Q}_H$ be the corresponding prime ideal of O_H. Thus, we have $P \subset Q_H \subset Q$ and $\mathfrak{P} \subset \mathfrak{Q}_H \subset \mathfrak{Q}$—in fact, $\mathfrak{Q} \cap O_H = \mathfrak{Q}_H$, and $\mathfrak{Q}_H \cap O_F = \mathfrak{P}$.

From the definitions, it is immediate that

$$\mathfrak{Z}_{Q/Q_H} = \mathfrak{Z}_{Q/P} \cap \mathfrak{H} \qquad \mathfrak{V}^{(r)}_{Q/Q_H} = \mathfrak{V}^{(r)}_{Q/P} \cap \mathfrak{H} \qquad (r = 0, 1, 2, \ldots)$$

If one takes $H = Z$, then $\mathfrak{H} = \mathfrak{G}(E/Z) = \mathfrak{Z}_{Q/P}$ and it follows that

$$\mathfrak{Z}_{Q/Q_Z} = \mathfrak{Z}_{Q/P} \qquad \mathfrak{V}^{(r)}_{Q/Q_Z} = \mathfrak{V}^{(r)}_{Q/P} \qquad (r = 0, 1, 2, \ldots)$$

Furthermore, if we take $H = V^{(i)}$ for any $i = 0, 1, 2, \ldots,$ then $\mathfrak{H} = \mathfrak{G}(E/V^{(i)}) = \mathfrak{V}^{(i)}_{Q/P}$, so that upon writing $Q_{V^{(i)}} = Q_i$ we have

$$\mathfrak{Z}_{Q/Q_i} = \mathfrak{V}^{(i)}_{Q/P} \qquad \mathfrak{V}^{(r)}_{Q/Q_i} = \mathfrak{V}^{(i)}_{Q/P} \qquad (r = 0, 1, 2, \ldots, i - 1)$$

$$\mathfrak{V}^{(r)}_{Q/Q_i} = \mathfrak{V}^{(r)}_{Q/P} \qquad (r = i, i + 1, i + 2, \ldots)$$

4-10-3. Proposition. Let H be any field between E and F; then

$$\mathfrak{Q}_H \text{ is a power of } \mathfrak{Q} \Leftrightarrow H \supset Z \Leftrightarrow \mathfrak{Z}_{Q/Q_H} = \mathfrak{G}\left(\frac{E}{H}\right)$$

Proof. $\mathfrak{Q}_H$ is a power of $\mathfrak{Q}$ (meaning that $\mathfrak{Q}_H O_E$ is a power of $\mathfrak{Q}$) $\Leftrightarrow Q_H$ has a unique extension to E, namely, $Q \Leftrightarrow \mathfrak{Z}_{Q/Q_H} = \mathfrak{G}(E/H)$ $= \mathfrak{H} \Leftrightarrow \mathfrak{Z}_{Q/P} \supset \mathfrak{H} \Leftrightarrow H \supset Z$.

This simple result enables us to connect $\mathfrak{P}$ and $\mathfrak{Q}$ via $\mathfrak{Q}_Z$; namely, we have the following result.

4-10-4. Proposition. We have

(i) $e\left(\frac{Q}{Q_Z}\right) = e\left(\frac{Q}{P}\right) \qquad f\left(\frac{Q}{Q_Z}\right) = f\left(\frac{Q}{P}\right) \qquad e\left(\frac{Q_Z}{P}\right) = f\left(\frac{Q_Z}{P}\right) = 1$

(ii) $\mathfrak{Q}_Z O_E = \mathfrak{Q}^{e(Q/P)} \qquad N_{Z\to F}(\mathfrak{Q}_Z) = \mathfrak{P} \qquad \mathfrak{P}O_Z = \mathfrak{Q}_Z\mathfrak{A}$

where $\mathfrak{A}$ is an integral Z ideal prime to $\mathfrak{Q}_Z$.

Proof: i. Let r be the number of extensions of P to E; so there are r prime ideals $\mathfrak{Q}_1 = \mathfrak{Q}, \mathfrak{Q}_2, \ldots, \mathfrak{Q}_r$ of O_E lying over $\mathfrak{P}$. In view of (2-5-5), we have

$$\mathfrak{P}O_E = (\mathfrak{Q}_1 \cdots \mathfrak{Q}_r)^{e(Q/P)}$$

and $f(Q_i/P) = f(Q/P)$ for $i = 1, \ldots, r$. This implies that

$$n = [E:F] = r\, e\left(\frac{Q}{P}\right) f\left(\frac{Q}{P}\right)$$

Furthermore, because Q is the unique extension of Q_Z to E, we have $[E:Z] = e(Q/Q_Z)f(Q/Q_Z)$. Combining these facts with $r = (\mathcal{G}:\mathcal{Z}) = n/[E:Z]$ yields

$$e\left(\frac{Q}{P}\right) f\left(\frac{Q}{P}\right) = e\left(\frac{Q}{Q_Z}\right) f\left(\frac{Q}{Q_Z}\right)$$

The desired conclusions now follow from the multiplicative character of e and f.

ii. The various assertions of (ii) are clear, since $\mathfrak{Q}_Z O_E = \mathfrak{Q}^{e(Q/Q_Z)}$, $N_{Z\to F}(\mathfrak{Q}_Z) = \mathfrak{P}^{f(Q_Z/P)}$, and the power of $\mathfrak{Q}_Z$ appearing in $\mathfrak{P}O_Z$ is $\mathfrak{Q}_Z^{e(Q_Z/P)}$. Because $\mathfrak{P}$ splits off the prime ideal factor $\mathfrak{Q}_Z$ of degree 1, Z is often referred to as the ***splitting field*** of $\mathfrak{Q}$ over F.

Suppose that $\sigma \in \mathcal{Z}$; so $\sigma O_E = O_E$, and $\sigma\mathfrak{Q} = \mathfrak{Q}$. Thus, σ determines an automorphism $\bar{\sigma}$ of the residue class field $\bar{E} = O_E/\mathfrak{Q}$ of E at Q, and $\bar{\sigma}$ leaves the residue class field $\bar{F} = O_F/\mathfrak{P}$ of F at P pointwise fixed; that is, $\bar{\sigma} \in \mathcal{G}(\bar{E}/\bar{F})$. The mapping $\sigma \to \bar{\sigma}$ may be used to study $\mathcal{Z}$ directly; however, we prefer to relate $\mathcal{Z}_{Q/P}$ and the various ramification groups to the local extension E_Q/F_P, which has been studied extensively in Sections 3-5 and 3-6.

4-10-5. Proposition. There is a canonical isomorphism

$$\mathcal{Z}_{Q/P} \approx \mathcal{G}\left(\frac{E_Q}{F_P}\right)$$

Proof. We may write $E_Q = F_P \cdot E$. Because E/F is a finite Galois extension, so is E_Q/F_P. An element $\sigma \in \mathcal{Z}$ is an automorphism of E such that $\sigma^*Q = Q = (\sigma^{-1})^*Q$. Therefore, σ may be extended uniquely to an automorphism $\tilde{\sigma}$ of the completion E_Q, and $\tilde{\sigma}$ is the identity on F_P because σ is the identity on F. The mapping $\sigma \to \tilde{\sigma}$ is clearly a monomorphism of $\mathcal{Z} \to \mathcal{G}(E_Q/F_P)$.

Suppose then that $\lambda \in \mathcal{G}(E_Q/F_P)$, and put $\sigma = \lambda|E$. Since σ leaves F pointwise fixed, it follows that $\sigma E = E$; so $\sigma \in \mathcal{G}(E/F)$. Furthermore, denoting the prime divisor and normalized valuation of E_Q by Q and ν_Q, we have $\nu_Q \circ \lambda = \nu_{\lambda^* Q} = \nu_Q \Rightarrow \nu_Q \circ \sigma = \nu_Q \Rightarrow \sigma^*Q = Q$; thus $\sigma \in \mathcal{Z}$, and $\lambda = \tilde{\sigma}$.

It should be noted that the previous map $\sigma \to \bar{\sigma}$ has not been lost. One considers the local residue class field extension $\bar{E}_Q/\bar{F}_P$, and when it is identified with $\bar{E}/\bar{F}$ (as is customary), it is clear that $\bar{\tilde{\sigma}} = \bar{\sigma}$.

As a rule *we shall not distinguish between σ and $\tilde{\sigma}$*. One consequence of this is that $\mathcal{G}(E_Q/F_P)$ is "identified" with the subgroup $\mathcal{Z}_{Q/P}$ of $\mathcal{G}(E/F)$. Our next step is to connect the ramification groups $\mathcal{V}^{(r)} = \mathcal{V}^{(r)}_{Q/P}$ with the local ramification groups $\mathcal{V}_r = \mathcal{V}_r(E_Q/F_P)$ of the extension E_Q/F_P.

4-10-6. Proposition. The canonical isomorphism between $\mathcal{Z}$ and $\mathcal{G}(E_Q/F_P)$ is such that

$$\mathcal{V}^{(r)}_{Q/P} \approx \mathcal{V}_r \qquad r = 0, 1, 2, \ldots$$

Proof. Denote the prime ideal of O_Q by $\mathfrak{Q}$, so $\mathfrak{Q} = \{\alpha \in E | \nu_Q(\alpha) \geq 1\}$ and denote the prime ideal of $\bar{O}_Q$ by $\bar{\mathfrak{Q}}$, so $\bar{\mathfrak{Q}} = \{\alpha \in E_Q | \nu_Q(\alpha) \geq 1\}$. For $\sigma \in \mathcal{Z} = \mathcal{G}(E_Q/F_P)$ we have

$$\begin{aligned}
\sigma \in \mathcal{V}_r &\Leftrightarrow (\sigma - 1)\bar{O}_Q \subset \bar{\mathfrak{Q}}^{r+1} \\
&\Rightarrow (\sigma - 1)O_Q \subset \mathfrak{Q}^{r+1} \qquad (\bar{O}_Q \cap E = O_Q,\ \bar{\mathfrak{Q}}^{r+1} \cap E = \mathfrak{Q}^{r+1}) \\
&\Rightarrow (\sigma - 1)O_E \subset \mathfrak{O}^{r+1} \qquad (\mathfrak{Q}^{r+1} \cap O_E = \mathfrak{O}^{r+1}) \\
&\Leftrightarrow \sigma \in \mathcal{V}^{(r)}_{Q/P}
\end{aligned}$$

To complete the proof, observe that by the strong approximation theorem [see item 1 of (4-1-5)] O_E is dense in $\bar{O}_Q$ and $\mathfrak{O}^{r+1}$ is dense in $\bar{\mathfrak{Q}}^{r+1}$. Since $\sigma - 1$ is a continuous map of $\bar{O}_Q \to \bar{O}_Q$ and $\bar{\mathfrak{Q}}^{r+1}$ is closed, it follows that

$$(\sigma - 1)O_E \subset \mathfrak{O}^{r+1} \Rightarrow (\sigma - 1)\bar{O}_Q \subset \bar{\mathfrak{Q}}^{r+1}$$

Thus we may also identify $\mathcal{V}^{(r)}_{Q/P}$ and $\mathcal{V}_r$.

4-10-7. Remark. All the group-theoretic results of the local theory now carry over, for example, (3-6-1), (3-6-4), (3-6-5), (3-6-6), and (3-7-25).

It should likewise be noted that fields also behave properly under our passage from the global to the local. To make this precise, let H be any field between Z and E. Its completion H_{Q_H} may be taken to lie between F_P and E_Q. From (4-10-3) we know that $\mathcal{Z}_{Q/Q_H} = \mathcal{G}(E/H)$. Now, the canonical isomorphism maps $\mathcal{G}(E/Z) = \mathcal{Z}_{Q/P}$ onto $\mathcal{G}(E_Q/F_P)$, and when it is restricted to $\mathcal{Z}_{Q/Q_H}$ we see that $\mathcal{G}(E/H) = \mathcal{Z}_{Q/Q_H}$ is mapped isomorphically onto the group $\{\sigma \in \mathcal{G}(E_Q/F_P) \,|\sigma|H = \text{identity}\}$ $= \{\sigma \in \mathcal{G}(E_Q/F_P) \,|\sigma|H_{Q_H} = \text{identity}\} = \mathcal{G}(E_Q/H_{Q_H})$. We conclude that the restriction to $\mathcal{Z}_{Q/Q_H}$ is identical with the canonical isomorphism for the extension E/H.

Suppose that $H = Z$; then taking the image of $\mathcal{Z}_{Q/Q_Z} = \mathcal{G}(E/Z) = \mathcal{Z}_{Q/P}$ under the canonical isomorphism, we see that $\mathcal{G}(E_Q/Z_{Q_Z}) = \mathcal{G}(E_Q/F_P)$, so that $Z_{Q_Z} = F_P$. Of course, this fact is also an immediate consequence of $e(Q_Z/P) = f(Q_Z/P) = 1$.

Suppose next that $H = V^{(i)}$ for $i = 0, 1, 2, \ldots$. Since $\mathcal{V}^{(i)} = \mathcal{G}(E/V^{(i)}) = \mathcal{Z}_{Q/Q_i}$ we conclude as above that $\mathcal{G}(E_Q/V_i) = \mathcal{V}_i = \mathcal{G}(E_Q/V^{(i)}_{Q_i})$. This means that the completion $V^{(i)}_{Q_i}$ of the ith ramification field $V^{(i)}$ is precisely $V_i = V_i(E_Q/F_P)$, the ith ramification field of the local extension E_Q/F_P.

Under our convention [introduced in the proof of (4-10-5)] that $E_Q = F_P \cdot E$ we have $\mathcal{G}(E_Q/F_P) \approx \mathcal{G}(E/E \cap F_P)$, so that $Z = E \cap F_P$ $= E \cap Z_{Q_Z}$. In the same way, $H = E \cap H_{Q_H}$ for any intermediate field H. These assertions should be taken with a grain of salt, especially in concrete situations, for then it is advisable to assume at the start that $E \cap F_P = F$ and to recall that $E_Q = F_P(\mu E)$ for some composition map μ.

Some further consequences of (4-10-6) are given in the next two results.

4-10-8. Proposition. We have

(i) $$[E:T] = (\mathfrak{J}_{Q/P}:\{1\}) = e\left(\frac{Q}{P}\right)$$

(ii) $$[T:Z] = (\mathcal{Z}_{Q/P}:\mathfrak{J}_{Q/P}) = f\left(\frac{Q}{P}\right)$$

(iii) $[Z:F] = r$ (number of extensions of P to E)

(iv) $$e\left(\frac{Q}{Q_T}\right) = e\left(\frac{Q}{P}\right) \qquad e\left(\frac{Q_T}{Q_Z}\right) = e\left(\frac{Q_Z}{P}\right) = 1$$

(v) $$f\left(\frac{Q}{Q_T}\right) = f\left(\frac{Q_Z}{P}\right) = 1 \qquad f\left(\frac{Q_T}{Q_Z}\right) = f\left(\frac{Q}{P}\right)$$

(vi) $$\mathfrak{Q}_T O_E = \mathfrak{Q}^{e(Q/P)} = \mathfrak{Q}_Z O_E \qquad \mathfrak{Q}_Z O_T = \mathfrak{Q}_T$$

(vii) $$\bar{E} = \bar{T} \qquad [\bar{T}:\bar{Z}] = f\left(\frac{Q}{P}\right) \qquad \bar{Z} = \bar{F}$$

(viii) $$N_{E\to T}\mathfrak{Q} = \mathfrak{Q}_T \qquad N_{T\to Z}\mathfrak{Q}_T = \mathfrak{Q}_Z^{f(Q/P)}$$

Proof. Let $\mathfrak{J}(E_Q/F_P) = \mathfrak{V}_0$ be the inertia group of E_Q/F_P, so that $[E:T] = (\mathfrak{J}_{Q/P}:\{1\}) = (\mathfrak{J}(E_Q/F_P):\{1\}) = e(Q/P)$ and also $[T:Z] = (\mathfrak{Z}_{Q/P}:\mathfrak{J}_{Q/P}) = (\mathfrak{G}(E_Q/F_P):\mathfrak{J}(E_Q/F_P)) = f(Q/P)$. This proves (i) and (ii). Since $[E:F] = re(Q/P)f(Q/P)$, (iii) is immediate. From the case $i = 0$ of (4-10-2) we know that $\mathfrak{Z}_{Q/Q_T} = \mathfrak{J}_{Q/P}$ and $\mathfrak{J}_{Q/Q_T} = \mathfrak{J}_{Q/P}$. Because [by (i) and (ii)] $(\mathfrak{J}_{Q/Q_T}:\{1\}) = e(Q/Q_T)$ and $(\mathfrak{Z}_{Q/Q_T}:\mathfrak{J}_{Q/Q_T}) = f(Q/Q_T)$, (iv) is obvious and (v) follows from (4-10-4). The rest is trivial.

It may also be noted that for any $V^{(i)}$ we have $f(Q_i/P) = f(Q/P)$, $f(Q/Q_i) = 1$, and $e(Q_i/P) = e(Q_i/Q_T)$.

4-10-9. *Proposition.* Let H be a field between F and E; then

(i) $$\mathfrak{Q}_H \text{ is unramified over } F \Leftrightarrow H \subset T \Leftrightarrow \mathfrak{H} \supset \mathfrak{J}_{Q/P}$$

(ii) $$e\left(\frac{Q_H}{P}\right) = f\left(\frac{Q_H}{P}\right) = 1 \Leftrightarrow H \subset Z \Leftrightarrow \mathfrak{H} \supset \mathfrak{Z}_{Q/P}$$

Proof

(i) $e\left(\frac{Q_H}{P}\right) = 1 \Leftrightarrow e\left(\frac{Q}{Q_H}\right) = e\left(\frac{Q}{P}\right) \Leftrightarrow \mathfrak{J}_{Q/Q_H} = \mathfrak{J}_{Q/P} \Leftrightarrow \mathfrak{J}_{Q/P} \subset \mathfrak{H} \Leftrightarrow$ $H \subset T$

(ii) Suppose that $e(Q_H/P) = 1$—that is, $H \subset T$. Then

$$f\left(\frac{Q_H}{P}\right) = 1 \Leftrightarrow f\left(\frac{Q}{Q_H}\right) = f\left(\frac{Q}{P}\right) \Leftrightarrow (\mathfrak{Z}_{Q/Q_H}:\mathfrak{J}_{Q/Q_H}) = f\left(\frac{Q}{P}\right)$$
$$\Leftrightarrow (\mathfrak{Z}_{Q/P} \cap \mathfrak{H}:\mathfrak{J}_{Q/P} \cap \mathfrak{H}) = f\left(\frac{Q}{P}\right)$$
$$\Leftrightarrow (\mathfrak{Z}_{Q/P} \cap \mathfrak{H}:\mathfrak{J}_{Q/P}) = (\mathfrak{Z}_{Q/P}:\mathfrak{J}_{Q/P}) \Leftrightarrow \mathfrak{Z}_{Q/P} \cap \mathfrak{H} = \mathfrak{Z}_{Q/P}$$
$$\Leftrightarrow \mathfrak{Z}_{Q/P} \subset \mathfrak{H} \Leftrightarrow H \subset Z$$

4-10-10. *Exercise.* Suppose that the Galois extension E/F is the splitting field of the subextension H/F; then the prime divisor P of F is unramified in H if and only if it is unramified in E.

Consider the isomorphism and identifications

$$\frac{\mathfrak{Z}_{Q/P}}{\mathfrak{J}_{Q/P}} = \frac{\mathfrak{G}(E_Q/F_P)}{\mathfrak{J}(E_Q/F_P)} \approx \mathfrak{G}\left(\frac{\bar{E}_Q}{\bar{F}_P}\right) = \mathfrak{G}\left(\frac{\bar{E}}{\bar{F}}\right)$$

Suppose that the residue class field $\bar{F}_P = \bar{F}$ is finite with q elements; so $q = \mathfrak{N}\mathfrak{P} = \mathfrak{N}P = \#(\bar{F})$ is a power of the characteristic p of $\bar{F}$.

Then $\mathfrak{Z}_{Q/P}/\mathfrak{J}_{Q/P}$ is a cyclic group of order $f(Q/P)$, and there exist elements $\sigma \in \mathfrak{Z}_{Q/P}$ such that $\bar{\sigma}$ is the canonical generator of $\mathcal{G}(\bar{E}/\bar{F})$; in other words, $\bar{\sigma}: x \to x^q$ for all $x \in \bar{E}$. These elements σ have, therefore, the property that

$$(*) \qquad \sigma\alpha \equiv \alpha^q \pmod{\mathfrak{Q}} \qquad \text{for all } \alpha \in O_E$$

Of course, such a σ is uniquely determined $\Leftrightarrow \mathfrak{J}_{Q/P} = \{1\} \Leftrightarrow E_Q/F_P$ is unramified $\Leftrightarrow \mathfrak{Q}$ is unramified over F; and then

$$\mathfrak{Z}_{Q/P} = \mathcal{G}\left(\frac{E_Q}{F_P}\right) \approx \mathcal{G}\left(\frac{\bar{E}_Q}{\bar{F}_P}\right) = \mathcal{G}\left(\frac{\bar{E}}{\bar{F}}\right)$$

In this case, we denote the unique σ by $\left[\frac{E/F}{\mathfrak{Q}}\right]$ or $\left[\frac{E/F}{Q}\right]$ and call it the ***Frobenius automorphism*** (or ***Frobenius symbol***) of $\mathfrak{Q}$ over F. Clearly, $\left[\frac{E/F}{\mathfrak{Q}}\right]$ is characterized by the property $(*)$—that is, for $\sigma \in \mathfrak{Z}_{Q/P}$

$$\sigma = \left[\frac{E/F}{\mathfrak{Q}}\right] \Leftrightarrow \sigma\alpha \equiv \alpha^{\mathfrak{N}\mathfrak{P}} \pmod{\mathfrak{Q}} \text{ for all } \alpha \in O_E$$

As a matter of fact, this equivalence holds for $\sigma \in \mathcal{G}(E/F)$ because if σ satisfies $(*)$, then $\sigma \in \mathfrak{Z}_{Q/P}$. [To see this, observe that, if $\alpha \in \mathfrak{Q}$, then $\alpha \in O_E$, $\nu_Q(\sigma\alpha - \alpha^q) \geq 1$, $\nu_Q(\alpha) \geq 1$, and $\nu_Q(\alpha^q) \geq 1$; so that $\nu_Q(\sigma\alpha) \geq 1$ and $\sigma\alpha \in \mathfrak{Q}$.]

Now, in the local situation we have [see (3-5-5)] the Frobenius automorphism $[E_Q/F_P] \in \mathcal{G}(E_Q/F_P)$. Since both

$$\left[\frac{E/F}{\mathfrak{Q}}\right] \in \mathfrak{Z}_{Q/P} \qquad \text{and} \qquad \left[\frac{E_Q}{F_P}\right] \in \mathcal{G}\left(\frac{E_Q}{F_P}\right)$$

correspond to the same automorphism of the residue class field extension $\bar{E}/\bar{F} = \bar{E}_Q/\bar{F}_P$ it follows that upon identification of $\mathfrak{Z}_{Q/P}$ and $\mathcal{G}(E_Q/F_P)$, we have

$$\left[\frac{E/F}{\mathfrak{Q}}\right] = \left[\frac{E_Q}{F_P}\right]$$

This means, in particular, that, for $\sigma \in \mathfrak{Z}_{Q/P} = \mathcal{G}(E_Q/F_P)$,

$$\sigma\alpha \equiv \alpha^q \pmod{\mathfrak{Q}} \quad \forall \alpha \in O_E \Leftrightarrow \sigma\alpha \equiv \alpha^q \pmod{\bar{\mathfrak{Q}}} \quad \forall \alpha \in \bar{O}_Q$$

This assertion may also be proved in a straightforward manner.

4-10-11. Proposition. Suppose that the prime ideal $\mathfrak{Q}$ of the Galois extension E/F is unramified over F and that the residue class field $\bar{F}$ is finite with $q = \mathfrak{N}\mathfrak{P}$ elements; then the Frobenius automorphism $\left[\frac{E/F}{\mathfrak{Q}}\right]$ has the following properties:

i. $\left[\frac{E/F}{\mathfrak{Q}}\right]$ is a generator of the cyclic group $\mathfrak{Z}_{Q/P}$ of order $f(Q/P)$. It is characterized by the fact that, for $\sigma \in \mathcal{G}(E/F)$,

$$\sigma = \left[\frac{E/F}{\mathfrak{Q}}\right] \Leftrightarrow \sigma\alpha \equiv \alpha^{\mathfrak{N}\mathfrak{P}} \pmod{\mathfrak{Q}} \text{ for all } \alpha \in O_E$$

ii. For any $\tau \in \mathcal{G}(E/F)$,

$$\left[\frac{E/F}{\tau\mathfrak{Q}}\right] = \tau\left[\frac{E/F}{\mathfrak{Q}}\right]\tau^{-1}$$

iii. If H is any intermediate field, then

$$\left[\frac{E/H}{\mathfrak{Q}}\right] = \left[\frac{E/F}{\mathfrak{Q}}\right]^{f(Q_H/P)}$$

iv. If, moreover, H/F is a Galois extension, then

$$\left[\frac{E/F}{\mathfrak{Q}}\right]\bigg| H = \left[\frac{H/F}{\mathfrak{Q}_H}\right]$$

and

$$\left[\frac{E/F}{\mathfrak{Q}}\right] \in \mathcal{H} \Leftrightarrow f\left(\frac{Q_H}{P}\right) = 1$$

v. Let K be any finite extension of F contained in a fixed algebraic closure Ω of E. Let Q' be any extension of Q to KE, and let P' be its restriction to K. Then $\mathfrak{Q}'$ is unramified over K, and

$$\left[\frac{KE/K}{\mathfrak{Q}'}\right]\bigg| E = \left[\frac{E/F}{\mathfrak{Q}}\right]^{f(P'/P)}$$

Proof. i. This has already been done.

ii. Since $\mathfrak{Q}$ is unramified over F, so is $\tau\mathfrak{Q}$; and

$$\left[\frac{E/F}{\mathfrak{Q}}\right](\tau^{-1}\alpha) \equiv (\tau^{-1}\alpha)^q \pmod{\mathfrak{Q}} \qquad \text{for all } \alpha \in O_E$$

implies that

$$\left(\tau\left[\frac{E/F}{\mathfrak{Q}}\right]\tau^{-1}\right)(\alpha) \equiv \alpha^q \pmod{\tau\mathfrak{Q}} \qquad \text{for all } \alpha \in O_E$$

iii. $\mathfrak{Q}$ is unramified over H, and

$$\left[\frac{E/F}{\mathfrak{Q}}\right]^{f(Q_H/P)}(\alpha) \equiv \alpha^{qf(Q_H/P)} = \alpha^{\mathfrak{N}\mathfrak{Q}_H} \pmod{\mathfrak{Q}} \qquad \text{for all } \alpha \in O_E$$

iv. Note first that $\mathfrak{Q}_H$ is unramified over F, that

$$\sigma = \left[\frac{E/F}{\mathfrak{Q}}\right]\Big| H \in \mathcal{G}\left(\frac{H}{F}\right)$$

and that

$$\left[\frac{E/F}{\mathfrak{Q}}\right]\alpha \equiv \alpha^q \pmod{\mathfrak{Q}} \qquad \text{for all } \alpha \in O_E$$

implies that $\sigma\alpha \equiv \alpha^q \pmod{\mathfrak{Q}_H = \mathfrak{Q} \cap O_H}$ for all $\alpha \in O_H$. Moreover,

$$\left[\frac{E/F}{\mathfrak{Q}}\right] \in \mathfrak{H} \Leftrightarrow \sigma = 1 \Leftrightarrow \left[\frac{H/F}{\mathfrak{Q}_H}\right] = 1 \Leftrightarrow f\left(\frac{Q_H}{P}\right) = 1$$

v. Consider $\sigma \in Z_{Q'/P'}$. If σ is restricted to E, we may view $\sigma \in \mathcal{G}(E/F)$. Since $\sigma\mathfrak{Q}$ lies below $\sigma\mathfrak{Q}' = \mathfrak{Q}'$, it follows that $\sigma\mathfrak{Q} = \mathfrak{Q}$ and $\sigma \in Z_{Q/P}$. Now,

$$\begin{aligned}\sigma \in \mathfrak{I}_{Q'/P'} &\Rightarrow \sigma\alpha \equiv \alpha \pmod{\mathfrak{Q}'} \quad \forall \alpha \in O_{KE}\\ &\Rightarrow \sigma\alpha \equiv \alpha \pmod{\mathfrak{Q}} \quad \forall \alpha \in O_E\\ &\Rightarrow \sigma \in \mathfrak{I}_{Q/P} = \{1\}\\ &\Rightarrow \sigma = \text{identity on } KE\end{aligned}$$

Thus, $\mathfrak{I}_{Q'/P'} = \{1\}$ and $\mathfrak{Q}'$ is unramified over K.

Furthermore, both

$$\tau_1 = \left[\frac{KE/K}{\mathfrak{Q}'}\right]\Big| E \qquad \text{and} \qquad \tau_2 = \left[\frac{E/F}{\mathfrak{Q}}\right]^{f(P'/P)}$$

belong to $Z_{Q/P}$. Since $\tau_1\alpha \equiv \alpha^{\mathfrak{N}\mathfrak{P}'} \pmod{\mathfrak{Q}}$ for all $\alpha \in O_E$ and also $\tau_2\alpha \equiv (\alpha^{\mathfrak{N}\mathfrak{P}})^{f(P'/P)} = \alpha^{\mathfrak{N}\mathfrak{P}'} \pmod{\mathfrak{Q}}$ for all $\alpha \in O_E$, we conclude that $\tau_1 = \tau_2$.

We may leave it to the reader to provide hypotheses and a proof for a statement of type

$$\left[\frac{E_1E_2/F}{\mathfrak{Q}}\right] = \left[\frac{E_1/F}{\mathfrak{Q}_1}\right]\left[\frac{E_2/F}{\mathfrak{Q}_2}\right]$$

4-10-12. Remark. Suppose that the prime ideal $\mathfrak{P}$ of F is unramified in E. According to (4-10-11ii), we may associate with $\mathfrak{P}$ a full class of conjugate elements in $\mathcal{G} = \mathcal{G}(E/F)$—namely, if $\mathfrak{Q}$ is a prime ideal of E lying over F, then take the class

$$\left\{\tau\left[\frac{E/F}{\mathfrak{Q}}\right]\tau^{-1}\,\Big|\, \tau \in \mathcal{G}\right\}$$

This class, which is independent of the choice of $\mathfrak{Q}$ lying over $\mathfrak{P}$, is called the ***Artin class*** or ***Artin symbol*** and is denoted by $\left(\frac{E/F}{\mathfrak{P}}\right)$ or

$\left(\frac{E}{\mathfrak{P}}\right)$. It was Artin who recognized the significance of the map

$$\mathfrak{P} \rightarrow \left(\frac{E}{\mathfrak{P}}\right)$$

for class field theory and the higher reciprocity laws.

The Artin symbol has properties analogous to those of the Frobenius symbol as described in (4-10-11). For example, (iv) becomes

$$\left(\frac{E}{\mathfrak{P}}\right)\bigg| H = \left(\frac{H}{\mathfrak{P}}\right)$$

while (iii) becomes

$$\left(\frac{E/H}{\mathfrak{Q}_H}\right) \subset \left(\frac{E/F}{\mathfrak{P}}\right)^{f(Q_H/P)}$$

We may note that $(E/\mathfrak{P}) = 1$ if and only if $\mathfrak{P}$ splits completely in E; even more, if H/F is a Galois subextension, then $\mathfrak{P}$ splits completely in H if and only if $\left(\frac{E}{\mathfrak{P}}\right) \subset \mathcal{H}$.

Suppose in addition that E/F is abelian. Then

$$\left[\frac{E/F}{\tau\mathfrak{Q}}\right] = \left[\frac{E/F}{\mathfrak{Q}}\right] \qquad \text{for all } \tau \in \mathcal{G}$$

and the Artin class consists of a single element which may be denoted by

$$\left[\frac{E}{\mathfrak{P}}\right] = \left[\frac{E/F}{\mathfrak{P}}\right] = \left(\frac{E/F}{\mathfrak{P}}\right)$$

Since $\mathfrak{P}O_E = \mathfrak{Q}_1 \cdots \mathfrak{Q}_r = \bigcap_1^r \mathfrak{Q}_i$, it is permissible and customary to write

$$\left[\frac{E}{\mathfrak{P}}\right]\alpha \equiv \alpha^{\mathfrak{N}\mathfrak{P}} \pmod{\mathfrak{P}} \qquad \text{for all } \alpha \in O_E$$

EXERCISES

4-1. Suppose that $\mathfrak{A}$ and $\mathfrak{B}$ are ideals of the ordinary arithmetic field $\{F, \mathcal{S}\}$. If $\mathfrak{B}$ is integral, there exists an element a divisible by $\mathfrak{A}$ such that $(a)/\mathfrak{A}$ is prime to $\mathfrak{B}$. If $\mathfrak{A} \subset \mathfrak{B}$, then the number of ideals between them is finite and there exists $b \in \mathfrak{B}$ such that $\mathfrak{B} = \mathfrak{A} + bO$. If $\mathfrak{A}$ is not divisible by any of $\mathfrak{B}_1, \ldots, \mathfrak{B}_n$, then $\mathfrak{A} \not\subset \cup \mathfrak{B}_i$. If $\mathfrak{A}_1, \ldots, \mathfrak{A}_n$ are integral ideals which are relatively prime in pairs and $a_1, \ldots, a_n \in O$, then the simultaneous congruences $x \equiv a_i \pmod{\mathfrak{A}_i}$ have a solution in O.

4-2. In $\{F, \mathcal{S}\}$ consider an integral ideal $\mathfrak{M} = \mathfrak{P}_1^{m_1} \cdots \mathfrak{P}_r^{m_r}$ with $r \geq 1$,

$m_i \geq 1$, and $\{P_1, \ldots, P_r\} \neq \mathcal{S}$. The following conditions on an element $a \in F$ are equivalent:

a. $a = b/c$, where $b, c \in O$ and $(c, \mathfrak{M}) = 1$.

b. $\nu_{P_i}(a) \geq 0$ for $i = 1, \ldots, r$.

c. $a = b/c$, where $b, c \in O$ and $c \equiv 1 \pmod{\mathfrak{M}}$.

The set of all such elements is an over-ring O' of O; it is called the ***ring of*** $\mathfrak{M}$ ***integers.*** O' is a principal ideal domain, and for each $i = 1, \ldots, r$ there exists $\pi_i \in O$ such that $\mathfrak{P}_i' = \mathfrak{P}_i O' = \pi_i O'$ [notation as in (4-5-5)]. If $\mathfrak{A} \in \mathfrak{J}\{\mathcal{S}\}$, then

$$\mathfrak{A}' = \{b/c | b \in \mathfrak{A}, c \in O, (c, \mathfrak{M}) = 1\}$$

4-3. With the notation as in (4-5-5) consider $O = O\{\mathcal{S}\} \prec O' = O\{\mathcal{S}'\}$. If $\mathfrak{A}$ is an integral $\mathcal{S}$ ideal and $P \in \mathcal{S}'$ whenever $\nu_P(\mathfrak{A}) > 0$, then $O/\mathfrak{A} \approx O'/\mathfrak{A}'$.

4-4. Suppose that $\mathfrak{A}$ and $\mathfrak{B}$ are ideals of $\{F, \mathcal{S}\}$ and that $\mathfrak{B}$ is integral; then the additive groups $O/\mathfrak{B}$ and $\mathfrak{A}/\mathfrak{A}\mathfrak{B}$ are isomorphic. This may be used to show that the counting norm $\mathfrak{N}$ is multiplicative.

In Exercises 4-5 through 4-7 we are concerned with an ordinary arithmetic field $\{F, \mathcal{S}(F)\}$ and an extension $\{E, \mathcal{S}(E)\}$ such that $[E:F] = n$.

4-5. *a.* Given $\mathfrak{A} \in \mathfrak{J}\{\mathcal{S}(E)\}$, what are $\mathfrak{A} \cap F$ and $(\mathfrak{A} \cap F)O_E$?

b. Describe $N_{E\to F}(\mathfrak{A})$ in terms of the various $N_{E\to F}(\mathfrak{A}O_P)$, $P \in \mathcal{S}(F)$, and also in terms of the local norms $N_{E_Q \to F_P}$.

c. If $\mathfrak{P}$ is a prime ideal of F and $\alpha \in O_E$ is prime to $\mathfrak{P}O_E$, then $N_{E\to F}(\alpha)$ is prime to $\mathfrak{P}$.

4-6. Suppose that O_F is a principal ideal domain; then O_E is contained in a finitely generated O_F module $\Leftrightarrow$ O_E is a finitely generated O_F module $\Leftrightarrow$ there exists an integral basis for O_E over O_F.

4-7. Suppose that $\mathcal{S}(F)$ consists of a single prime divisor P and that we write $\mathcal{S}(E) = \{Q_1, \ldots, Q_r\}$. Note that O_F is a principal ideal domain, $\mathfrak{P} = \mathcal{P}$, and $\mathfrak{P}O_E = \mathfrak{Q}_1^{e(Q_1/P)} \cdots \mathfrak{Q}_r^{e(Q_r/P)}$.

a. If $\mathfrak{A}$ is an integral O_E ideal containing $\mathcal{P}O_E$, then $O_E/\mathfrak{A}$ may be viewed as a vector space over $O_F/\mathcal{P}$. If $\mathfrak{B}$ is also such an integral O_E ideal and $\mathfrak{A}\mathfrak{B} \supset \mathcal{P}O_E$, then

$$\dim\left(\frac{O_E}{\mathfrak{A}\mathfrak{B}}\right) = \dim\left(\frac{O_E}{\mathfrak{A}}\right) + \dim\left(\frac{O_E}{\mathfrak{B}}\right)$$

It follows that

$$\dim\left(\frac{O_E}{\mathcal{P}O_E}\right) = \sum_1^r e\left(\frac{Q_i}{P}\right) f\left(\frac{Q_i}{P}\right)$$

b. If O_E has an integral basis over O_F—say, $\omega_1, \ldots, \omega_n$—then $\dim(O_E/\mathcal{P}O_E) = n$ and $\omega_1, \ldots, \omega_n$ are representatives of a basis. Conversely, if $\dim(O_E/\mathcal{P}O_E) = n$ and $\omega_1, \ldots, \omega_n \in O_E$ are representatives of a basis, then $\omega_1, \ldots, \omega_n$ form an integral basis for O_E over O_F.

c. We have $g_i = g(Q_i/P) = 1$ for $i = 1, \ldots, r \Leftrightarrow \sum_1^r n_i = n \Leftrightarrow \sum_1^r e_i f_i = n \Leftrightarrow$ $\dim(O_E/\mathcal{P}O_E) = n \Leftrightarrow$ there exists an integral basis for O_E over $O_F \Leftrightarrow O_E$ is a finitely generated module over O_F. One may note, in particular, that, if E/F is separable, there exists an integral basis for O_E over O_F.

4-8. The algebraic closure Ω of $\mathbf{Q}$ is known as the field of all algebraic numbers, and the integral closure O_Ω of $\mathbf{Z}$ in Ω is known as the ring of all algebraic integers. Ω is the quotient field of O_Ω. If F is any algebraic number field, then $O_\Omega \cap F = O_F$, the ring of integers of F, and $O_\Omega = \bigcup O_F$ where F runs over all algebraic number fields. The domain O_Ω has no irreducible elements. Using the fact [to be proved in (5-3-11)] that an algebraic number field has finite class number, it is easy to show that any two algebraic integers have a greatest common divisor. More precisely, given $\alpha \neq 0$, $\beta \neq 0 \in O_\Omega$, there exist δ, ξ, $\eta \in O_\Omega$ such that $\delta|\alpha$, $\delta|\beta$, and $\delta = \xi\alpha + \eta\beta$.

4-9. *a.* Consider the quadratic extension $E/F = \mathbf{Q}(\sqrt{m})/\mathbf{Q}$ with $m \in \mathbf{Z}$, m square-free. Show that $\{1, \sqrt{m}\}$ is not an integral basis $\Leftrightarrow \Delta_{E/F} = m \Leftrightarrow \dfrac{1+\sqrt{m}}{2}$ is integral $\Leftrightarrow \dfrac{1-m}{4} \in \mathbf{Z} \Leftrightarrow m \equiv 1 \pmod 4 \Leftrightarrow \left\{1, \dfrac{1+\sqrt{m}}{2}\right\}$ is an integral basis, and that $\{1, \sqrt{m}\}$ is an integral basis $\Leftrightarrow \Delta_{E/F} = 4m \Leftrightarrow m \equiv 2$ or $3 \pmod 4$. [*Hint:* Use (4-8-15).]

b. Is $\{\frac{1}{2}(3 + 7\sqrt{5}), \frac{1}{2}(-1 - 3\sqrt{5})\}$ an integral basis for $\mathbf{Q}(\sqrt{5})$?

c. Show that $\mathbf{Q}(\sqrt{-23})$ and $\mathbf{Q}(\sqrt{-89})$ do not have class number 1.

d. Find an infinite number of units in the ring of integers of $\mathbf{Q}(\sqrt{2})$.

4-10. Prove the following assertions in $\mathbf{Q}(\sqrt{-5})$:

a. $3 + \sqrt{-5}$ and $8 + \sqrt{-5}$ are relatively prime.

b. The least common multiple of $\mathfrak{A} = (2, 1 + \sqrt{-5})$ and $\mathfrak{B} = (3, 1 + \sqrt{-5})$ is $(1 + \sqrt{-5})$.

c. Find an $\alpha \in \mathfrak{A}$ such that $(\alpha)/\mathfrak{A}$ is relatively prime to $\mathfrak{B}$.

d. For a, $b \in \mathbf{Z}$, $(a, b + \sqrt{-5}) = \mathbf{Z}a \oplus \mathbf{Z}(b + \sqrt{-5}) \Leftrightarrow N(b + \sqrt{-5})$ is divisible by a.

e. $(23, 8 + \sqrt{-5})$ is a prime ideal.

f. $(2, 1 + \sqrt{-5})$ is not principal.

g. $(21, 9 + 3\sqrt{-5}, -2 + 4\sqrt{-5})$ is principal.

h. Find the greatest common divisor of $(3 - \sqrt{-5})$ and $(1 + 2\sqrt{-5})$, and show that it is a prime ideal.

4-11. Consider the fifth degree field $E = \mathbf{Q}(\alpha)$, where α is a root of the irreducible polynomial $f(x) = x^5 + 20x + 16$.

a. Show that $\Delta(1, \alpha, \alpha^2, \alpha^3, \alpha^4) = (5^6)(16^4)$, so that $\{1, \alpha, \alpha^2, \alpha^3, \alpha^4\}$ is an integral basis at all primes p except possibly 2 or 5.

b. 3 remains prime in E.

c. 5 has a unique extension with $e = 5$ and $f = 1$.

d. 7 has three extensions with $e_1 = e_2 = e_3 = 1$, $f_1 = f_2 = 1$, $f_3 = 3$.

e. 2 has two extensions with $e_1 = 4$, $e_2 = 1$, $f_1 = f_2 = 1$. [*Hint:* One way to do this is to put $\beta = \alpha^2/2$, $\gamma = (\alpha^3 + 2\alpha)/4$. The $\mathbf{Z}$ module $M = \mathbf{Z} + \mathbf{Z}\alpha + \mathbf{Z}\beta + \mathbf{Z}\gamma + \mathbf{Z}\beta^2$ is then a ring containing $\mathbf{Z}[\alpha]$. Moreover, $M \subset O_E$ because it is finitely generated over $\mathbf{Z}$. The field equation for $\gamma - 1$ is a polynomial with a Newton polygon which gives the desired information.]

4-12. Is there an ideal in $\mathbf{Q}(\sqrt{2})$ equal to the principal prime ideal $(1 + i)$ of $\mathbf{Q}(i)$?

4-13. When E/F is a Galois extension, an ideal $\mathfrak{A} \in \mathfrak{J}\{\mathfrak{S}(E)\}$ is said to be ***invariant*** when $\sigma\mathfrak{A} = \mathfrak{A}$ for all $\sigma \in \mathcal{G}(E/F)$. If $\mathfrak{A}$ is invariant and $n = [E:F]$, then $\mathfrak{A}^n \in \mathfrak{J}\{\mathfrak{S}(F)\}$. The different $\mathfrak{D}_{E/F}$ is an invariant ideal, and

$$N_{E\to F}(\mathfrak{D}_{E/F}) = \mathfrak{D}^n_{E/F} = \Delta_{E/F}$$

5

Global Fields

If one wishes to study the units and the class number of an algebraic number field, it is necessary to make use of the archimedean primes. For this and many other reasons one is led to consider an algebraic number field with all its prime divisors—or, better still, all possible completions of an algebraic number field. Of course, the theory of ordinary arithmetic fields is not lost in the process. We shall go one step further and make the formulation sufficiently general to include the function fields for which class field theory holds.

5-1. Global Fields and the Product Formula

This chapter is devoted to the study of global fields. By a ***global field*** we mean a field which is either (1) an algebraic number field, or (2) an algebraic function field in one variable over a finite constant field.

In this section we fix some notation and isolate certain simple properties of global fields which will enable us to treat both classes of global fields simultaneously.

First let us note that the statement that F is a global field of type 2 means that there exist in F a finite subfield k and an element x transcendental over k such that $[F:k(x)] < \infty$. Of course, a subfield $k(x)$ with these properties is not unique; however, any such subfield $k(x)$ will be called a ***rational subfield*** of F (or simply a ***rational field***) and will be denoted by R. Naturally, when F is a number field, there is a unique ***rational field*** R, namely, $R = \mathbf{Q}$.

Given a global field F, we denote the set of all prime divisors of F (excluding the trivial one) by $\mathfrak{M}(F)$. Suppose that R is a rational subfield; then $\mathfrak{M}(F)$ is the set of extensions to F of the primes in $\mathfrak{M}(R)$ and $\mathfrak{M}(R)$ is the set of restrictions to R of the primes in $\mathfrak{M}(F)$. If $R = \mathbf{Q}$, then $\mathfrak{M}(R)$ consists of the archimedean prime ∞, which is called the ***infinite prime,*** and a countably infinite collection of discrete primes, which are known as ***finite primes.*** If $R = k(x)$, then, because k is finite, $\mathfrak{M}(R)$ is the set of all prime divisors of R which are trivial on k. Thus, according to (1-4-5), $\mathfrak{M}(R)$ consists of a countably infinite collection of discrete prime divisors; one of these (which one depends on the choice of the generator x of R over k) is called the ***infinite prime*** ∞, while the others (these correspond to the irreducible polynomials in $k[x]$) are called the ***finite primes.*** In either case, we put $\mathfrak{M}_\infty(R) = \{\infty\}$ and let $\mathfrak{M}_0(R)$ denote the set of all finite primes of R. The same notation may be carried over to the global field F. The elements of the finite set $\mathfrak{M}_\infty(F) = \{P \in \mathfrak{M}(F) | P \supset \infty\}$ are known as the ***infinite primes*** of F, and the elements of the countably infinite set $\mathfrak{M}_0(F)$ of primes lying over $\mathfrak{M}_0(R)$ are known as the ***finite primes*** of F. In particular, we have a disjoint union $\mathfrak{M}(F) = \mathfrak{M}_0(F) \cup \mathfrak{M}_\infty(F)$. The fact that $\mathfrak{M}_0(F)$ and $\mathfrak{M}_\infty(F)$ are not really unique for function fields will cause no difficulty, because $R = k(x)$ will usually be fixed.

For almost all $P \in \mathfrak{M}(F)$ we know that P is discrete and that the residue class field is finite with $\mathfrak{N}P$ elements. For archimedean P, it is convenient to put $\mathfrak{N}P = 1$. It is clear that, for any $M > 0$, $\{P \in \mathfrak{M}(F) | \mathfrak{N}P \leq M\}$ is a finite set. A more significant observation is that $\{F, \mathfrak{M}_0(F)\}$ is an ordinary arithmetic field; more will be said about this in Section 5-2.

Some rather deep properties of global fields are connected with the existence of a product formula. Here we shall simply make precise what is meant by the product formula and show that a global field has a product formula.

Suppose that F is an arbitrary field and that $\mathfrak{M}(F)$ is a collection of prime divisors of F. We say that $\{F, \mathfrak{M}(F)\}$ is a ***product formula field*** if there exists for each $P \in \mathfrak{M}(F)$ a valuation $\varphi_P \in P$ such that, for any $a \neq 0 \in F$, we have

(i) $$\varphi_P(a) \leq 1 \text{ for almost all } P \in \mathfrak{M}(F)$$

(ii) $$\prod_{P \in \mathfrak{M}(F)} \varphi_P(a) = 1$$

The product in (ii) is well defined because condition (i) is equivalent to the requirement that $\varphi_P(a) = 1$ for almost all $P \in \mathfrak{M}(F)$.

5-1-1. Proposition. Suppose that $\{F, \mathfrak{M}(F)\}$ is a product formula field. Let E/F be a finite extension, and let $\mathfrak{M}(E)$ denote the extensions to E of the primes in $\mathfrak{M}(F)$; then $\{E, \mathfrak{M}(E)\}$ is a product formula field.

Proof. The product formula in F is given in terms of $\{\varphi_P | P \in \mathfrak{M}(F)\}$. For each $Q \in \mathfrak{M}(E)$ choose φ_Q in the manner prescribed in (2-5-1). Then, under the assumption that (i) holds, we have, for any $\alpha \neq 0 \in E$,

$$\begin{aligned}\prod_{Q \in \mathfrak{M}(E)} \varphi_Q(\alpha) &= \prod_{P \in \mathfrak{M}(F)} \Big(\prod_{Q \supset P} \varphi_Q(\alpha)\Big) \\ &= \prod_{P \in \mathfrak{M}(F)} \varphi_P[N_{E \to F}(\alpha)] \\ &= 1\end{aligned}$$

It remains to verify (i). If $P \in \mathfrak{M}(F)$ is archimedean, then $F \supset \mathbf{Q}$ and $\varphi_P(2) > 1$; therefore, $\mathfrak{M}(F)$ [and consequently $\mathfrak{M}(E)$ too] can include only a finite number of archimedean primes. Since we are concerned with almost all primes in $\mathfrak{M}(E)$, the archimedean primes may be excluded. For $\alpha \neq 0 \in E$, consider its minimum polynomial

$$f(\alpha, F) = x^m + a_1 x^{m-1} + \cdots + a_m$$

From the product formula in F, it follows that for almost all $P \in \mathfrak{M}(F)$ we have $\varphi_P(a_i) \leq 1$ for $i = 1, \ldots, m$. For such P, α is integral over $O_P = \{a \in F | \varphi_P(a) \leq 1\}$. According to (2-5-4), this means that $\alpha \in \bigcap_{Q \supset P} O_Q$, so that $\varphi_Q(\alpha) \leq 1$ for almost all $Q \in \mathfrak{M}(E)$.

5-1-2. Proposition. If F is a global field, then $\{F, \mathfrak{M}(F)\}$ is a product formula field. Moreover, the valuations φ_P that appear in the product formula may be taken to be of form

$\varphi_P(a) = \lvert a \rvert$	P real archimedean
$\varphi_P(a) = \lvert a \rvert^2$	P complex archimedean
$\varphi_P(a) = (\mathfrak{N}P)^{-\nu_P(a)}$	P discrete

Proof. If F is a rational field, the assertions follow, from Section 1-4. For an arbitrary global field, one uses (5-1-1) to move up from a rational subfield. To check that the φ_P's then have the proper form, one uses (2-5-1) and the fact that all g's are 1 (by separability and Exercise 5-1).

As a rule, our proofs will depend only on the form of the φ_P's and on the product formula; that is, it will not be necessary to go back to the definition of a global field. Underlying this is the fact (proved in [3]) that, if $\{F, \mathfrak{M}(F)\}$ is any product formula field and there exists in $\mathfrak{M}(F)$ a prime which is either archimedean or discrete with finite residue class field, then F is a global field and $\mathfrak{M}(F)$ is the set of all prime divisors of F.

5-2. Adèles, Idèles, Divisors, and Ideals

Let F be a global field, and consider the product space $\prod_{P \in \mathfrak{M}(F)} F_P$. With componentwise algebraic operations and the product topology, it is a topological ring with identity. By making use of the various canonical imbeddings $\mu_P : F \to F_P$, F may be imbedded isomorphically in ΠF_P along the diagonal, and, upon identification of F with its image, we have $1 \in F \subset \Pi F_P$. Because emphasis will be placed on the completions F_P, the prime divisor of F_P will be denoted by P and objects associated with F_P will be indexed by P. Thus, for example, when P is nonarchimedean, ν_P will denote the normalized exponential valuation of F_P and $\bar{F}_P$ will denote the residue class field; we shall use O_P, $\mathcal{P}$, and U_P for the ring of integers, prime ideal, and group of units of F_P, respectively.

Clearly, ΠF_P contains all information about F that its prime divisors have to offer; however, it is too big for the treatment of arithmetical questions in that its elements do not behave enough like field elements. The search for a subset of ΠF_P of proper size leads to the following definition: Let $\mathbf{F}$ denote the subset of ΠF_P consisting of all functions f on $\mathfrak{M}(F)$ such that

(i) $\qquad f(P) \in F_P$ for all $P \in \mathfrak{M}(F)$

(ii) $\qquad f(P) \in O_P$ for almost all $P \in \mathfrak{M}(F)$

It will be convenient to denote such a function by a (or by any of the other symbols usually reserved for elements of F) and to let $a_P \in F_P$ be the value of the function at any $P \in \mathfrak{M}(F)$. An element $a \in \mathbf{F}$ is called a ***valuation vector***, or ***adèle*** of F—it is distinguished by the property that $\varphi_P(a_P) \leq 1$ for almost all $P \in \mathfrak{M}(F)$. Of course, the algebraic operations in $\mathbf{F}$ are those induced from ΠF_P, and $\mathbf{F}$ is clearly a commutative ring with identity. It is called the ***adèle ring,*** or the ***ring of valuation vectors,*** of F. The discussion of topological questions in $\mathbf{F}$ is deferred until Section 5-5.

If $a \in F$, then under the isomorphic imbedding of F in ΠF_P, a is identified with the element of $\mathbf{F}$ (naturally, this element is also denoted by a) for which $a_P = \mu_P a = a \in F_P$ for every $P \in \mathfrak{M}(F)$. We have $1 \in F \subset \mathbf{F} \subset \Pi F_P$. The elements of F are then called ***principal valuation vectors,*** or simply ***principal adèles,*** of F.

The units of the ring $\mathbf{F}$ are called ***idèles*** of F. The multiplicative group of units is denoted by $\mathbf{F}^*$ and is called the ***idèle group*** of F. It is immediate from the definitions that, if $a \in \mathbf{F}$, then

$$a \in \mathbf{F}^* \Leftrightarrow \begin{cases} \text{(i)} \ a_P \in F_P^* \text{ for all } P \in \mathfrak{M}(F) \\ \text{(ii)} \ a_P \in U_P \text{ for almost all } P \in \mathfrak{M}(F) \end{cases}$$

We have $1 \in F^* \subset \mathbf{F}^*$; that is, F^* is a subgroup of $\mathbf{F}^*$. In fact, if $a \in F^*$, then both a and a^{-1} belong to O_P for almost all P; hence, $a = \mu_P a \in U_P$ for almost all P, and $a \in \mathbf{F}^*$. The elements of F^* (when viewed in $\mathbf{F}^*$) are called the ***principal idèles*** of F; the factor group $\mathbf{F}^*/F^*$ is known as the ***idèle class group*** of F.

It should be remarked that idèles could have been defined directly by taking a multiplicative approach parallel to that for $\mathbf{F}$—that is, by replacing ΠF_P by ΠF_P^*. In fact, both situations may be subsumed under a single general principle; namely, suppose that to every P in some indexing set $\mathfrak{M}$ there is associated an algebraic object A_P and that for almost all $P \in \mathfrak{M}$ we have in addition an algebraic object $B_P \subset A_P$. We may then form the ***restricted direct product*** $\Pi' A_P$—that is, the functions on $\mathfrak{M}$ which take values in A_P for all P and in B_P for almost all P. In particular, if $B_P = A_P$ for all $P \in \mathfrak{M}$, then the restricted direct product is simply the direct product ΠA_P, while if B_P consists solely of the identity element of A_P for all $P \in \mathfrak{M}$, then $\Pi' A_P$ is the algebraic direct sum $\Sigma \oplus A_P$.

Historically, the idèles were introduced first—by Chevalley, in 1936. Some fifteen years then elapsed before the adèles were defined and exploited.

Returning to the global field F, we let S denote any finite subset of $\mathfrak{M}(F)$ such that $S \supset \mathfrak{M}_\infty(F)$; the collection of all such sets S will be denoted by Σ. Consider

$$\begin{aligned} \mathbf{F}_S &= \prod_{P \in S} F_P \times \prod_{P \notin S} O_P \\ &= \{a \in \mathbf{F} \mid \varphi_P(a_P) \leq 1 \quad \forall P \notin S\} \end{aligned}$$

This is a subring of $\mathbf{F}$, containing 1; it is called the ***ring of S integers of $\mathbf{F}$***. Note that $S \subset S'$ implies $\mathbf{F}_S \subset \mathbf{F}_{S'}$ and that

$$\mathbf{F} = \bigcup_{S \in \Sigma} \mathbf{F}_S$$

The group of units of the ring $\mathbf{F}_S$ is known as the ***group of S idèles of F*** and is denoted by $\mathbf{F}_S^*$. It is immediate that

$$\begin{aligned} \mathbf{F}_S^* &= \prod_{P \in S} F_P^* \times \prod_{P \notin S} U_P \\ &= \{a \in \mathbf{F}^* \mid \varphi_P(a_P) = 1 \quad \forall P \notin S\} \end{aligned}$$

and

$$\mathbf{F}^* = \bigcup_{S \in \Sigma} \mathbf{F}_S^*$$

Of course, $S \subset S'$ implies $\mathbf{F}_S^* \subset \mathbf{F}_{S'}^*$. The objects F_S and F_S^* are defined by

$$F_S = F \cap \mathbf{F}_S = \{a \in F | \varphi_P(a) \leq 1 \quad \forall P \notin S\}$$

$$F_S^* = F^* \cap \mathbf{F}_S^* = \{a \in F^* | \varphi_P(a) = 1 \quad \forall P \notin S\}$$

F_S is a subring of F, containing the identity, and is called the ***ring of*** S ***integers of*** F; F_S^* is the group of units of F_S and is called the ***group of*** S ***units of*** F. Clearly,

$$F = \bigcup_S F_S \qquad \text{and} \qquad F^* = \bigcup_S F_S^*$$

It should be noted that F_S and F_S^* have already been encountered. More precisely, if $\mathcal{S}$ denotes the complement of S in $\mathfrak{M}(F)$, then $\{F, \mathcal{S}\}$ is an OAF for which F_S is the ring of $\mathcal{S}$ integers and F_S^* is the group of $\mathcal{S}$ units.

As was done in the discussion of ordinary arithmetic fields, let us form the divisor group $\mathbf{D}_S = \mathbf{D}\{\mathcal{S}\}$. Its elements, which we call the S ***divisors of*** F, are of form

$$A = \prod_{P \notin S} P^{\nu_P(A)}$$

with $\nu_P(A) = 0$ for almost all $P \in \mathcal{S}$. It is easy to provide the connection between idèles and S divisors; for this, it is convenient to extend our valuations to $\mathbf{F}$. Thus, when P is discrete, we put $\nu_P(a) = \nu_P(a_P)$ for $a \in \mathbf{F}$, and similarly for any $P \in \mathfrak{M}(F)$ we put $\varphi_P(a) = \varphi_P(a_P)$.

Given any $a \in \mathbf{F}^*$, we may associate with it the element

$$A = \prod_{P \notin S} P^{\nu_P(a)} \in \mathbf{D}_S$$

The mapping $a \to A$ is clearly a homomorphism of $\mathbf{F}^*$ onto $\mathbf{D}_S$ with kernel $\mathbf{F}_S^*$. In other words, we have isomorphic groups

$$\frac{\mathbf{F}^*}{\mathbf{F}_S^*} \approx \mathbf{D}_S$$

Since $F^* \subset \mathbf{F}^*$, the homomorphism above may be restricted to F^*. The image of F^* is known as the group of ***principal*** S ***divisors of*** F; we denote it by $\mathbf{D}_S^* = \mathbf{D}^*\{\mathcal{S}\}$. The kernel is now $F^* \cap \mathbf{F}_S^* = F_S^*$, and we have the isomorphism

$$\frac{F^*}{F_S^*} \approx \mathbf{D}_S^*$$

which has already been encountered in Chapter 4.

Under the homomorphism of $\mathbf{F}^*$ onto $\mathbf{D}_S$, the complete inverse image of $\mathbf{D}_S^*$ is precisely $F^*\mathbf{F}_S^* = \{ab|a \in F^*, b \in \mathbf{F}_S^*\}$, so that

$$\frac{F^*\mathbf{F}_S^*}{\mathbf{F}_S^*} \approx \mathbf{D}_S^*$$

Of course, by a standard isomorphism theorem we have

$$\mathbf{D}_S^* \approx \frac{F^*}{F_S^*} = \frac{F^*}{F^* \cap \mathbf{F}_S^*} \approx \frac{F^*\mathbf{F}_S^*}{\mathbf{F}_S^*}$$

and it follows that

$$\frac{\mathbf{F}^*}{F^*\mathbf{F}_S^*} \approx \frac{\mathbf{D}_S}{\mathbf{D}_S^*}$$

where the right side is known as the group of S ***divisor classes of*** F.

Divisors are formal objects and not very exciting in themselves. Things become more interesting when we introduce ideals of the adèle ring. For any $a \in \mathbf{F}^*$ consider the $\mathbf{F}_S$ module $a\mathbf{F}_S = \{ab|b \in \mathbf{F}_S\}$. It is immediate that

$$a\mathbf{F}_S = \prod_{P \in S} F_P \times \prod_{P \notin S} \mathcal{P}^{\nu_P(a)}$$

Although it may appear somewhat artificial, it turns out to be very convenient to call such sets $a\mathbf{F}_S$ the S ***ideals of*** $\mathbf{F}$. One observes that any set of form

$$\prod_{P \in S} F_P \times \prod_{P \notin S} \mathcal{P}^{m_P}$$

with $m_P = 0$ for almost all P is an S ideal of $\mathbf{F}$. The collection of all S ideals of $\mathbf{F}$ will be denoted by $\mathbf{I}_S$. By a ***principal*** S ***ideal of*** $\mathbf{F}$ we shall mean an S ideal which can be expressed in the form $a\mathbf{F}_S$ with $a \in F^*$; the set of all these will be denoted by $\mathbf{I}_S^*$.

We have already treated the product $F^*\mathbf{F}_S^* = \{ab|a \in F^*, b \in \mathbf{F}_S^*\}$. This definition is the usual one for the product of the multiplicative groups F^* and $\mathbf{F}_S^*$ in the ring $\mathbf{F}$. On the other hand, if M_1 and M_2 are additive subgroups of $\mathbf{F}$, then by their product one means

$$M_1M_2 = \left\{\sum_i m_i^{(1)}m_i^{(2)} | m_i^{(1)} \in M_1, m_i^{(2)} \in M_2\right\}$$

It is clear that the product of two S ideals of $\mathbf{F}$ is an S ideal—in fact, if $a, b \in \mathbf{F}^*$, then

$$(a\mathbf{F}_S)(b\mathbf{F}_S) = ab\mathbf{F}_S$$

It follows that $\mathbf{I}_S$ is a group under this multiplication—its identity element is $\mathbf{F}_S = 1\mathbf{F}_S$ and $(a\mathbf{F}_S)^{-1} = a^{-1}\mathbf{F}_S$. Furthermore, $\mathbf{I}_S^*$ is a subgroup of $\mathbf{I}_S$. The factor group $\mathbf{I}_S/\mathbf{I}_S^*$ is called the ***group of S ideal classes of* F.**

The choice of definition for S ideals makes it trivial to relate them to the idèles. One simply uses the homomorphism of $\mathbf{F}^*$ onto $\mathbf{I}_S$ given by $a \to a\mathbf{F}_S$. Since $\mathbf{F}_S$ is the identity of $\mathbf{I}_S$, the kernel of this homomorphism is $\mathbf{F}_S^*$ and

$$\frac{\mathbf{F}^*}{\mathbf{F}_S^*} \approx \mathbf{I}_S$$

Consequently, $\mathbf{D}_S$ and $\mathbf{I}_S$ are isomorphic; that is,

$$\mathbf{D}_S \approx \frac{\mathbf{F}^*}{\mathbf{F}_S^*} \approx \mathbf{I}_S$$

and the isomorphism may be given explicitly by tracing the isomorphisms out of which it is composed. Thus, if $A \in \mathbf{D}_S$ is given and one chooses any $a \in \mathbf{F}^*$ such that $\nu_P(a) = \nu_P(A)$ for all $P \notin S$, then $a\mathbf{F}_S \in \mathbf{I}_S$ ($a\mathbf{F}_S$ is independent of the choice of such an $a \in \mathbf{F}^*$) and

$$A = \prod_{P \notin S} P^{\nu_P(A)} \to a\mathbf{F}_S = \prod_{P \in S} F_P \times \prod_{P \notin S} \mathcal{P}^{\nu_P(A)}$$

The inverse isomorphism is given by

$$a\mathbf{F}_S \to \prod_{P \notin S} P^{\nu_P(a)}$$

(note that it is well defined).

Of course, we also have the induced isomorphisms

$$\mathbf{D}_S^* \approx \frac{F^*}{F_S^*} \approx \mathbf{I}_S^* \qquad \text{and} \qquad \frac{\mathbf{D}_S}{\mathbf{D}_S^*} \approx \frac{\mathbf{F}^*}{F^*\mathbf{F}_S^*} \approx \frac{\mathbf{I}_S}{\mathbf{I}_S^*}$$

Furthermore, the isomorphisms between $\mathbf{I}_S$ and $\mathbf{D}_S$ preserve joins and meets—that is, if $a\mathbf{F}_S \leftrightarrow A$ and $b\mathbf{F}_S \leftrightarrow B$, then

$$a\mathbf{F}_S + b\mathbf{F}_S \leftrightarrow A + B \qquad \text{and} \qquad a\mathbf{F}_S \cap b\mathbf{F}_S \leftrightarrow A \cap B$$

In fact, these correspondences follow from

$$\begin{aligned} a\mathbf{F}_S + b\mathbf{F}_S &= \{\alpha + \beta \mid \alpha \in a\mathbf{F}_S,\ \beta \in b\mathbf{F}_S\} \\ &= \prod_{P \in S} F_P \times \prod_{P \notin S} \mathcal{P}^{\nu_P(a)} + \prod_{P \in S} F_P \times \prod_{P \notin S} \mathcal{P}^{\nu_P(b)} \\ &= \prod_{P \in S} F_P \times \prod_{P \notin S} (\mathcal{P}^{\nu_P(a)} + \mathcal{P}^{\nu_P(b)}) \\ &= \prod_{P \in S} F_P \times \prod_{P \notin S} \mathcal{P}^{\min\{\nu_P(a),\, \nu_P(b)\}} \end{aligned}$$

and

$$\begin{aligned} a\mathbf{F}_S \cap b\mathbf{F}_S &= \prod_{P \in S} F_P \times \prod_{P \notin S} (\mathcal{P}^{\nu_P(a)} \cap \mathcal{P}^{\nu_P(b)}) \\ &= \prod_{P \in S} F_P \times \prod_{P \notin S} \mathcal{P}^{\max\{\nu_P(a),\, \nu_P(b)\}} \end{aligned}$$

Thus far, we have not made use of the fact that $\{F, \mathcal{S}\}$ is an OAF. When this is done and we write $\mathfrak{J}_S = \mathfrak{J}\{\mathcal{S}\}$ and $\mathfrak{J}_S^* = \mathfrak{J}^*\{\mathcal{S}\}$, then combining (4-3-2) with the discussion above yields the following proposition.

5-2-1. Proposition. If F is a global field, then for any $S \in \Sigma$ the groups $\mathfrak{J}_S$ and $\mathbf{I}_S$ are lattice-isomorphic. We have

$$\begin{aligned} \mathfrak{J}_S &\approx \mathbf{D}_S \approx \frac{\mathbf{F}^*}{\mathbf{F}_S^*} \approx \mathbf{I}_S \\ \mathfrak{J}_S^* &\approx \mathbf{D}_S^* \approx \frac{F^*}{F_S^*} \approx \mathbf{I}_S^* \\ \frac{\mathfrak{J}_S}{\mathfrak{J}_S^*} &\approx \frac{\mathbf{D}_S}{\mathbf{D}_S^*} \approx \frac{\mathbf{F}^*}{F^*\mathbf{F}_S^*} \approx \frac{\mathbf{I}_S}{\mathbf{I}_S^*} \end{aligned}$$

It turns out that the two (inverse) isomorphisms between $\mathfrak{J}_S$ and $\mathbf{I}_S$ have natural algebraic descriptions; this will be shown in (5-2-5).

5-2-2. Proposition. If $a, b \in \mathbf{F}^*$, then

$$b\mathbf{F}_S \subset a\mathbf{F}_S \Leftrightarrow \nu_P(a) \le \nu_P(b) \quad \forall P \notin S$$

Moreover, in such a case

$$(a\mathbf{F}_S : b\mathbf{F}_S) = \prod_{P \notin S} (\mathfrak{N}P)^{-\nu_P(a/b)} = \prod_{P \notin S} \varphi_P\left(\frac{a}{b}\right)$$

Proof. Since

$$a\mathbf{F}_S = \prod_{P \in S} F_P \times \prod_{P \notin S} \mathcal{P}^{\nu_P(a)} \qquad \text{and} \qquad b\mathbf{F}_S = \prod_{P \in S} F_P \times \prod_{P \notin S} \mathcal{P}^{\nu_P(b)}$$

the first statement is clear, and then

$$\begin{aligned}(a\mathbf{F}_S : b\mathbf{F}_S) &= \prod_{P \notin S} (\mathcal{P}^{\nu_P(a)} : \mathcal{P}^{\nu_P(b)}) \\ &= \prod_{P \notin S} (\mathfrak{N}P)^{\nu_P(b) - \nu_P(a)} \\ &= \prod_{P \notin S} (\mathfrak{N}P)^{-\nu_P(a/b)}\end{aligned}$$

5-2-3. Proposition. The set $\mathfrak{a} \subset \mathbf{F}$ is an S ideal of $\mathbf{F} \Leftrightarrow$ the following conditions are satisfied:

i. $\mathfrak{a}$ is an $\mathbf{F}_S$ module.

ii. There exists $b \neq 0 \in F_{\mathfrak{M}_\infty(F)}$ such that

$$b\mathbf{F}_S \subset \mathfrak{a} \subset \frac{1}{b}\mathbf{F}_S$$

Proof. $\Rightarrow$: Suppose that

$$\mathfrak{a} = a\mathbf{F}_S = \prod_{P \in S} F_P \times \prod_{P \notin S} \mathcal{P}^{\nu_P(a)}$$

with $a \in \mathbf{F}^*$; then (i) is trivial. Furthermore, we note that $\nu_P(a) = 0$ for almost all $P \in \mathfrak{M}_0(F)$. Since $\{F, \mathfrak{M}_0(F)\}$ is an OAF, there exists $b \neq 0 \in F$ such that $\nu_P(b) \geq |\nu_P(a)|$ for all $P \in \mathfrak{M}_0(F)$. This b suffices for the proof of (ii).

$\Leftarrow$: According to (ii) we know that

$$b\mathbf{F}_S = \prod_{P \in S} F_P \times \prod_{P \notin S} \mathcal{P}^{\nu_P(b)} \subset \mathfrak{a} \subset \prod_{P \in S} F_P \times \prod_{P \notin S} \mathcal{P}^{-\nu_P(b)} = \frac{1}{b}\mathbf{F}_S$$

For each $P \notin S$, let $\mathfrak{a}_P = \{a_P \in F_P | a \in \mathfrak{a}\}$. Since $\mathfrak{a}_P$ is a nonzero O_P module and $\{\nu_P(a_P) | a_P \in \mathfrak{a}_P\}$ is bounded below, it follows easily that $\mathfrak{a}_P$ is a fractional ideal of F_P. Therefore, $\mathfrak{a}_P = \mathcal{P}^{m_P}$, where $-\nu_P(b) \leq m_P \leq \nu_P(b)$ and $m_P = 0$ for almost all P. If we choose $a \in \mathbf{F}^*$ such that $\nu_P(a) = m_P$ for all $P \notin S$, then

$$\begin{aligned}\mathfrak{a} = \mathfrak{a}\mathbf{F}_S &= \Big\{\sum_i a_i b_i | a_i \in \mathfrak{a},\, b_i \in \mathbf{F}_S\Big\} \\ &= \prod_{P \in S} F_P \times \prod_{P \notin S} \mathcal{P}^{m_P} \\ &= a\mathbf{F}_S \in \mathbf{I}_S\end{aligned}$$

5-2-4. Remarks. 1. The argument employed above shows that an $\mathbf{F}_S$ module which is caught between two S ideals of $\mathbf{F}$ is itself an S ideal of $\mathbf{F}$.

2. Condition (ii) may be replaced by
iii. There exist $b, c \neq 0 \in F_{\mathfrak{M}_\infty(F)}$ such that

$$b\mathbf{F}_S \subset \mathfrak{a} \subset \frac{1}{c}\mathbf{F}_S$$

for then $bc\mathbf{F}_S \subset b\mathbf{F}_S \subset \mathfrak{a} \subset \frac{1}{c}\mathbf{F}_S = \frac{b}{bc}\mathbf{F}_S \subset \frac{1}{bc}\mathbf{F}_S$.

3. If R is a rational subfield of F, then from the connections between the valuations of F and R it is easy to see that in (ii) or (iii) the elements b and c may be taken from $R_{\mathfrak{M}_\infty(R)}$; in particular, when F is an algebraic number field, b and c may be taken from $\mathbf{Z}$.

5-2-5. Theorem. If F is a global field and $S \in \Sigma$, then:
i. The mapping

$$\mathfrak{A} \to \bar{\mathfrak{A}} = \mathfrak{A}\mathbf{F}_S$$

is a lattice isomorphism of $\mathfrak{J}_S$ onto $\mathbf{I}_S$.

ii. The inverse isomorphism is given by

$$\mathfrak{a} \to \mathfrak{a} \cap F$$

iii. These isomorphisms are identical with those of (5-2-1).

iv. If $\mathfrak{A} \in \mathfrak{J}_S$ and $\mathfrak{a} \in \mathbf{I}_S$, then

$$\mathfrak{A}\mathbf{F}_S \cap F = \mathfrak{A} \qquad \text{and} \qquad (\mathfrak{a} \cap F)\mathbf{F}_S = \mathfrak{a}$$

Furthermore, if $\mathfrak{A}, \mathfrak{B} \in \mathfrak{J}_S$ are such that $\mathfrak{B} \subset \mathfrak{A}$, then

v. $\bar{\mathfrak{A}} = \mathfrak{A}\mathbf{F}_S = \mathfrak{A} + \mathfrak{B}\mathbf{F}_S = \mathfrak{A} + \bar{\mathfrak{B}}$.

vi. $\mathfrak{A} \cap \bar{\mathfrak{B}} = \mathfrak{A} \cap \mathfrak{B}\mathbf{F}_S = \mathfrak{B}$.

vii. $\mathfrak{A}/\mathfrak{B} \approx \bar{\mathfrak{A}}/\bar{\mathfrak{B}} = \mathfrak{A}\mathbf{F}_S/\mathfrak{B}\mathbf{F}_S$ (as F_S modules).

viii. $(\mathfrak{A}:\mathfrak{B}) = \prod_{P \notin S} (\mathfrak{N}P)^{\nu_P(\mathfrak{B}/\mathfrak{A})}$.

ix. If we write $\mathfrak{a} = a\mathbf{F}_S$ with $a \in \mathbf{F}^*$ and $\mathfrak{a} \cap F = \mathfrak{A}$, then

$$\mathfrak{N}\mathfrak{A} = \prod_{P \notin S} \varphi_P\left(\frac{1}{a}\right)$$

Proof. If $\mathfrak{A} \in \mathfrak{J}_S$, then $\mathfrak{A}\mathbf{F}_S$ is an S ideal of $\mathbf{F}$; in fact, it is clear that

$$\mathfrak{A}\mathbf{F}_S = \prod_{P \in S} F_P \times \prod_{P \notin S} \mathcal{P}^{\nu_P(\mathfrak{A})}$$

Now let us make explicit the various isomorphisms [see (5-2-1)] that go into the lattice isomorphisms between $\mathfrak{J}_S$ and $\mathbf{I}_S$.

Given $\mathfrak{A} \in \mathfrak{J}_S$, the corresponding element of $\mathbf{D}_S$ is

$$A = \delta(\mathfrak{A}) = \prod_{P \notin S} P^{\nu_P(\mathfrak{A})}$$

and the element of $\mathbf{I}_S$ corresponding to $A = \delta(\mathfrak{A})$ is

$$\prod_{P \in S} F_P \times \prod_{P \notin S} \mathcal{P}^{\nu_P(\mathfrak{A})} = \mathfrak{A}\mathbf{F}_S = \bar{\mathfrak{A}}$$

As for the inverse isomorphism—given $\mathcal{A} = a\mathbf{F}_S \in \mathbf{I}_S$, the corresponding element of $\mathbf{D}_S$ is

$$A = \prod_{P \notin S} P^{\nu_P(a)}$$

and the element of $\mathfrak{J}_S$ corresponding to A is

$$\kappa(A) = \{b \in F \mid \nu_P(b) \geq \nu_P(a) \quad \forall P \in \mathcal{S}\} = a\mathbf{F}_S \cap F$$

These observations take care of the proofs of (i), (ii), (iii), and (iv).

Let us turn to the second part of the theorem. It is trivial that $\mathfrak{A} + \bar{\mathfrak{B}} \subset \bar{\mathfrak{A}}$. On the other hand, suppose that

$$a \in \bar{\mathfrak{A}} = \prod_{P \in S} F_P \times \prod_{P \notin S} \mathcal{P}^{\nu_P(\mathfrak{A})}$$

and let $S' = \{P_1, \ldots, P_r\} = \{P \in \mathcal{S} \mid \nu_P(\mathfrak{A}) \neq 0 \text{ or } \nu_P(\mathfrak{B}) \neq 0\}$. According to the strong approximation theorem [see (4-1-5)] in the ordinary arithmetic field $\{F, \mathcal{S}\}$, there exists $c \in F$ such that

$$\begin{aligned} \nu_{P_i}(c - a_{P_i}) &\geq \nu_{P_i}(\mathfrak{B}) \geq \nu_{P_i}(\mathfrak{A}) \qquad && i = 1, \ldots, r \\ \nu_P(c) &\geq 0 = \nu_P(\mathfrak{A}) && P \in \mathcal{S},\ P \notin S' \end{aligned}$$

So $c \in \mathfrak{A}$. Consider $b = c - a$. Since

$$\nu_{P_i}(b) = \nu_{P_i}(c - a_{P_i}) \geq \nu_{P_i}(\mathfrak{B}) \qquad i = 1, \ldots, r$$

and

$$\begin{aligned} \nu_P(b) &= \nu_P(c - a_P) \geq \min\{\nu_P(c), \nu_P(a_P)\} \\ &\geq 0 = \nu_P(\mathfrak{B}) \qquad && P \in \mathcal{S},\ P \notin S' \end{aligned}$$

it follows that

$$b \in \prod_{P \in S} F_P \times \prod_{P \notin S} \mathcal{P}^{\nu_P(\mathfrak{B})} = \bar{\mathfrak{B}}$$

Therefore, $\bar{\mathfrak{A}} = \mathfrak{A} + \bar{\mathfrak{B}}$, and (v) is proved.

The proof of (vi) is immediate from

$$\mathfrak{B} \subset \mathfrak{A} \cap \bar{\mathfrak{B}} \subset F \cap \bar{\mathfrak{B}} = \mathfrak{B}$$

The proof of (vii) consists of the observation that $\mathfrak{A}, \mathfrak{B}, \bar{\mathfrak{A}}, \bar{\mathfrak{B}}$ are F_S

modules and that

$$\frac{\bar{\mathfrak{A}}}{\bar{\mathfrak{B}}} = \frac{\mathfrak{A} + \bar{\mathfrak{B}}}{\bar{\mathfrak{B}}} \approx \frac{\mathfrak{A}}{\mathfrak{A} \cap \bar{\mathfrak{B}}} = \frac{\mathfrak{A}}{\mathfrak{B}}$$

Finally, it follows from

$$\bar{\mathfrak{A}} = \prod_{P \in S} F_P \times \prod_{P \notin S} \mathcal{P}^{\nu_P(\mathfrak{A})} \qquad \text{and} \qquad \bar{\mathfrak{B}} = \prod_{P \in S} F_P \times \prod_{P \notin S} \mathcal{P}^{\nu_P(\mathfrak{B})}$$

that

$$\begin{aligned}(\mathfrak{A}:\mathfrak{B}) &= \prod_{P \notin S} (\mathcal{P}^{\nu_P(\mathfrak{A})} : \mathcal{P}^{\nu_P(\mathfrak{B})}) \\ &= \prod_{P \notin S} (\mathfrak{N}P)^{\nu_P(\mathfrak{B}/\mathfrak{A})}\end{aligned}$$

The proof of (ix) is easy and may be left to the reader.

5-2-6. Remarks. 1. For each $a \neq 0 \in F_S$ the theorem asserts that $\mathbf{F}_S = F_S + a\mathbf{F}_S$, $F_S \cap a\mathbf{F}_S = aF_S$, and $\mathbf{F}_S/a\mathbf{F}_S \approx F_S/aF_S$ (as F_S modules).

2. The technique employed in the proof of (v) may be used to show that

$$\mathbf{F} = F + \mathbf{F}_S = F + a\mathbf{F}_S$$

for every $a \in \mathbf{F}^*$ and any $S \in \Sigma$. Note that this assertion is equivalent to the strong approximation theorem.

3. Among other things (5-2-5) justifies the introduction of idèles as a generalization of ordinary ideals. The correspondence between $\mathfrak{I}_S$ and $\mathbf{I}_S$ often makes it possible to give simple proofs of statements about F_S ideals of F by transferring them to S ideals of $\mathbf{F}$. For example, suppose that $\mathfrak{A} \in \mathfrak{I}_S$ is integral; so, by definition, $\mathfrak{N}\mathfrak{A}$ is the number of elements in the residue class ring $F_S/\mathfrak{A}$. From (viii) or (ix) it is immediate that

$$\mathfrak{N}\mathfrak{A} = \prod_{P \notin S} (\mathfrak{N}P)^{\nu_P(\mathfrak{A})}$$

Thus, if $\mathfrak{B} \in \mathfrak{I}_S$ is also integral, then

$$\mathfrak{N}(\mathfrak{A}\mathfrak{B}) = (\mathfrak{N}\mathfrak{A})(\mathfrak{N}\mathfrak{B})$$

In the same vein, if $\mathfrak{P}$ is the prime ideal of F_S corresponding to $P \in \mathcal{S}$, then

$$\frac{F_S}{\mathfrak{P}} \approx \frac{O_P}{\mathcal{P}} = \bar{F}_P$$

The discussion in this section should be considered as another way of viewing a global field as an OAF, since the primes of S really do not play any role. As a matter of fact, it is possible to develop the entire theory of ordinary arithmetic fields for global fields within the present framework and to give proofs that look different from those in Chapter 4.

5-3. Unit Theorem and Class Number

In this section we shall treat some truly global questions, that is, questions that cannot be reduced to local considerations.

Suppose that F is a global field. For any $\mathbf{a} \in \mathbf{F}^*$ we put

$$V(\mathbf{a}) = \prod_{P \in \mathfrak{M}(F)} \varphi_P(\mathbf{a})$$

and call it the ***volume*** of the idèle $\mathbf{a}$; the product is, of course, well defined. If, in addition, $\alpha \in F^*$, then the product formula implies that $V(\alpha\mathbf{a}) = V(\mathbf{a})$, so that V could be viewed as a function on $\mathbf{F}^*/F^*$.

For $\mathbf{a} \in \mathbf{F}^*$, we also put

$$\Pi_{\mathbf{a}} = \{\mathbf{b} \in \mathbf{F} | \varphi_P(\mathbf{b}) \leq \varphi_P(\mathbf{a}) \quad \forall P \in \mathfrak{M}(F)\}$$

and call it the ***parallelotope of size*** $\mathbf{a}$. Furthermore, we let $M(\mathbf{a})$ denote the number (possibly infinite) of elements of F contained in the parallelotope of size $\mathbf{a}$—that is,

$$M(\mathbf{a}) = \#(F \cap \Pi_{\mathbf{a}})$$

If, in addition, $\alpha \in F^*$, then it is easy to see that $M(\alpha\mathbf{a}) = M(\mathbf{a})$, so that M could be viewed as a function on $\mathbf{F}^*/F^*$.

Our treatment of both the units and the class number of a global field is based on a paper of Artin and Whaples [3] and does not require ideal theory or the Minkowski theory of lattices. The key step in this approach is to prove the next lemma, which provides a rough estimate of $M(\mathbf{a})$ and, in particular, says that it is finite.

5-3-1. Lemma. Suppose that F is a global field; then there exist positive constants C and D such that, for all $\mathbf{a} \in \mathbf{F}^*$,

$$CV(\mathbf{a}) < M(\mathbf{a}) \leq \max\{1, DV(\mathbf{a})\}$$

Proof. The easy part is to locate D. We know that $M(\mathbf{a}) \geq 1$, since $0 \in \Pi_{\mathbf{a}}$ for any $\mathbf{a} \in \mathbf{F}^*$. Consequently, suppose that $M(\mathbf{a}) > 1$, and let $\mathfrak{J}$ be any finite subset of $F \cap \Pi_{\mathbf{a}}$ such that $t = \#(\mathfrak{J}) > 1$. Fix any $P \in \mathfrak{M}_0(F)$, and note that there exists an integer $r \geq 0$ such that $(\mathfrak{N}P)^r < t \leq (\mathfrak{N}P)^{r+1}$. Since $t < \infty$, there exists an $\alpha_0 \neq 0 \in \mathfrak{J}$

such that

$$\varphi_P(\alpha_0) = \max_{\alpha \in \mathfrak{J}} \{\varphi_P(\alpha)\}$$

Thus, $\varphi_P(\alpha/\alpha_0) \leq 1$ for all $\alpha \in \mathfrak{J}$. We have, therefore, t distinct elements α/α_0 in O_P. Since $(O_P:\mathcal{P}^r) = (\mathfrak{N}P)^r$, the "pigeonhole principle" says that two of them (of which one may be 0) are in the same class of O_P modulo $\mathcal{P}^r$; that is, we have $\alpha_1, \alpha_2 \in \mathfrak{J}$ such that $(\alpha_1 - \alpha_2)/\alpha_0 \neq 0 \in \mathcal{P}^r$. In other words, $\nu_P[(\alpha_1 - \alpha_2)/\alpha_0] \geq r$, and $\varphi_P[(\alpha_1 - \alpha_2)/\alpha_0] \leq 1/(\mathfrak{N}P)^r \leq \mathfrak{N}P/t$, so that

$$\varphi_P(\alpha_1 - \alpha_2) \leq \left(\frac{\mathfrak{N}P}{t}\right) \varphi_P(a)$$

For any other nonarchimedean $P \in \mathfrak{M}(F)$,

$$\varphi_P(\alpha_1 - \alpha_2) \leq \max \{\varphi_P(\alpha_1), \varphi_P(\alpha_2)\} \leq \varphi_P(a)$$

while, for any archimedean $P \in \mathfrak{M}(F)$,

$$\varphi_P(\alpha_1 - \alpha_2) \leq 2 \max \{\varphi_P(\alpha_1, \varphi_P(\alpha_2)\} \leq 2\varphi_P(a) \qquad P \text{ real}$$
$$\varphi_P(\alpha_1 - \alpha_2) \leq 4\varphi_P(a) \qquad P \text{ complex}$$

Let r_1 be the number of real archimedean primes in $\mathfrak{M}(F)$, and let r_2 be the number of complex ones; then, denoting the fixed $P \in \mathfrak{M}_0(F)$ by P_0, we have

$$1 = \prod_{P \in \mathfrak{M}(F)} \varphi_P(\alpha_1 - \alpha_2) \leq 2^{r_1+2r_2} \frac{(\mathfrak{N}P_0)}{t} \prod_{P \in \mathfrak{M}(F)} \varphi_P(a)$$

and

$$t \leq 2^{r_1+2r_2}(\mathfrak{N}P_0)V(a)$$

From the way in which $\mathfrak{J}$ is chosen, it follows that, if we put $D = 2^{r_1+2r_2}(\mathfrak{N}P_0)$, then $M(a) \leq DV(a)$.

The proof of the existence of C is somewhat more complicated and seems to depend in an essential way on the structure of a rational subfield.

Let R be any rational subfield of F, and suppose that $[F:R] = n$. We fix $S = \mathfrak{M}_\infty(F)$ and then choose a basis $\{\alpha_1, \ldots, \alpha_n\}$ for F over R such that all $\alpha_i \in F_S$. Let

$$B = 4^n \max \{\varphi_P(\alpha_i) | P \in S, i = 1, \ldots, n\}$$

and choose the integer $m \geq 5$ so large that the open interval $(4B, mB)$ contains some power (positive, negative, or zero) of $\mathfrak{N}P$ for every nonarchimedean $P \in S$.

Consider any $a \in \mathbf{F}^*$; then, from the ordinary approximation theorem and the form of the φ_P's, it follows that there exists $\alpha \in F^*$ such that

$$4B \leq \varphi_P(\alpha a) \leq mB \qquad \forall P \in S$$

(If P is complex, then $\|\varphi_P\| = 4$; this accounts for the appearance of the integer 4 here and in the definition of B. When F is an algebraic number field, m may be taken equal to 5—in fact, $4 + \varepsilon$ would do.) By the strong approximation theorem we may choose $b \neq 0 \in R_{\{\infty\}} = R_{\mathfrak{M}_\infty(R)}$ such that $b\alpha a \in \mathbf{F}_S$—that is, $\varphi_P(b\alpha a) \leq 1$ for all $P \in \mathfrak{M}_0(F)$. Since $V(b\alpha a) = V(\alpha a) = V(a)$ and $M(b\alpha a) = M(a)$, we may assume at the start that the idèle a satisfies the conditions

$$\varphi_P(a) \leq 1 \qquad P \in \mathfrak{M}_0(F)$$

and

$$4B\varphi_P(b) \leq \varphi_P(a) \leq mB\varphi_P(b) \qquad P \in S = \mathfrak{M}_\infty(F)$$

for some $b \neq 0 \in R_{\{\infty\}}$

For such an $a \in \mathbf{F}^*$ and with C still at our disposal, we note that

$$CV(a) = C \prod_{P \in \mathfrak{M}(F)} \varphi_P(a) = \frac{C \prod_{P \in \mathfrak{M}_\infty(F)} \varphi_P(a)}{\prod_{P \in \mathfrak{M}_0(F)} \varphi_P(1/a)}$$

and that by (5-2-2) and (5-2-5)

$$\prod_{P \in \mathfrak{M}_0(F)} \varphi_P\left(\frac{1}{a}\right) = (\mathbf{F}_S : a\mathbf{F}_S) = (\mathbf{F}_S \cap F : a\mathbf{F}_S \cap F) = (F_S : a\mathbf{F}_S \cap F)$$

Call this integer N_a.

Now, let us construct a subset of F_S which contains more than $C \prod_{P \in \mathfrak{M}_\infty(F)} \varphi_P(a)$ elements. To be precise, the set we want is

$$G = \left\{ \beta \in F \;\middle|\; \begin{array}{ll} \varphi_P(\beta) \leq 1 & P \in \mathfrak{M}_0(F) \\ \varphi_P(\beta) \leq B\varphi_P(b) & P \in \mathfrak{M}_\infty(F) \end{array} \right\}$$

Surely G contains the set

$$G' = \left\{ c_1\alpha_1 + \cdots + c_n\alpha_n \in F \;\middle|\; \begin{array}{ll} \varphi_P(c_i) \leq 1 & P \in \mathfrak{M}_0(R) \\ \varphi_\infty(c_i) \leq \varphi_\infty(b) & \{\infty\} = \mathfrak{M}_\infty(R) \end{array} \right\}$$

Moreover, it is easy to see that

$$\#(G') > [\varphi_\infty(b)]^n$$

[of course, by the product formula $\varphi_\infty(b) \geq 1$]. In fact, if $R = \mathbf{Q}$, then $c_i \in \mathbf{Z}$ and $|c_i| \leq \varphi_\infty(b)$ for $i = 1, \ldots, n$; so there are $> \varphi_\infty(b)$ choices for each c_i. On the other hand, if $R = k(x)$, we write $\infty = Q = \mathfrak{M}_\infty(R)$ and put $q = \mathfrak{N}Q = \#(k)$. Each c_i is then a polynomial in x with $\varphi_\infty(c_i) = (\mathfrak{N}Q)^{\deg c_i} \leq \varphi_\infty(b) = (\mathfrak{N}Q)^{\deg b}$. But the number of polynomials of degree $\leq \deg b$ is $q^{1+\deg b}$; so there are $> \varphi_\infty(b)$ choices for each c_i.

If we put $C = (1/mB)^r$, where $r = \#\{\mathfrak{M}_\infty(F)\}$, then

$$\#(G) > [\varphi_\infty(b)]^n = \prod_{P\in\mathfrak{M}_\infty(F)} \varphi_P(b) \geq \prod_{P\in\mathfrak{M}_\infty(F)} \left(\frac{\varphi_P(a)}{mB}\right)$$

$$= C \prod_{P\in\mathfrak{M}_\infty(F)} \varphi_P(a)$$

Since $G \subset F_S$, one of the N_a cosets of $(a\mathbf{F}_S \cap F)$ in F_S contains (by the pigeonhole principle) more than

$$\frac{C \prod_{P\in\mathfrak{M}_\infty(F)} \varphi_P(a)}{N_a} = CV(a)$$

elements of G. Let these elements be $\{\beta_1, \ldots, \beta_t\} \subset G$, where $t > CV(a)$. The set $\{\beta_i - \beta_1 | i = 1, \ldots, t\}$ has t distinct elements, all of which belong to $a\mathbf{F}_S \cap F$. Therefore, we may conclude that

$$\varphi_P(\beta_i - \beta_1) \leq \varphi_P(a) \qquad P \in \mathfrak{M}_0(F)$$
$$\varphi_P(\beta_i - \beta_1) \leq 4B\varphi_P(b) \leq \varphi_P(a) \qquad P \in \mathfrak{M}_\infty(F)$$

In other words, $\{\beta_i - \beta_1\}$ is a subset of $\Pi_a \cap F$ containing more than $CV(a)$ elements. This completes the proof.

We shall need some simple consequences of this lemma; they deal with the existence of field elements satisfying certain conditions at all primes of F. It should be observed that, once a C has been found, then any smaller positive number will do; hence, we may assume that $C < 1$.

5-3-2. Corollary. Suppose that $a \in \mathbf{F}^*$ has $V(a) \geq 1/C$; then there exists $\beta \in F^*$ such that

$$1 \leq \varphi_P(\beta a) \leq V(a)$$

for all $P \in \mathfrak{M}(F)$.

Proof. We have $M(a) > CV(a) \geq 1$; so there exists $\beta' \neq 0 \in F \cap \Pi_a$. If $\beta = 1/\beta'$, then, for any $P \in \mathfrak{M}(F)$,

$$1 \leq \frac{\varphi_P(a)}{\varphi_P(\beta')} = \varphi_P(\beta a) = \frac{V(\beta a)}{\prod_{Q\neq P} \varphi_Q(\beta a)} \leq V(\beta a) = V(a)$$

5-3-3. Corollary. Suppose that $a \in \mathbf{F}^*$, and fix $Q \in \mathfrak{M}(F)$; then there exists $\beta \in F^*$ such that

$$\frac{C}{\mathfrak{N}Q} V(a) \leq \varphi_Q(\beta a) \leq V(a)$$
$$1 \leq \varphi_P(\beta a) \leq \frac{\mathfrak{N}Q}{C} \qquad \forall P \neq Q$$

Proof. We may modify only the Q component of a to get an idèle a' with

$$\frac{1}{C} \leq V(a') \leq \frac{\mathfrak{N}Q}{C}$$

Applying (5-3-2) to a', we get an element $\beta \in F^*$ such that

$$1 \leq \varphi_Q(\beta a') \leq V(a')$$
$$1 \leq \varphi_P(\beta a') = \varphi_P(\beta a) \leq V(a') \leq \frac{\mathfrak{N}Q}{C} \qquad \forall P \neq Q$$

Therefore,

$$\frac{C}{\mathfrak{N}Q} V(a) \leq \frac{V(a)}{V(a')} = \frac{\varphi_Q(a)}{\varphi_Q(a')} = \frac{\varphi_Q(\beta a)}{\varphi_Q(\beta a')} \leq \varphi_Q(\beta a)$$
$$= \frac{V(\beta a)}{\prod\limits_{P \neq Q} \varphi_P(\beta a)} \leq V(a)$$

Our next result formalizes a type of argument which occurs later.

5-3-4. Proposition. Suppose that $S \in \Sigma$ and that $M > 1$. Consider the set

$$\mathfrak{M} = \{a \in \mathbf{F}^* | 1 \leq \varphi_P(a) \leq M \quad \forall P \notin S\}$$

If $\mathfrak{M}'$ is any subset of $\mathfrak{M}$, then there exists a finite set of elements $a_1, \ldots, a_m$ from $\mathfrak{M}'$ with the following property: given $a \in \mathfrak{M}'$, there exists an a_i such that $\varphi_P(a) = \varphi_P(a_i)$ for all $P \notin S$.

Proof. Each $a \in \mathfrak{M}$ determines a function from the complement of S in $\mathfrak{M}(F)$ into the reals—namely,

$$a: P \rightarrow \varphi_P(a)$$

Consider the finite set $T = \{P \notin S | \mathfrak{N}P \leq M\}$. From the form of the functions φ_P we see that, for $a \in \mathfrak{M}$, $\varphi_P(a) = 1$ for $P \notin S \cup T$ and $\varphi_P(a)$ can take only a finite number of values for $P \in T$. Thus, the elements of $\mathfrak{M}$ determine only a finite number of distinct functions.

We are now in a position to determine the structure of F_S^*—the group of S units of F.

Consider first the subset

$$U_F = \{\alpha \in F^* | \varphi_P(\alpha) = 1 \quad \forall P \in \mathfrak{M}(F)\}$$

of F_S^*. It is a multiplicative group, called the group of ***absolute units*** of F. Since U_F is contained in the parallelotope of size 1, it follows from the easy part of (5-3-1) that it is a finite group. Consequently, U_F is a finite cyclic group, namely, the group of all roots of unity in F. The next object of study is, therefore, F_S^*/U_F.

Let $S = \{P_1, \ldots, P_s\}$, and let $\mathbf{R}^{s-1}$ represent euclidean space of dimension $s - 1$. We define a homomorphism

$$\Psi_S : F_S^* \to \mathbf{R}^{s-1}$$

by putting, for ε in F_S^*,

$$\Psi_S(\varepsilon) = (\log \varphi_{P_1}(\varepsilon), \ldots, \log \varphi_{P_{s-1}}(\varepsilon))$$

(It will turn out that it does not matter which element of S is excluded in the definition of Ψ_S.) By the product formula,

$$\log \varphi_{P_s}(\varepsilon) = - \sum_1^{s-1} \log \varphi_{P_i}(\varepsilon)$$

so that, if $\Psi_S(\varepsilon)$ is known, then we know $\varphi_P(\varepsilon)$ for all $P \in \mathfrak{M}(F)$. [Note that, when $s = 1$, we put $\Psi_S(F_S^*) = (0)$. In this case, the product formula says that $\varphi_{P_1}(\varepsilon) = 1$, so that $F_S^* = U_F$.] It is clear that the kernel of Ψ_S is U_F, and we get a monomorphism, also denoted by Ψ_S, of $F_S^*/U_F \to \mathbf{R}^{s-1}$. This moves our considerations into euclidean space, and we have the following lemma.

5-3-5. Lemma. Any bounded set in $\mathbf{R}^{s-1}$ contains only a finite number of elements of $\Psi_S(F_S^*/U_F) = \Psi_S(F_S^*)$.

Proof. It suffices to consider a rectangular neighborhood N of the origin in $\mathbf{R}^{s-1}$ whose sides are $[-M_i, +M_i]$ $(i = 1, \ldots, s-1)$. If $\Psi_S(\varepsilon) \in N$, then

$$-M_i \le \log \varphi_{P_i}(\varepsilon) \le M_i \qquad i = 1, \ldots, s-1$$

$$- \sum_1^{s-1} M_i \le \log \varphi_{P_s}(\varepsilon) \le \sum_1^{s-1} M_i$$

It is no restriction to require that, for $i = 1, \ldots, s-1$, M_i be such that e^{M_i} is a value taken on by φ_{P_i} and that

$$M_s \ge \sum_1^{s-1} M_i$$

be such that e^{M_i} is a value of φ_{P_i}. This means that there exists an idèle a with

$$\varphi_{P_i}(a) = e^{M_i} \qquad i = 1, \ldots, s$$
$$\varphi_P(a) = 1 \qquad P \notin S$$

Since $F \cap \Pi_a$ is finite and since $\Psi_S(F_S^*) \cap N \subset \Psi_S(F^* \cap \Pi_a)$ (with the obvious definition of Ψ_S on F^*), the assertion of the lemma holds.

This result says that $\Psi_S(F_S^*)$ is a discrete group in $\mathbf{R}^{s-1}$. Now, if an additive subgroup G of a euclidean space $\mathbf{R}^{s-1}$ is discrete, then it is said to be a ***lattice;*** the subspace of $\mathbf{R}^{s-1}$ spanned by G will be denoted by $[G]$. Our next result provides a standard algebraic characterization of lattices and is the only fact about them needed for the discussion here.

5-3-6. Lemma. Let G be an additive subgroup of $\mathbf{R}^{s-1}$; then G is a lattice $\Leftrightarrow$ there exist $v_1, \ldots, v_m \in G$ which are linearly independent over $\mathbf{R}$ and such that $G = \mathbf{Z}v_1 \oplus \cdots \oplus \mathbf{Z}v_m$.

The elements $v_1, \ldots, v_m$ are then said to be a ***lattice basis*** for G, and m is called the ***dimension*** of the lattice; of course, we know that $[G] = \mathbf{R}v_1 \oplus \cdots \oplus \mathbf{R}v_m$.

Proof. $\Leftarrow$: This part is obvious.

$\Rightarrow$: We proceed by induction on $m = \dim [G]$. If $m = 0$, everything is trivial. If $m = 1$, we may write $[G] = \mathbf{R}w$, $w \in G$. Every $v \in G$ is of form rw, $r \in \mathbf{R}$. If we let $v_1 \in G$ be one with minimal $r > 0$, then $G = \mathbf{Z}v_1$. Consider then any $m > 1$. Let us write

$$[G] = \mathbf{R}v_1 \oplus \cdots \oplus \mathbf{R}v_m \qquad v_i \in G$$

and put

$$V = \mathbf{R}v_1 \oplus \cdots \oplus \mathbf{R}v_{m-1}$$

Since $G \cap V$ is discrete, we may (in virtue of the induction hypothesis) assume at the start that $v_1, \ldots, v_{m-1}$ are a lattice basis for $G \cap V$; in particular,

$$G \cap V = \mathbf{Z}v_1 \oplus \cdots \oplus \mathbf{Z}v_{m-1} \qquad v_i \in G$$

If any $v \in G$ is given, then, by subtracting off suitable integral multiples of $v_1, \ldots, v_m$, we can get a unique $v' \in G$ of form

$$v' = \sum_1^m a_i v_i \qquad 0 \leq a_i < 1,\ i = 1, \ldots, m$$

Since all such v' belong to a bounded region, there are only a finite number of distinct ones. Choose one for which $a_m \neq 0$ is minimal, and call it

$$v'_m = \sum_1^m a'_i v_i$$

We have

$$G \supset \mathbf{Z}v_1 \oplus \cdots \oplus \mathbf{Z}v_{m-1} \oplus \mathbf{Z}v'_m$$

Now, given $v \in G$, by subtracting suitable integral multiples of v'_m, $v_1, \ldots, v_{m-1}$ we can get an element $v'' \in G$ of form

$$v'' = b_m v_m + \sum_1^{m-1} b_i v_i$$

with $0 \leq b_m < a'_m$ and $0 \leq b_i < 1$ for $i = 1, \ldots, m-1$. Therefore, $b_m = 0$, $v'' \in G \cap V$ and $v'' = 0$; so $v \in \mathbf{Z}v_1 \oplus \cdots \oplus \mathbf{Z}v_{m-1} \oplus \mathbf{Z}v'_m$, and the proof is complete.

The next step is the heart of the matter—namely, to show that the lattice $\Psi_S(F_S^*)$ has maximal dimension $s - 1$. The proof of this fact rests on the difficult part of (5-3-1).

5-3-7. Lemma. There exists a constant M_S with the following property: Given $Q \in S$ and $a \in \mathbf{F}_S^*$, there exists an ε in F_S^* such that

$$\varphi_P(\varepsilon a) \leq M_S \qquad \forall P \in S,\ P \neq Q$$

Proof. As in (5-3-3), choose $\beta \in F^*$ so that

$$1 \leq \varphi_P(\beta a) \leq \frac{\mathfrak{M}Q}{C} \qquad \forall P \neq Q$$

Since $a \in \mathbf{F}_S^*$, we have

$$1 \leq \varphi_P(\beta) \leq \frac{\mathfrak{M}Q}{C} \qquad \forall P \notin S$$

Let

$$\mathfrak{M} = \left\{ b \in \mathbf{F}^* \;\middle|\; 1 \leq \varphi_P(b) \leq \frac{\mathfrak{M}Q}{C} \quad \forall P \notin S \right\}$$

and, as in (5-3-4), let $\{\beta_1, \ldots, \beta_m\} \subset \mathfrak{M}' = \mathfrak{M} \cap F$ be a full set of representatives for $\mathfrak{M}'$. In other words, for any β chosen as above, there exists some β_i with

$$\varphi_P(\beta) = \varphi_P(\beta_i) \qquad \forall P \notin S$$

Since $\varepsilon = \beta/\beta_i \in F_S^*$ and

$$\varphi_P(\varepsilon a) \leq \left(\frac{\mathfrak{M}Q}{C}\right)\left(\frac{1}{\varphi_P(\beta_i)}\right) \qquad \forall P \neq Q$$

we may take

$$M_S = \max_{\substack{P,Q \in S \\ i=1,\ldots,m}} \left\{ \frac{\mathfrak{N}Q}{C\varphi_P(\beta_i)} \right\}$$

5-3-8. Corollary. Suppose that $S = \{P_1, \ldots, P_s\}$; then, for each $i = 1, \ldots, s$, there exists an element ε_i in F_S^* such that

$$\varphi_{P_j}(\varepsilon_i) < 1 \qquad j \neq i$$

Proof. Select an $a \in \mathbf{F}_S^*$ with $\varphi_P(a) > M_S$ for every $P \in S$; then apply (5-3-7).

The S units, whose existence has just been proved, enable us to bring matters to a conclusion. By the product formula

$$\varphi_{P_i}(\varepsilon_i) > 1 \qquad i = 1, \ldots, s$$

Let us put

$$a_{ij} = \log \varphi_{P_j}(\varepsilon_i) \qquad i, j = 1, \ldots, s-1$$

so that $a_{ii} > 0$, $a_{ij} < 0$ for $i \neq j$ and

$$\sum_{\nu=1}^{s-1} a_{i\nu} = \sum_{\nu=1}^{s-1} \log \varphi_{P_\nu}(\varepsilon_i) = -\log \varphi_{P_s}(\varepsilon_i) > 0$$

To show that the lattice $\Psi_S(F_S^*)$ has dimension $s-1$, it suffices to show that the $s-1$ vectors $\Psi_S(\varepsilon_i) = (a_{i1}, \ldots, a_{i\,s-1})$ of $\mathbf{R}^{s-1}$ are linearly independent, and this is the content of the following lemma.

5-3-9. Lemma (Minkowski). Let (a_{ij}) be an $n \times n$ real matrix such that

(i) $$a_{ij} < 0 \qquad i \neq j$$

(ii) $$\sum_{\nu=1}^{n} a_{i\nu} > 0 \qquad i = 1, \ldots, n$$

Then (a_{ij}) is nonsingular.

Proof (Artin). Suppose (a_{ij}) is singular; then the columns are linearly dependent, and the system of equations

$$\sum_{\nu=1}^{n} a_{i\nu} x_\nu = 0 \qquad i = 1, \ldots, n$$

has a nontrivial solution. Let $(x_1, \ldots, x_n)$ be such a solution, and suppose $|x_j| = \max_{i=1,\ldots,n} \{|x_i|\}$. Multiplying the solution by -1, if necessary, we may assume that $x_j > 0$. Now, consider the jth equation; so

$$\begin{aligned}\sum_{\nu=1}^{n} a_{j\nu}x_\nu &= a_{jj}x_j + \sum_{\nu\neq j} a_{j\nu}x_\nu \\ &\geq a_{jj}x_j + \sum_{\nu\neq j} a_{j\nu}x_j \\ &= x_j\left(\sum_{\nu=1}^{n} a_{j\nu}\right) > 0\end{aligned}$$

which is the desired contradiction.

5-3-10. Unit theorem (Dirichlet-Hasse-Chevalley). If F is a global field, then for any $S \in \Sigma$ the group of S units of F is the direct product of $s - 1$ infinite cyclic groups (where s is the number of elements in S) and the finite cyclic group U_F (whose order is even) of all roots of unity in F.

Proof. Choose $\eta_1, \ldots, \eta_{s-1} \in F_S^*$ such that $\Psi_S(\eta_1), \ldots, \Psi_S(\eta_{s-1})$ form a lattice basis for $\Psi_S(F_S^*)$. It we let (η_i) denote the infinite cyclic group generated by η_i, then it follows easily that

$$F_S^* = U_F \times (\eta_1) \times \cdots \times (\eta_{s-1}) \qquad \text{(direct product)}$$

The set $\{\eta_1, \ldots, \eta_{s-1}\}$ is called a set of ***fundamental S units.***

If ζ is a generator of U_F and $m = \#(U_F)$ is odd, then $-\zeta \in U_F$ and $(-\zeta)^m = 1$, a contradiction; so m must be even.

As a by-product of the proof of the unit theorem, we can show that the class number is finite; thus the units and the class number are intimately related.

5-3-11. Theorem. If $\mathbf{F}$ is a global field, then for any $S \in \Sigma$ the S class number of F is finite; more precisely, the groups

$$\frac{\mathbf{F}^*}{F^*\mathbf{F}_S^*} \approx \frac{\mathbf{D}_S}{\mathbf{D}_S^*} \approx \frac{\mathbf{I}_S}{\mathbf{I}_S^*} \approx \frac{\mathfrak{J}_S}{\mathfrak{J}_S^*}$$

are finite. Moreover, for sufficiently large S, the S class number is 1.

Proof. Fix any $Q \in S$, and let

$$\mathfrak{M} = \left\{a \in \mathbf{F}^* \;\middle|\; 1 \leq \varphi_P(a) \leq \frac{\mathfrak{N}Q}{C} \quad \forall P \notin S\right\}$$

As in (5-3-4), let $a_1, \ldots, a_m \in \mathfrak{M}$ be a full set of representatives for $\mathfrak{M}$. Now, given any $a \in \mathbf{F}^*$, there exists, by (5-3-3), an element $\beta \in F^*$ such that

$$1 \leq \varphi_P(\beta a) \leq \frac{\mathfrak{N}Q}{C} \qquad \forall P \neq Q$$

In particular, there is an a_i for which

$$\varphi_P(\beta a) = \varphi_P(a_i) \qquad \forall P \notin S$$

Therefore, $\beta a/a_i \in \mathbf{F}_S^*$ and $a \in a_i F^* \mathbf{F}_S^*$, which implies that $(\mathbf{F}^* : F^* \mathbf{F}_S^*) < \infty$.

Moreover, if S contains all P for which $\mathfrak{N}P \leq \mathfrak{N}Q/C$, then $\mathfrak{M} \subset \mathbf{F}_S^*$ and

$$\mathbf{F}^* = F^* \mathbf{F}_S^*$$

The discussion in this section has been rather formal and unmotivated; its connection with the classical approach will be described in Section 5-4. Examples will be given there and in Chapters 6 and 7.

5-3-12. Exercise. Suppose that F is an algebraic number field and that $S = \mathfrak{M}_\infty(F) = \{P_1, \ldots, P_{r+1}\}$. If $\varepsilon_1, \ldots, \varepsilon_r$ is a fundamental system of S units of F, the absolute value of the determinant

$$\begin{vmatrix} \log \varphi_{P_1}(\varepsilon_1) & \cdots & \log \varphi_{P_r}(\varepsilon_1) \\ \cdots & \cdots & \cdots \\ \log \varphi_{P_1}(\varepsilon_r) & \cdots & \log \varphi_{P_r}(\varepsilon_r) \end{vmatrix}$$

is denoted by $\mathfrak{R}$ and is known as the ***regulator*** of F. The regulator is >0; it is independent of the ordering of S and of the choice of fundamental units.

5-4. Class Number of an Algebraic Number Field

The classical proof of the finiteness of class number for an algebraic number field (by this we mean, of course, the S class number when S consists of the archimedean primes) is rather direct; we shall sketch it. Throughout this section, F will denote an arbitrary but fixed algebraic number field with $[F:\mathbf{Q}] = n$.

5-4-1. Exercise. 1. There exists a constant B (depending on F) such that every ideal $\mathfrak{A}$ of F contains an element $\alpha \neq 0$ with

$$|N_{F\to\mathbf{Q}}(\alpha)| \leq B(\mathfrak{N}\mathfrak{A})$$

(*Hint:* It suffices to assume that $\mathfrak{A}$ is integral. Let $\{\omega_1, \ldots, \omega_n\}$ be a basis for F over $\mathbf{Q}$ whose elements come from $O_F = F_S$, where $S = \mathfrak{M}_\infty(F)$, and (since the ω_j come from a field of degree n) let $\omega_j^{(i)}$ $(i, j = 1, 2, \ldots, n)$ denote the ith conjugate of ω_j. Writing any element of F in the form $\xi = \Sigma a_j \omega_j$, $a_j \in \mathbf{Q}$, we have $|N\xi| \leq$

$B\{\max_j |a_j|\}^n$, where

$$B = \prod_{i=1}^{n} \sum_{j=1}^{n} |\omega_j^{(i)}|$$

Let each a_j range over the integers in the interval $[0, (\mathfrak{N}\mathfrak{A})^{1/n}]$, and apply the pigeonhole principle.)

2. Every ideal class contains an integral ideal $\mathfrak{A}$ with

$$\mathfrak{N}\mathfrak{A} \leq B$$

(*Hint:* Consider an ideal $\mathfrak{B}$ belonging to the inverse of the class in question, and choose $\beta \neq 0 \in \mathfrak{B}$ as above; then $\mathfrak{A} = \beta\mathfrak{B}^{-1}$ satisfies the requirements.)

3. For any integer $m > 0$ there are only a finite number of integral ideals $\mathfrak{A}$ of F with $\mathfrak{N}\mathfrak{A} = m$; hence the class number is finite.

From the proof of (5-4-1) it is clear that part 1 is the key to the finiteness of class number. By using (5-3-1), it is possible to give an even simpler proof of part 1 and in this way to gain further insight into (5-3-1) and its origins. In fact, suppose that $\mathfrak{A}$ is an ideal of the algebraic number field F. Choose $a \in \mathbf{F}^*$ such that $\nu_P(a) = \nu_P(\mathfrak{A})$ for all $P \in \mathfrak{M}_0(F)$, and $V(a) = 1/C$, where C is the constant appearing in (5-3-1). Now (5-3-1) says, in particular, that there exists $\alpha \neq 0 \in F \cap \Pi_a$. This means that $\alpha \neq 0 \in \mathfrak{A}$ and $\varphi_P(\alpha) \leq \varphi_P(a)$ for all $P \in \mathfrak{M}_\infty(F)$. Since

$$|N\alpha| = \prod_{P \in \mathfrak{M}_\infty} \varphi_P(\alpha) \qquad \text{and} \qquad \mathfrak{N}\mathfrak{A} = \prod_{P \in \mathfrak{M}_0} \varphi_P\left(\frac{1}{a}\right)$$

it follows that

$$|N\alpha| \leq \prod_{P \in \mathfrak{M}_\infty} \varphi_P(a) = \frac{V(a)}{\prod_{P \in \mathfrak{M}_0} \varphi_P(a)} = \left(\frac{1}{C}\right)(\mathfrak{N}\mathfrak{A})$$

Therefore, part 1 of (5-4-1) holds, with $B = 1/C$.

Suppose further that $a \in \mathbf{F}^*$ is chosen with the additional requirement that

$$\varphi_P(a) = \left[\left(\frac{\mathfrak{N}\mathfrak{A}}{C}\right)^{\delta_P}\right]^{1/n}$$

for all $P \in \mathfrak{M}_\infty(F)$, where δ_P is 1 when P is real and 2 when P is complex. Then (because of the form of φ_P for complex P) it follows that for our $\alpha \neq 0 \in \mathfrak{A}$ we have

$$|\alpha^{(i)}| \leq \left(\frac{\mathfrak{N}\mathfrak{A}}{C}\right)^{1/n} \qquad i = 1, \ldots, n$$

This stronger version of $|N\alpha| \leq \left(\frac{1}{C}\right)(\mathfrak{N}\mathfrak{A})$ may also be proved directly; more precisely, we have the following exercise.

5-4-2. Exercise. There exists a constant A (depending on F) such that every ideal $\mathfrak{A}$ of F contains an element $\alpha \neq 0$ with

$$|\alpha^{(i)}| \leq A(\mathfrak{N}\mathfrak{A})^{1/n} \qquad i = 1, \ldots, n$$

Hint: Proceed as in item 1 of (5-4-1), and take $A = \max\limits_i \left\{\sum\limits_j |\omega_j^{(i)}|\right\}$.

We have mentioned (5-4-2) only because it is used in the classical proof of the Dirichlet unit theorem in order to get at the assertion of (5-3-8)—that is, in order to prove the existence of a sufficient number of ***independent units*** (see Exercise 5-7) where units are said to be independent when their images under Ψ_S are linearly independent. Otherwise the proof of the unit theorem developed in Section 5-3 is essentially identical with the classical one.

The constant B of (5-4-1) suffices for theoretical work; however, for the concrete problem of finding the class number of a given field, it is desirable to take B [as it appears in item 2 of (5-4-1)] as small as possible. The method for treating this question is due to Minkowski and involves the study of lattices and convex bodies in euclidean space; first, it is necessary to fix the notation carefully.

Let us write $\mathfrak{M}_\infty(F) = \{P_1, \ldots, P_{r_1+r_2}\}$, where $P_1, \ldots, P_{r_1}$ are real (so that $F_{P_i} = \mathbf{R}$ for $i = 1, \ldots, r_1$) and $P_{r_1+1}, \ldots, P_{r_1+r_2}$ are complex (so that $F_{P_i} = \mathbf{C}$ for $i = r_1 + 1, \ldots, r_1 + r_2$). The composition maps $\mu_1, \ldots, \mu_n$ which comprise $\Gamma(\mathbf{Q}, F \to \mathbf{C})$ may be ordered [see (2-5-2)] so that μ_i is associated with P_i for $1 \leq i \leq r_1 + r_2$ and $\mu_{r_1+r_2+j} = \bar{\mu}_{r_1+j}$ is associated with P_{r_1+j} for $j = 1, \ldots, r_2$. Of course, $\bar{\mu}_i = \mu_i$ for $i = 1, \ldots, r_1$. Note that the conjugates of an element $\alpha \in F$ may be indexed so that

$$\mu_i(\alpha) = \alpha^{(i)} \qquad i = 1, \ldots, n$$

Consider the "infinite part of **F**"—that is,

$$F_\infty = \sum_1^{r_1+r_2} \oplus F_{P_i} = \mathbf{R}^{r_1} \times \mathbf{C}^{r_2}$$

Thus F is imbedded in F_∞ via the map

$$\alpha \to (\alpha^{(1)}, \ldots, \alpha^{(r_1+r_2)})$$

If we write $\mathbf{C} = \mathbf{R} \times \mathbf{R}$, then F_∞ may be viewed as euclidean n-space with the usual measure and F is imbedded (additively) in $\mathbf{R}^n = F_\infty$ by

$$\alpha \to (\alpha^{(1)}, \ldots, \alpha^{(r_1)}, u_\alpha^{(1)}, v_\alpha^{(1)}, \ldots, u_\alpha^{(r_2)}, v_\alpha^{(r_2)})$$

where $\alpha^{(r_1+j)} = u_\alpha^{(j)} + iv_\alpha^{(j)}, u_\alpha^{(j)}, v_\alpha^{(j)} \in \mathbf{R}$ $(j = 1, \ldots, r_2)$. The various ways of viewing F_∞ will be used interchangeably.

5-4-3. Proposition. Every ideal $\mathfrak{A}$ of F is a lattice of maximal dimension in $\mathbf{R}^n$.

Proof. Choose $a \in \mathbf{F}^*$ such that $\nu_P(a) = \nu_P(\mathfrak{A})$ for all $P \in \mathfrak{M}_0(F)$ and $V(a) < 1$. It follows from the product formula that there exists a neighborhood of 0 in $\sum_1^{r_1+r_2} F_{P_i}$ which contains only the zero element of $\mathfrak{A}$. Since $\mathfrak{A}$ is an additive subgroup of $\mathbf{R}^n$, it is discrete and, hence, a lattice. Furthermore, $\mathfrak{A}$ surely contains a basis for $F/\mathbf{Q}$—say, $\{\omega_1, \ldots, \omega_n\}$. By separability, we know that det $(\omega_j^{(i)}) \neq 0$ [see (3-7-7) and (3-7-8)]. This implies that the n rows of the matrix $(\omega_j^{(i)})$ $(i = 1, \ldots, r_1 + r_2, j = 1, \ldots, n)$ are linearly independent over $\mathbf{R}$. In other words, $\{\omega_1, \ldots, \omega_n\}$ is a basis for F_∞ over $\mathbf{R}$, and the lattice $\mathfrak{A}$ has maximal dimension.

Suppose that $\mathfrak{L}$ is a lattice in $\mathbf{R}^n$ with lattice basis $\{\omega_1, \ldots, \omega_n\}$. The set of all elements in $\mathbf{R}^n$ of form

$$\sum_1^n x_i\omega_i \qquad 0 \le x_i < 1, i = 1, \ldots, n$$

is known as a ***fundamental domain of*** $\mathfrak{L}$. The volume or measure of a fundamental domain is then the absolute value of the determinant of the matrix gotten by expressing the ω_i in terms of the canonical basis of $\mathbf{R}^n$. Any other lattice basis of $\mathfrak{L}$ is related to this one by a matrix with coefficients from $\mathbf{Z}$ and determinant ± 1. Therefore, all fundamental domains of $\mathfrak{L}$ have the same measure, and we may speak of the measure $m(\mathfrak{L})$ of the lattice. It may be remarked that for any $\omega_1, \ldots, \omega_n \in \mathbf{R}^n$, if $\mathfrak{L} = \mathbf{Z}\omega_1 + \cdots + \mathbf{Z}\omega_n$ is discrete, then it is a lattice of dimension $s \le n$ and we can talk about its measure $m(\mathfrak{L})$; of course, $m(\mathfrak{L}) \neq 0 \Leftrightarrow s = n$.

5-4-4. Proposition. If $\mathfrak{A}$ is an ideal of F, then

$$m(\mathfrak{A}) = \frac{1}{2^{r_2}} (\mathfrak{N}\mathfrak{A}) \sqrt{|\Delta_F|}$$

Proof. Let $\{\omega_1, \ldots, \omega_n\}$ be a lattice basis for $\mathfrak{A}$; so $m(\mathfrak{A})$ is the absolute value of the determinant of the matrix

$$\begin{pmatrix} \omega_1^{(1)}\omega_1^{(2)} & \cdots & \omega_1^{(r_1)}u_1^{(1)}v_1^{(1)} & \cdots & u_1^{(r_2)}v_1^{(r_2)} \\ \cdot\ \cdot & \cdots\cdot & \cdot\ \ \cdot & \cdots\cdot & \cdot \\ \omega_n^{(1)}\omega_n^{(2)} & \cdots & \omega_n^{(r_1)}u_n^{(1)}v_n^{(1)} & \cdots & u_n^{(r_2)}v_n^{(r_2)} \end{pmatrix}$$

Now, let us rewrite this (with obvious choice of notation) as a matrix of column vectors and perform some elementary operations on the columns, namely,

$$\begin{aligned}
&(W^{(1)}, \ldots, W^{(r_1)}, U^{(1)}, V^{(1)}, \ldots, U^{(r_2)}, V^{(r_2)}) \to \\
&(W^{(1)}, \ldots, W^{(r_1)}, U^{(1)} + iV^{(1)}, V^{(1)}, \ldots, U^{(r_2)} + iV^{(r_2)}, V^{(r_2)}) \to \\
&(W^{(1)}, \ldots, W^{(r_1)}, U^{(1)} + iV^{(1)}, -2iV^{(1)}, \ldots, U^{(r_2)} + iV^{(r_2)}, -2iV^{(r_2)}) \to \\
&(W^{(1)}, \ldots, W^{(r_1)}, U^{(1)} + iV^{(1)}, U^{(1)} - iV^{(1)}, \ldots, U^{(r_2)} + iV^{(r_2)}, U^{(r_2)} - iV^{(r_2)}) \to \\
&(W^{(1)}, \ldots, W^{(r_1)}, W^{(r_1+1)}, \ldots, W^{(r_1+r_2)}, \ldots, W^{(n)})
\end{aligned}$$

The factors introduced by these operations are such that

$$m(\mathfrak{A}) = \frac{1}{2^{r_2}} \sqrt{|\Delta(\omega_1, \ldots, \omega_n)|}$$

Then, by (4-8-10),

$$m(\mathfrak{A}) = \frac{1}{2^{r_2}} (\mathfrak{N}\mathfrak{A}) \sqrt{|\Delta_F|}$$

Incidentally, one can give an alternative proof of (5-4-3) in the following manner: According to the discussion of the definition of discriminant in Section 4-8, the ideal $\mathfrak{A}$ has a basis of n elements, say, $\mathfrak{A} = \mathbf{Z}\alpha_1 \oplus \cdots \oplus \mathbf{Z}\alpha_n$, and [as in (5-4-4)] the measure of

$$\left\{\sum_1^n x_i\alpha_i \middle| 0 \le x_i < 1\right\}$$

is not zero. Thus, $\mathfrak{A}$ is indeed a lattice of maximal dimension.

How big does a "nice" set about the origin in $\mathbf{R}^n$ have to be in comparison with the measure of a lattice $\mathfrak{L}$ in order that it should contain a nonzero element of $\mathfrak{L}$? The answer is given by the following proposition.

5-4-5. Proposition (Minkowski). Suppose that $\mathfrak{L}$ is a lattice of maximal dimension in $\mathbf{R}^n$ and that M is a compact, centrally symmetric, convex set. If

$$m(M) \ge 2^n m(\mathfrak{L})$$

then

$$M \cap \mathfrak{L} \ne (0)$$

Proof. The proof will be given in several small steps, but first we note that "centrally symmetric" means that $\theta \in M$ implies $-\theta \in M$ and "convexity" means that, if $\theta_1, \theta_2 \in M$, then the line segment $\{x\theta_1 + (1-x)\theta_2 | 0 \leq x \leq 1\}$ is contained in M.

Assert. If M is a measurable set in $\mathbf{R}^n$ with $m(M) > m(\mathfrak{L})$, then there exist $\theta_1, \theta_2 \in M$, $\theta_1 \neq \theta_2$, such that $\theta_1 - \theta_2 \in \mathfrak{L}$.

Proof. Fix a lattice basis $\{\omega_1, \ldots, \omega_n\}$ for $\mathfrak{L}$, and use it also as a basis for $\mathbf{R}^n$. Thus, $\mathfrak{L}$ consists of all n-tuples $(x_1, \ldots, x_n)$ with $x_i \in \mathbf{Z}$ $(i = 1, \ldots, n)$. For each such n-tuple, consider the set

$$L(x_1, \ldots, x_n) = \left\{\sum_1^n y_i\omega_i | x_i \leq y_i < x_i + 1, i = 1, \ldots, n\right\}$$

It is clear that $L(x_1, \ldots, x_n)$ is a translate of the fundamental domain $L(0, \ldots, 0)$ of $\mathfrak{L}$ [namely, by the vector $(x_1, \ldots, x_n)$], that it is measurable [with measure $m(\mathfrak{L})$], and that all the sets $L(x_1, \ldots, x_n)$ provide a countable disjoint covering of $\mathbf{R}^n$. Take each of the measurable sets $M \cap L(x_1, \ldots, x_n)$, and translate it by the vector $(-x_1, \ldots, -x_n)$ to get it inside $L(0, \ldots, 0)$; call the resulting set $M(x_1, \ldots, x_n)$. If the sets $M(x_1, \ldots, x_n)$ are all disjoint, then

$$\begin{aligned} m(M) &= \sum m(M \cap L(x_1, \ldots, x_n)) \\ &= \sum m(M(x_1, \ldots, x_n)) \\ &\leq m(\mathfrak{L}) \end{aligned}$$

a contradiction. Therefore, two of the sets $M(x_1, \ldots, x_n)$ must intersect, and the assertion follows immediately.

Assert. If M is convex, centrally symmetric, and $m(M) > 2^n m(\mathfrak{L})$, then $M \cap \mathfrak{L} \neq (0)$.

Proof. Consider $\frac{1}{2}M = \{\theta/2 | \theta \in M\}$, so that $m(\frac{1}{2}M) > m(\mathfrak{L})$. Therefore, there exist $\theta_1/2, \theta_2/2 \in \frac{1}{2}M$ with $\theta_1 \neq \theta_2$ and $(\theta_1 - \theta_2)/2 \in \mathfrak{L}$. In particular, $\theta_1, \theta_2 \in M$; so, by symmetry, $-\theta_2 \in M$, and, by convexity, $(\theta_1 - \theta_2)/2 \in M$.

We have proved (5-4-5) except for the case where $m(M) = 2^n m(\mathfrak{L})$ [and, for this, compactness is essential, as may be seen by taking $\mathfrak{L} = \mathbf{Z} \times \mathbf{Z}$ in $\mathbf{R}^2$ and M as the square with vertices at $(\pm 1, \pm 1)$]. In this situation, we observe that for any $\varepsilon > 0$ the set

$$M_{1+\varepsilon} = (1 + \varepsilon)M$$

is compact and $m(M_{1+\varepsilon}) > 2^n m(\mathfrak{L})$. Therefore, $M_{1+\varepsilon} \cap \mathfrak{L} \neq (0)$; and because $M_{1+\varepsilon}$ is bounded, $M_{1+\varepsilon} \cap \mathfrak{L}$ is a finite set. It follows that there is a nonzero lattice vector belonging to $\bigcap M_{1+\varepsilon} = M$. (Note where the fact that M is closed is used.) This completes the proof.

For $\alpha \in F$ we are interested—because our objective is part 1 of (5-4-1)—in controlling

$$|N\alpha| = \prod_{1}^{r_1+r_2} \varphi_{P_i}(\alpha) = |\alpha^{(1)}| \cdots |\alpha^{(r_1)}|\, |\alpha^{(r_1+1)}|^2 \cdots |\alpha^{(r_1+r_2)}|^2 = \prod_{1}^{n} |\alpha^{(i)}|$$

Therefore, let us define a norm in $\mathbf{R}^n = \mathbf{R}^{r_1} \times \mathbf{C}^{r_2}$. Consider the element $z = (z_1, \ldots, z_{r_1+r_2})$ with $z_1, \ldots, z_{r_1} \in \mathbf{R}$ and $z_{r_1+1}, \ldots, z_{r_1+r_2} \in \mathbf{C}$; for $j = 1, \ldots, r_2$ we put $z_{r_1+r_2+j} = \bar{z}_{r_1+j}$ and then define

$$Nz = \prod_{1}^{n} z_i$$

Because sets of form

$$\left\{ z \;\middle|\; |Nz| = \prod_{1}^{n} |z_i| \le t \right\}$$

need not be compact or convex, it is necessary to find nice sets inside them. To do this, let us take, for every $t > 0$,

$$M_t = \left\{ z \in \mathbf{R}^n = \mathbf{R}^{r_1} \times \mathbf{C}^{r_2} \;\middle|\; \sum_{1}^{n} |z_i| \le t \right\}$$

It is easy to check that M_t is closed, bounded, centrally symmetric, and convex. Moreover,

$$M_t \subset \left\{ z \;\middle|\; |Nz| \le \left(\frac{t}{n}\right)^n \right\}$$

since for $z \in M_t$

$$|Nz|^{1/n} = (|z_1| \cdots |z_n|)^{1/n} \le \frac{|z_1| + \cdots + |z_n|}{n} \le \frac{t}{n}$$

Thus, for example, if $n = 2$, $r_1 = 2$, $r_2 = 0$, then M_t is the square with vertices $(\pm t, 0)$, $(0, \pm t)$ plus its interior. Because the area defined by the corresponding norm condition is that between the hyperbolas $xy = \pm t^2/4$, it is clear that M_t is the largest convex body that will work in all cases.

5-4-6. Proposition. The measure of M_t is given by

$$m(M_t) = \frac{2^{r_1 - r_2}\pi^{r_2} t^n}{n!}$$

Proof. If $n = 1$, then $r_1 = 1$, $r_2 = 0$, and $m(M_t) = 2t$. Suppose next that $n = 2$; if $r_1 = 2$, $r_2 = 0$, then $m(M_t) = 2t^2$, while if $r_1 = 0$, $r_2 = 1$, then M_t is the circle of radius $t/2$ and $m(M_t) = \pi t^2/4$. Now, we proceed by induction on $n = r_1 + 2r_2$, and suppose the assertion is true for all integers $< n$. If $r_1 > 0$, then

$$m(M_t) = 2\int_0^t \frac{2^{r_1-1-r_2}\pi^{r_2}}{(n-1)!}(t-\tau)^{n-1}\,d\tau$$

and if $r_2 > 0$, then

$$m(M_t) = \int_0^{t/2} \frac{2^{r_1-r_2+1}\pi^{r_2-1}}{(n-2)!}(t-2\tau)^{n-2}(2\pi\tau)\,d\tau$$

These integrals are easily evaluated, thus completing the proof.

5-4-7. Proposition. If we put

$$B = \left(\frac{4}{\pi}\right)^{r_2}\frac{n!}{n^n}\sqrt{|\Delta_F|}$$

and call it the ***Minkowski bound,*** then every ideal class of F contains an integral ideal with norm $\leq B$.

Proof. We prove part 1 of (5-4-1) for this B. Let $\mathfrak{A}$ be any ideal of F. Select t such that $m(M_t) = 2^n m(\mathfrak{A})$—that is,

$$\frac{2^{r_1-r_2}\pi^{r_2}t^n}{n!} = 2^{r_1+r_2}\mathfrak{N}(\mathfrak{A})\sqrt{|\Delta_F|}$$

Then there exists $\alpha \neq 0 \in \mathfrak{A} \cap M_t$, and consequently

$$|N\alpha| \leq \frac{t^n}{n^n} = B(\mathfrak{N}\mathfrak{A})$$

The Minkowski bound is the best general bound known; however, it can be improved upon for certain classes of fields.

5-4-8. Examples. Consider the cubic field with discriminant -503 discussed in (4-8-20). Here we have $B < 7$. According to (4-9-4) all its prime ideals of norm < 7 are principal. Therefore, every integral ideal of norm < 7 is principal, and we conclude that the class number is 1. The other three cubic fields considered in (4-8-20) have discriminants -23, -31, and -44, respectively. They all have class number 1 because B is less than 2. Similar applications of the Minkowski bound will be given in Section 6-4.

5-4-9. Exercise. Show that the cubic field $F = \mathbf{Q}(\alpha)$, where α is a root of $x^3 - 17x + 31$, has discriminant $(-5)(1259)$ and class number 1.

5-4-10. Theorem (Minkowski). The discriminant of an algebraic number field cannot be ± 1—that is, $|\Delta| > 1$. In particular, there exist no unramified extensions of $\mathbf{Q}$.

Proof. According to (5-4-7) every ideal class contains an integral ideal $\mathfrak{A}$ with $1 \leq \mathfrak{N}\mathfrak{A} \leq (4/\pi)^{r_2}(n!/n^n)\sqrt{|\Delta|}$, so that

$$\sqrt{|\Delta|} \geq \left(\frac{\pi}{4}\right)^{r_2} \frac{n^n}{n!} \geq \left(\frac{\pi}{4}\right)^{n/2} \frac{n^n}{n!} = s_n$$

Now $s_2 = \pi/2 > 1$, and the sequence $\{s_n\}$ is monotonic increasing since $s_{n+1}/s_n = \sqrt{\pi/4}\,(1 + 1/n)^n$. Hence $|\Delta| > 1$.

5-4-11. Exercise. Suppose that $F = \mathbf{Q}(\sqrt{-5})$ and $E = F(\sqrt{-1})$; show that the extension E/F is unramified. It may be pointed out that because F has class number 2 (as will be observed in Sec. 6-4), class field theory predicts the existence of a maximal unramified abelian extension of F of degree 2 (it is known as the Hilbert class field of F); we have exhibited it.

It is an unsolved problem to decide whether or not there exist infinite towers of Hilbert class fields; in other words, if we start with F, take its Hilbert class field, and keep repeating this operation, does the tower of fields thus constructed have a top? Now, it is easy to see that $|\Delta_F| \to \infty$ as $\deg F = n \to \infty$, and it has been conjectured that there exists a function $f(n)$ with $f(n) \to \infty$ as $n = \deg F \to \infty$ and such that $|\Delta_F| > f(n)^n$. This would provide a negative answer to the "tower problem" because of the following reasoning. Suppose that E/F is unramified, so that $\mathbf{\Delta}_{E/F} = O_F$. From (4-8-12) it follows that $|\Delta_{E/\mathbf{Q}}| = |\Delta_{F/\mathbf{Q}}|^{[E:F]}$, so that

$$|\Delta_E|^{1/[E:\mathbf{Q}]} = |\Delta_F|^{1/[F:\mathbf{Q}]}$$

which is contradicted by $|\Delta_F|^{1/n} \to \infty$.

5-4-12. Remark. Suppose that A is any ideal class of F, and for every $t > 0$ let $Z(t, A)$ be the number of integral ideals of norm $\leq t$ belonging to A. It can be proved (by analytic techniques) that $\lim_{t\to\infty} [Z(t, A)/t]$ exists. In fact, this limit is independent of the choice of the ideal class and is equal to $(2^{r_1+r_2}\pi^{r_2}\mathcal{R})/(w\sqrt{|\Delta|})$, where $\mathcal{R}$ is the regulator and w is the number of roots of unity in F. This result is known as the ***Dirichlet-Dedekind density theorem.*** In particular, if $Z(t)$ is the number of integral ideals of F of norm $\leq t$, then

$$\lim_{t\to\infty} \frac{Z(t)}{t} = h\,\frac{2^{r_1+r_2}\pi^{r_2}\mathcal{R}}{w\sqrt{|\Delta|}}$$

This limit may also be evaluated in terms of the ζ-function of F, and because the ζ-function depends on the prime ideals of F, it is possible (when the prime ideals are known) to get information about the class number h. For a discussion of these questions, see [7].

5-5. Topological Considerations

In this section we introduce rather natural and useful topologies for the algebraic objects associated with a global field F. As usual, S denotes an arbitrary finite subset of $\mathfrak{M}(F)$ such that $S \supset \mathfrak{M}_\infty(F)$.

For every $P \in \mathfrak{M}(F)$, F_P is a locally compact field with the second axiom of countability, and, for every $P \notin S$, O_P is a compact, open subring of F_P with the second axiom of countability. Therefore, under the product topology

$$\mathbf{F}_S = \prod_{P \in S} F_P \times \prod_{P \notin S} O_P$$

is a locally compact ring with the second axiom of countability. In a sense, $\mathbf{F}_S$ takes up a large part of $\mathbf{F}$; hence in putting a topology on $\mathbf{F}$ it is natural to require that $\mathbf{F}_S$ should be an open set. Let us then topologize $\mathbf{F}$ according to this requirement—of course, we also require that the induced topology on $\mathbf{F}_S$ coincide with the product topology that we already have. This means that, if $\{N_i | i = 1, 2, \ldots\}$ is a fundamental sequence of neighborhoods of 0 in $\mathbf{F}_S$, we take it also as a fundamental system of neighborhoods of 0 in $\mathbf{F}$. Naturally, the N_i may be taken to be "nice" rectangular sets, and it is then easy to verify the standard criteria [see N. Bourbaki's "Topologie Générale," Groupes Topologiques, Chapter III (Hermann et Cie, Paris, 1940)] which guarantee that $\mathbf{F}$ is a topological ring. Since $\mathbf{F}_S$ is locally compact and open (and closed) in $\mathbf{F}$, it follows that $\mathbf{F}$ is locally compact. Furthermore, it is not hard to see that the index of $\mathbf{F}_S$ in $\mathbf{F}$ (as additive groups) is countably infinite; consequently, $\mathbf{F}$ satisfies the second axiom of countability.

The ***restricted direct product topology*** which we have just defined for $\mathbf{F}$ seems to depend on the choice of S; however, this is not the case. To see this, one has only to consider any two sets S_1 and S_2 in Σ and to observe that one may select a single system of neighborhoods of 0 which is fundamental for both $\mathbf{F}_{S_1}$ and $\mathbf{F}_{S_2}$. Thus the S_1 topology coincides with the S_2 topology.

It is easy to describe a fundamental system of neighborhoods of 0 in $\mathbf{F}$ explicitly in terms of the valuations. Fix, for convenience,

$S_\infty = \mathfrak{M}_\infty(F)$. For each $S \supset S_\infty$ and each ε with $0 < \varepsilon \leq 1$, we put

$$N(\varepsilon, S) = \left\{ a \in \mathbf{F} \,\middle|\, \begin{array}{ll} \varphi_P(a_P) < \varepsilon & P \in S \\ \varphi_P(a_P) \leq 1 & P \notin S \end{array} \right\}$$

The set of all such $N(\varepsilon, S)$ is a fundamental system of neighborhoods of 0 in $\mathbf{F}_{S_\infty}$ and hence in $\mathbf{F}$, and from this set a countable fundamental system may be extracted. Of course, one could start with the sets $N(\varepsilon, S)$, take them to be a fundamental system of neighborhoods of 0 in $\mathbf{F}$, and then show that $\mathbf{F}$ has the properties described above.

For each $P \in \mathfrak{M}(F)$ the field F_P is imbedded in $\mathbf{F}$ via the map $a \to (0, \ldots, 0, a_P = a, 0, \ldots, 0)$, and the topology then induced on F_P from $\mathbf{F}$ is clearly the same as the original topology of F_P. Each such map therefore provides an imbedding of F in $\mathbf{F}$. Another imbedding of F in $\mathbf{F}$, and this is the one we really care about, is given by $a \to (a, \ldots, a)$—that is, along the diagonal.

We may leave it to the reader to show that a subset of $\mathbf{F}$ has compact closure if and only if it is contained in a set of form ΠA_P where A_P is a compact subset of F_P for all $P \in \mathfrak{M}(F)$ and $A_P = O_P$ for almost all P.

Let us now consider the idèle group $\mathbf{F}^*$. As a subset of $\mathbf{F}$ it has an induced topology, and since $\mathbf{F}$ is a topological ring, multiplication in $\mathbf{F}^*$ is continuous. However, the inverse map $a \to a^{-1}$ in $\mathbf{F}^*$ is not continuous in this topology, so that $\mathbf{F}^*$ is not a topological group. In addition, it is not hard to see that $\mathbf{F}^*$ is not complete in the induced topology. Thus, it is desirable to have another topology for $\mathbf{F}^*$, and this is accomplished by taking (as for $\mathbf{F}$) the restricted direct product topology.

For every $P \in \mathfrak{M}(F)$, F_P^* is a locally compact group with the second axiom of countability; and for every $P \notin S$, U_P is a compact group with second axiom of countability. Under the product topology,

$$\mathbf{F}_S^* = \prod_{P \in S} F_P^* \times \prod_{P \notin S} U_P$$

is then a locally compact group with the second axiom of countability. We take any fundamental system of neighborhoods of 1 in $\mathbf{F}_S^*$ and require that it be a fundamental system of neighborhoods of 1 in $\mathbf{F}^*$; in other words, we topologize $\mathbf{F}^*$ so that $\mathbf{F}_S^*$ is open and so that the induced topology on $\mathbf{F}_S^*$ coincides with the given product topology. Then $\mathbf{F}^*$ is a locally compact group, and because (as is easily checked) $(\mathbf{F}^*:\mathbf{F}_S^*)$ is countably infinite, it follows that $\mathbf{F}^*$ satisfies the second axiom of countability. This topology is independent of the choice of S used in its definition. A fundamental system of neighborhoods of

1 in $\mathbf{F}^*_{S_\infty}$, and hence in $\mathbf{F}^*$, is given by

$$N'(\varepsilon, S) = \left\{ a \in \mathbf{F}^* \,\middle|\, \begin{matrix} \varphi_P(a_P - 1) < \varepsilon & P \in S \\ \varphi_P(a_P) = 1 & P \notin S \end{matrix} \right\}$$

where S runs over Σ and ε is arbitrary with $0 < \varepsilon \leq 1$. For every $P \in \mathfrak{M}(F)$, the map $a \to (1, \ldots, 1, a_P = a, 1, \ldots, 1)$ provides an imbedding of F^*_P in $\mathbf{F}^*$, and the induced topology on F^*_P coincides with the original topology.

The discussion for $\mathbf{F}^*$ has paralleled exactly the discussion for $\mathbf{F}$. Obviously one can set up a theory of restricted direct products and their topologies which includes both situations. Incidentally, it is easy to check that the restricted direct product topology of $\mathbf{F}^*$ is stronger than the topology induced from $\mathbf{F}$.

5-5-1. Proposition. Suppose that F is a global field; then F is discrete in $\mathbf{F}$, and F^* is discrete in $\mathbf{F}^*$.

Proof. Let $S = \mathfrak{M}_\infty(F)$; then application of the product formula shows that

$$F \cap N(1, S) = \{0\} \qquad \text{and} \qquad F^* \cap N'(1, S) = \{1\}$$

In Section 5-3 we introduced the volume

$$V(a) = \prod_{P \in \mathfrak{M}(F)} \varphi_P(a)$$

of an idèle a. Since φ_P is continuous, the function $V : \mathbf{F}^*_S \to \mathbf{R}_{>0}$ is continuous for any choice of $S \in \Sigma$; and because $\mathbf{F}^*_S$ is open in $\mathbf{F}^*$, we conclude that

$$V : \mathbf{F}^* \to \mathbf{R}_{>0}$$

is a continuous homomorphism. The kernel, which we denote by ${}^o\mathbf{F}^*$, is a closed subgroup of $\mathbf{F}^*$. In particular, ${}^o\mathbf{F}^*$ is a locally compact group with the second axiom of countability. The product formula implies that

$$\{1\} \subset F^* \subset {}^o\mathbf{F}^* \subset \mathbf{F}^*$$

Our next result indicates that both ${}^o\mathbf{F}^*$ and the restricted direct product topology on $\mathbf{F}^*$ have deep arithmetical significance.

5-5-2. Theorem. Suppose that F is a global field, and let

$$N = \{a \in \mathbf{F}^* | \varphi_P(a) = 1 \quad \forall P \in \mathfrak{M}(F)\}$$

then, for any $S \in \Sigma$:

i. The group $\dfrac{{}^o\mathbf{F}^*}{F^*(\mathbf{F}^*_S \cap {}^o\mathbf{F}^*)}$ is compact if and only if the S class number of F is finite.

ii. The group $\dfrac{F^*(\mathbf{F}_S^* \cap {}^\circ\mathbf{F}^*)}{F^*N}$ is compact if and only if the Dirichlet unit theorem, with respect to S, holds in F.

iii. The group ${}^\circ\mathbf{F}^*/F^*$ is compact if and only if the Dirichlet S unit theorem holds and the S class number is finite.

Proof. For convenience we list some elementary facts about topological groups which are needed for the proof; details may be found in L. Pontrjagin's "Topological Groups" (Princeton University Press, Princeton, N.J., 1946).

Suppose that G is a topological group with a closed subgroup H and a closed normal subgroup N:

1. If G satisfies the second axiom of countability, then so do H and G/N.
2. If G is locally compact, then so are H and G/N.
3. G is compact $\Leftrightarrow$ both N and G/N are compact.
4. If G and G' are both locally compact with the second axiom of countability, then a continuous homomorphism of G onto G' is open; and if N is the kernel, then we have a topological isomorphism $G/N \approx G'$.
5. If G has the second axiom of countability and N is compact, then HN is a closed subgroup of G.
6. If G is locally compact with the second axiom of countability and HN is closed, then there is a topological isomorphism $HN/N \approx H/H \cap N$.
7. Suppose that G is locally compact with the second axiom of countability, H is normal, $G = NH$, and $N \cap H = \{e\}$; then we have topological isomorphisms $G \approx H \times N$ (where the direct product $H \times N$ has the product topology) and $G/N \approx H$.

Consider the chain of groups

$$ {}^\circ\mathbf{F} \supset F^*(\mathbf{F}_S^* \cap {}^\circ\mathbf{F}^*) \supset F^*N \supset F^* $$

${}^\circ\mathbf{F}^*$ is a locally compact abelian group with the second axiom of countability. F^* is discrete, hence closed, in $\mathbf{F}^*$ and in ${}^\circ\mathbf{F}^*$. N is contained in $\mathbf{F}_{S_\infty}^*$ and has the product topology; hence, N is a compact group, and F^*N is a closed subgroup of ${}^\circ\mathbf{F}^*$. Finally, $\mathbf{F}_S^* \cap {}^\circ\mathbf{F}^*$ is open in ${}^\circ\mathbf{F}^*$, so that $F^*(\mathbf{F}_S^* \cap {}^\circ\mathbf{F}^*)$ is an open and closed subgroup of ${}^\circ\mathbf{F}^*$. It follows that all the groups of the chain, and all factor groups that arise from them, are locally compact with the second axiom of countability.

Using the fact that F^*N/F^* is topologically isomorphic to the compact group $N/(F^* \cap N)$, we note that

$$\frac{{}^o\mathbf{F}^*}{F^*} \text{ is compact} \Leftrightarrow \begin{cases} \text{both } \dfrac{F^*(\mathbf{F}_S^* \cap {}^o\mathbf{F}^*)}{F^*} \text{ and} \\ \dfrac{{}^o\mathbf{F}^*}{F^*(\mathbf{F}_S^* \cap {}^o\mathbf{F}^*)} \text{ are compact} \end{cases}$$

$$\Leftrightarrow \begin{cases} \dfrac{F^*N}{F^*}, \dfrac{F^*(\mathbf{F}_S^* \cap {}^o\mathbf{F}^*)}{F^*N}, \text{ and} \\ \dfrac{{}^o\mathbf{F}^*}{F^*(\mathbf{F}_S^* \cap {}^o\mathbf{F}^*)} \text{ are all compact} \end{cases}$$

$$\Leftrightarrow \begin{cases} \dfrac{{}^o\mathbf{F}^*}{F^*(\mathbf{F}_S^* \cap {}^o\mathbf{F}^*)} \text{ and} \\ \dfrac{F^*(\mathbf{F}_S^* \cap {}^o\mathbf{F}^*)}{F^*N} \text{ are compact} \end{cases}$$

Let us examine these two factor groups. Because ${}^o\mathbf{F}^*/F^*(\mathbf{F}_S^* \cap {}^o\mathbf{F}^*)$ is discrete, we know that it is compact if and only if it is finite. Furthermore, by making use of the fact that, if A, B, C are multiplicative abelian groups with $A \subset C$, then $A(B \cap C) = AB \cap C$, we observe that

$$\frac{{}^o\mathbf{F}^*}{F^*(\mathbf{F}_S^* \cap {}^o\mathbf{F}^*)} = \frac{{}^o\mathbf{F}^*}{F^*\mathbf{F}_S^* \cap {}^o\mathbf{F}^*} \approx \frac{{}^o\mathbf{F}^*\mathbf{F}_S^*}{F^*\mathbf{F}_S^*} \subset \frac{\mathbf{F}^*}{F^*\mathbf{F}_S^*} \approx \frac{\mathbf{I}_S}{\mathbf{I}_S^*}$$

Now, it is not hard to see that the group $\mathbf{F}^*/{}^o\mathbf{F}^*\mathbf{F}_S^*$ is finite; this is trivial when F is an algebraic number field and involves looking at the image of $\mathbf{F}^*$ under the homomorphism $V:\mathbf{F}^* \to \mathbf{R}_{>0}$ when F is a global function field. Consequently,

$$\frac{{}^o\mathbf{F}^*}{F^*(\mathbf{F}_S^* \cap {}^o\mathbf{F}^*)} \text{ is finite} \Leftrightarrow \frac{\mathbf{F}^*}{F^*\mathbf{F}_S^*} \text{ is finite}$$

This proves (i).

As for (ii), we shall give the proof only for the case of a number field and leave the necessary modifications for the function field case as an exercise for the reader. Because $N \subset \mathbf{F}_S^* \cap {}^o\mathbf{F}^*$, we have

$$\begin{aligned} \frac{F^*(\mathbf{F}_S^* \cap {}^o\mathbf{F}^*)}{F^*N} &= \frac{(F^*N)(\mathbf{F}_S^* \cap {}^o\mathbf{F}^*)}{F^*N} \\ &\approx \frac{\mathbf{F}_S^* \cap {}^o\mathbf{F}^*}{(F^*N) \cap (\mathbf{F}_S^* \cap {}^o\mathbf{F}^*)} \\ &= \frac{\mathbf{F}_S^* \cap {}^o\mathbf{F}^*}{N(\mathbf{F}_S^* \cap {}^o\mathbf{F}^* \cap F^*)} \\ &= \frac{\mathbf{F}_S^* \cap {}^o\mathbf{F}^*}{NF_S^*} \end{aligned}$$

Incidentally,
$$\frac{NF_S^*}{N} \approx \frac{F_S^*}{F_S^* \cap N} = \frac{F_S^*}{U_F}$$

where, by definition, $U_F = \{a \in F^* | \varphi_P(a) = 1 \quad \forall P \in \mathfrak{M}(F)\}$. The group F_S^*/U_F is discrete; in fact, F^* is discrete, and so is F_S^*. Since U_F is a discrete subset of the compact set N, we see that U_F is finite. Therefore, U_F is the group of all roots of unity in F.

Let $S = \{P_1, \ldots, P_s\}$ be ordered so that $\{P_1, \ldots, P_t\} = \mathfrak{M}_\infty(F)$ $(1 \leq t \leq s)$. Putting for $a \in \mathbf{F}_S^*$

$$\chi_S(a) = (\log \varphi_{P_2}(a), \ldots, \log \varphi_{P_s}(a))$$

we see that

$$\chi_S : \mathbf{F}_S^* \to \mathbf{R}^{s-1}$$

is a continuous homomorphism. In view of the product formula, there is a topological isomorphism also denoted by χ_S:

$$\frac{\mathbf{F}_S^* \cap {}^\circ\mathbf{F}^*}{N} \approx \mathbf{R}^{t-1} \oplus \mathbf{Z}^{s-t}$$

Let $\chi_S(F_S^*) = \chi_S(F_S^*/U_F) = L$; then L is a discrete subgroup of $\mathbf{R}^{s-1} = \mathbf{R}^{t-1} \oplus \mathbf{R}^{s-t}$—that is, L is a lattice in $\mathbf{R}^{s-1}$. The Dirichlet unit theorem for S holds $\Leftrightarrow L$ is a lattice of dimension $s - 1 \Leftrightarrow \mathbf{R}^{s-1}/L$ is compact $\Leftrightarrow (\mathbf{R}^{t-1} \oplus \mathbf{R}^{s-t})/L$ is compact $\Leftrightarrow [F^*(\mathbf{F}_S^* \cap {}^\circ\mathbf{F}^*)]/F^*N$ is compact. The last equivalence follows from

$$\frac{\mathbf{F}_S^* \cap {}^\circ\mathbf{F}^*}{NF_S^*} \approx \frac{(\mathbf{F}_S^* \cap {}^\circ\mathbf{F}^*)/N}{NF_S^*/N} \approx \frac{\mathbf{R}^{t-1} \oplus \mathbf{Z}^{s-t}}{L}$$

This completes the proof.

In virtue of (5-3-10) and (5-3-11) we conclude that ${}^\circ\mathbf{F}^*/F^*$ is compact. On the other hand, if a direct proof of the compactness of ${}^\circ\mathbf{F}^*/F^*$ is given, then [since the proof of (5-5-2) is independent of Section 5-3] we have alternative proofs of the unit theorem and the finiteness of the class number.

5-5-3. Proposition. If F is a global field, then ${}^\circ\mathbf{F}^*/F^*$ is compact.

Proof. Consider any $a \in {}^\circ\mathbf{F}^*$. It suffices to show that there exists a compact set from which we may select a representative of the class in ${}^\circ\mathbf{F}^*/F^*$ to which a belongs.

Fix $Q \in \mathfrak{M}_\infty(F)$. By (5-3-3) there exists $b \in F^*$ such that

$$\frac{C}{\mathfrak{N}Q} \leq \varphi_Q(ba) \leq 1$$

$$1 \leq \varphi_P(ba) \leq \frac{\mathfrak{N}Q}{C} \qquad \forall P \neq Q$$

Now consider the set

$$\mathfrak{M} = \left\{ c \in \mathbf{F}^* \middle| 1 \leq \varphi_P(c) \leq \frac{\mathfrak{N}Q}{C} \quad \forall P \notin \mathfrak{M}_\infty(F) \right\}$$

and as in (5-3-4) let $\{c_1, \ldots, c_m\} \subset \mathfrak{M}$ be a full set of representatives for $\mathfrak{M}$. Thus, there exists $i \in \{1, \ldots, m\}$ such that

$$\varphi_P(c_i) = \varphi_P(ba) \qquad \forall P \notin \mathfrak{M}_\infty(F)$$

Of course, it is no restriction to assume that $\varphi_P(c_i) = 1$ for $P \in \mathfrak{M}_\infty(F)$ and $i = 1, \ldots, m$.

If we put

$$X_Q = \{x \in F_Q^* | C/\mathfrak{N}Q \leq \varphi_Q(x) \leq 1\}$$
$$X_P = \{x \in F_P^* | 1 \leq \varphi_P(x) \leq \mathfrak{N}Q/C\} \qquad P \in \mathfrak{M}_\infty(F),\ P \neq Q$$

then it is immediate that the set

$$X = \prod_{P \in \mathfrak{M}_\infty(F)} X_P \times \prod_{P \notin \mathfrak{M}_\infty(F)} U_P$$

is compact, and therefore so is the set

$$Y = \bigcup_{i=1}^{m} c_i X$$

One checks easily that $d = ba/c_i \in X$, and since $a = b^{-1}c_i d$, the element $c_i d \in Y$ is a representative for the class to which a belongs. This completes the proof.

We may also give a more topological proof of (5-5-3), as is done in [9].

5-5-4. Exercise. i. Let F be an algebraic number field and let $P \in \mathfrak{M}(F)$ be arbitrary; so F_P^+ is a locally compact abelian group. Let χ be any nontrivial character of F_P^+. For each $\alpha \in F_P^+$ define a character χ_α of F_P^+ by

$$\chi_\alpha(\beta) = \chi(\alpha\beta) \qquad \beta \in F_P^+$$

Then the correspondence of $\alpha \leftrightarrow \chi_\alpha$ is a topological isomorphism between F_P^+ and its character group $\hat{F}_P^+$.

ii. Let us construct a canonical character χ^P of F_P^+. It suffices to construct a nontrivial continuous additive map Λ_P of F_P^+ into the group of reals (mod 1)—for then $\chi^P:\beta \to e^{2\pi i \Lambda_P(\beta)}$ is a nontrivial character of F_P^+. Let $p \in \mathfrak{M}(\mathbf{Q})$ lie below P. If λ_p is a nontrivial continuous additive map of $\mathbf{Q}_p^+$ into the group of reals (mod 1), then we may take $\Lambda_P = \lambda_p \circ S_{F_P \to \mathbf{Q}_p}$. Thus it suffices to produce λ_p.

If $p = \infty$, then for $x \in \mathbf{Q}_p^+$ we put

$$\lambda_p(x) \equiv -x \pmod 1$$

and if $p \neq \infty$, then for $x \in \mathbf{Q}_p^+$ we put

$$\lambda_p(x) = \text{principal part of } x$$

By this we mean that if $x = \Sigma a_i p^i$ with $i \geq r$ and $a_i \in \{0,1, \ldots ,p - 1\}$ is the p-adic expansion of x, then

$$\lambda_p(x) = \sum_{i \leq -1} a_i p^i \qquad \text{when } r < 0$$
$$\lambda_p(x) = 0 \qquad \text{when } r \geq 0$$

All this shows that F_P^+ is its own character group when we identify the character $\chi_\alpha^P : \beta \rightarrow e^{2\pi i \Lambda_P(\alpha\beta)}$ with the element α.

iii. It may be noted in passing that when $p \neq \infty$, λ_p is characterized (mod 1) by the following two properties. (1) $\lambda_p(x)$ is a rational number with only a p power in the denominator. (2) $\lambda_p(x) - x$ is a p-adic integer. Note also that if $x \in \mathbf{Q}$, then

$$\sum_{p \in \mathfrak{M}(\mathbf{Q})} \lambda_p(x) \equiv 0 \pmod 1$$

iv. For $P \in \mathfrak{M}_0(F)$, let

$$O_P^\perp = \{\chi \in \hat{F}_P^+ | \chi(O_P) = 1\}$$

Because O_P is compact and open in F_P^+, it follows that $O_P^\perp$ is compact and open in $\hat{F}_P^+$. On the other hand, it is easy to see that χ_α^P is trivial on O_P if and only if $\alpha \in \mathfrak{D}_{F_P/\mathbf{Q}_p}^{-1}$. Consequently,

$$O_P^\perp = \{\chi_\alpha^P | \alpha \in O_P\} \text{ for almost all } P$$

so that, for such P, $O_P^\perp = O_P$ under our identification of $\hat{F}_P^+$ and F_P^+.

v. For $\chi \in \hat{\mathbf{F}}$ and $P \in \mathfrak{M}(F)$ put $\chi_P = \chi | F_P$, so that $\chi_P \in \hat{F}_P^+$. Show that $\chi_P \in O_P^\perp$ for almost all P. The mapping

$$\chi \rightarrow \prod_{P \in \mathfrak{M}(F)} \chi_P$$

is a topological isomorphism of $\hat{\mathbf{F}}$ onto the restricted direct product of the locally compact abelian groups $\hat{F}_P^+$ with respect to the compact open subgroups $O_P^\perp$. Of course, in virtue of (iv), this restricted direct product may be identified with $\mathbf{F}$.

vi. The locally compact abelian group $\mathbf{F}$ is self-dual in a natural way. In fact, if for $a \in \mathbf{F}$, we define $\chi_a \in \hat{\mathbf{F}}$ by

$$\chi_a(b) = \prod_{P \in \mathfrak{M}(F)} \chi^P_{a_P}(b_P) \qquad b \in \mathbf{F}$$

then $a \to \chi_a$ is a topological isomorphism of $\mathbf{F}$ onto $\hat{\mathbf{F}}$. If we define an additive function Λ on $\mathbf{F}$ by

$$\Lambda(c) = \sum_{P \in \mathfrak{M}(F)} \Lambda_P(c_P) \qquad c \in \mathbf{F}$$

then we have

$$\chi_a(b) = e^{2\pi i \Lambda(ab)} \qquad b \in \mathbf{F}$$

and the dual pairing of $\mathbf{F}$ with itself,

$$(a,b) = e^{2\pi i \Lambda(ab)}$$

5-6. Relative Theory

Suppose that F is a global field and that E is a finite extension with $[E:F] = n$; we shall describe certain mappings between the objects associated with F and those associated with E. Of course, when E and F are viewed as ordinary arithmetic fields, these mappings should be related to the maps discussed in Chapter 4. By working within the framework of adèles and idèles, one can give simple proofs for the results of the relative theory in Chapter 4.

We begin by defining the inclusion map

$$i_{\mathbf{F}\to\mathbf{E}} : \mathbf{F} \to \mathbf{E}$$

To do this, consider any $a \in \mathbf{F}$, and let $i_{\mathbf{F}\to\mathbf{E}}(a)$ be the element of $\mathbf{E}$ whose coordinate at any $Q \in \mathfrak{M}(E)$ is a_P, where $P \in \mathfrak{M}(F)$ is the prime lying below Q. In other words,

$$(i_{\mathbf{F}\to\mathbf{E}}(a))_Q = a_P \qquad Q \supset P$$

Strictly speaking, a_P should be replaced by the image of a_P under the canonical imbedding of F_P in E_Q. It is clear that $i_{\mathbf{F}\to\mathbf{E}}$ is a continuous ring isomorphism of $\mathbf{F}$ into $\mathbf{E}$, so that one may identify and write $\mathbf{F} \subset \mathbf{E}$. If $F \subset E \subset K$, then $i_{\mathbf{F}\to\mathbf{K}} = i_{\mathbf{E}\to\mathbf{K}} \circ i_{\mathbf{F}\to\mathbf{E}}$. In connection with the imbeddings $F \subset \mathbf{F}$ and $E \subset \mathbf{E}$, we note that $i_{\mathbf{F}\to\mathbf{E}}$ is compatible with the inclusion map $i_{F\to E}: F \to E$—that is, for $a \in F$ the adèles $i_{F\to E}(a)$ and $i_{\mathbf{F}\to\mathbf{E}}(a)$ are the same. We observe also that $_{\mathbf{F}\to\mathbf{E}}$

takes idèles into idèles, and it follows that

$$i_{\mathbf{F}\to\mathbf{E}} : \mathbf{F}^* \to \mathbf{E}^*$$

is a monomorphism which is continuous in the idèle topologies, and so we may write $\mathbf{F}^* \subset \mathbf{E}^*$. Furthermore, this inclusion map is transitive and compatible with the imbeddings of F^* in E^* and F_P^* in E_Q^*.

Let us fix any set $S \in \Sigma(F)$ and write also

$$S = S^E = \{Q \in \mathfrak{M}(E) | Q \supset P,\ P \in S\} \in \Sigma(E)$$

Obviously, the preceding remarks apply when the objects are taken with respect to S; for example, $F_S \subset E_S$, $\mathbf{F}_S \subset \mathbf{E}_S$, and these imbeddings are compatible.

5-6-1. Remark. The map $i_{\mathbf{F}\to\mathbf{E}}$ may be carried over to S ideals. More precisely, consider an arbitrary S ideal $\mathfrak{a}$ of $\mathbf{F}$, that is, $\mathfrak{a} \in \mathbf{I}_S(F)$. Choose $a \in \mathbf{F}^*$ such that $\mathfrak{a} = a\mathbf{F}_S$, and put

$$i_{\mathbf{F}\to\mathbf{E}}(\mathfrak{a}) = (i_{\mathbf{F}\to\mathbf{E}}(a))\mathbf{E}_S = a\mathbf{E}_S = \mathfrak{a}\mathbf{E}_S$$

This is clearly an element of $\mathbf{I}_S(E)$ which does not depend on the choice of a. We observe that

$$i_{\mathbf{F}\to\mathbf{E}}(\mathfrak{a}) = \prod_{Q \in S} E_Q \times \prod_{Q \notin S} \mathfrak{Q}^{\nu_Q(a)}$$

It is easy to check that

$$i_{\mathbf{F}\to\mathbf{E}} : \mathbf{I}_S(F) \to \mathbf{I}_S(E)$$

is a monomorphism [which carries $\mathbf{I}_S^*(F)$ into $\mathbf{I}_S^*(E)$] whose effect is undone by intersecting with $\mathbf{F}$—that is to say,

$$\mathfrak{a}\mathbf{E}_S \cap \mathbf{F} = \mathfrak{a}$$

Moreover, this mapping corresponds to the inclusion map

$$i_{F\to E} : \mathfrak{I}_S(F) \to \mathfrak{I}_S(E)$$

in the sense that the obvious commutativities hold in the diagram

$$\begin{array}{ccc} \mathbf{I}_S(F) & \xrightarrow{i_{\mathbf{F}\to\mathbf{E}}} & \mathbf{I}_S(E) \\ \updownarrow & & \updownarrow \\ \mathfrak{I}_S(F) & \xrightarrow{i_{F\to E}} & \mathfrak{I}_S(E) \end{array}$$

In fact, if $\mathfrak{A} \in \mathfrak{I}_S(F)$, then $\mathfrak{a} = \mathfrak{A}\mathbf{F}_S \in \mathbf{I}_S(F)$ (with $\mathfrak{a} \cap F = \mathfrak{A}$) corresponds to it and we have

$$\begin{aligned}(i_{F\to E}(\mathfrak{A}))\mathbf{E}_S &= (\mathfrak{A}E_S)\mathbf{E}_S = \mathfrak{A}\mathbf{E}_S \\ &= (\mathfrak{A}\mathbf{F}_S)\mathbf{E}_S = \mathfrak{a}\mathbf{E}_S \\ &= i_{\mathbf{F}\to\mathbf{E}}(\mathfrak{a})\end{aligned}$$

and also (starting at $\mathfrak{a}$)

$$\begin{aligned}i_{F\to E}(\mathfrak{a} \cap F) &= \mathfrak{A}E_S \\ &= \mathfrak{A}\mathbf{E}_S \cap E \\ &= \mathfrak{a}\mathbf{E}_S \cap E \\ &= i_{\mathbf{F}\to\mathbf{E}}(\mathfrak{a}) \cap E\end{aligned}$$

It may be noted that, in addition, $\mathfrak{A}\mathbf{E}_S \cap F = \mathfrak{A}$.

On the other hand, when there is no relative theory of ordinary arithmetic fields at our disposal, the map $i_{\mathbf{F}\to\mathbf{E}}$ may be used to describe a map of $\mathfrak{J}_S(F) \to \mathfrak{J}_S(E)$. For this, one goes around the diagram and requires commutativity; thus,

$$\mathfrak{A} \to \mathfrak{A}\mathbf{F}_S \to (\mathfrak{A}\mathbf{F}_S)\mathbf{E}_S = \mathfrak{A}\mathbf{E}_S \to \mathfrak{A}\mathbf{E}_S \cap E = \mathfrak{A}E_S$$

so that $\mathfrak{A} \to \mathfrak{A}E_S$ is the desired map and $\mathfrak{A}E_S \cap F = \mathfrak{A}$. Moreover, it is easy to give the explicit form of $\mathfrak{A}E_S$. For this, it suffices to consider a prime ideal $\mathfrak{P}_0$ of F_S; $\mathfrak{P}_0$ corresponds to

$$\mathfrak{P}_0\mathbf{F}_S = \prod_{P \in S} F_P \times \prod_{\substack{P \notin S \\ P \neq P_0}} O_P \times \mathcal{P}_0 \in \mathbf{I}_S(F)$$

(such an object may be referred to as a ***prime*** S ***ideal of*** $\mathbf{F}$). Then

$$i_{\mathbf{F}\to\mathbf{E}}(\mathfrak{P}_0\mathbf{F}_S) = \mathfrak{P}_0\mathbf{E}_S = \prod_{Q \in S} E_Q \times \prod_{\substack{Q \notin S \\ Q \not\supset P_0}} O_Q \times \prod_{Q_i \supset P_0} \mathcal{Q}_i^{e(Q_i/P_0)}$$

and

$$\mathfrak{P}_0E_S = \mathfrak{P}_0\mathbf{E}_S \cap E = \prod_{Q_i \supset P_0} \mathfrak{Q}_i^{e(Q_i/P_0)}$$

Next, let us define the norm

$$N_{\mathbf{E}\to\mathbf{F}} : \mathbf{E} \to \mathbf{F}$$

Given $\alpha \in \mathbf{E}$, let $N_{\mathbf{E}\to\mathbf{F}}(\alpha)$ be the element of $\mathbf{F}$ which for each $P \in \mathfrak{M}(F)$ has P component

$$(N_{\mathbf{E}\to\mathbf{F}}(\alpha))_P = \prod_{Q \supset P} N_{E_Q \to F_P}(\alpha_Q) \in F_P$$

It is immediate that this norm is transitive, that it is compatible with $N_{E\to F}: E \to F$, and that, if $\alpha \in \mathbf{F}$, then $N_{\mathbf{E}\to\mathbf{F}}(\alpha) = \alpha^n$. Moreover, we have

$$N_{\mathbf{E}\to\mathbf{F}}: \mathbf{E}^* \to \mathbf{F}^*$$

and this is a continuous homomorphism with respect to the idèle topologies. To carry the norm over to the S ideals, consider any $\mathfrak{a} \in \mathbf{I}_S(E)$, write it in the form $\mathfrak{a} = \alpha\mathbf{E}_S$, $\alpha \in \mathbf{E}^*$, and put

$$N_{\mathbf{E}\to\mathbf{F}}(\mathfrak{a}) = (N_{\mathbf{E}\to\mathbf{F}}(\alpha))\mathbf{F}_S$$

It is easy to check that this is a well-defined homomorphism which is transitive and compatible with the norm $N_{E\to F}:\mathfrak{J}_S(E) \to \mathfrak{J}_S(F)$ (as defined in Section 4-7)—that is, for $\mathfrak{A} \in \mathfrak{J}_S(E)$,

$$(N_{E\to F}(\mathfrak{A}))\mathbf{F}_S = N_{\mathbf{E}\to\mathbf{F}}(\mathfrak{A}\mathbf{E}_S)$$

5-6-2. Exercise. Suppose that nothing is known about the relative theory of ordinary arithmetic fields.

1. Show how $N_{\mathbf{E}\to\mathbf{F}}$ may be used to define a homomorphism $N_{E\to F}:\mathfrak{J}_S(E) \to \mathfrak{J}_S(F)$.
2. Find $N_{E\to F}$ explicitly; it suffices to show that, if $Q \supset P$, then $N_{E\to F}(\mathfrak{Q}) = \mathfrak{P}^{f(Q/P)}$.
3. If $Q_1, \ldots, Q_r$ are the extensions of P, then

$$\sum_1^r e\left(\frac{Q_i}{P}\right) f\left(\frac{Q_i}{P}\right) = n$$

(*Hint:* Define the notion of a ***prime idèle,*** and make use of $i_{\mathbf{F}\to\mathbf{E}}$.)

In order to define the trace map

$$S_{\mathbf{E}\to\mathbf{F}} : \mathbf{E} \to \mathbf{F}$$

we assume that E/F is separable. Consider any $\alpha \in \mathbf{E}$, and let $S_{\mathbf{E}\to\mathbf{F}}(\alpha)$ be the element of $\mathbf{F}$ whose P component [for every $P \in \mathfrak{M}(F)$] is

$$(S_{\mathbf{E}\to\mathbf{F}}(\alpha))_P = \sum_{Q\supset P} S_{E_Q\to F_P}(\alpha_Q) \in F_P$$

It may be left to the reader to check such statements as the following:

1. $S_{\mathbf{E}\to\mathbf{F}}$ is additive, transitive, and continuous.
2. $S_{\mathbf{E}\to\mathbf{F}}(\alpha) = n\alpha$ when $\alpha \in \mathbf{F}$.
3. $S_{\mathbf{E}\to\mathbf{F}}(a\alpha) = aS_{\mathbf{E}\to\mathbf{F}}(\alpha)$ when $a \in \mathbf{F}$, $\alpha \in \mathbf{E}$.
4. $S_{\mathbf{E}\to\mathbf{F}}$ is compatible with $S_{E\to F}:E \to F$.
5. If $\mathfrak{a} \in \mathbf{I}_S(E)$ and $\mathfrak{b} \in \mathbf{I}_S(F)$, then $S_{\mathbf{E}\to\mathbf{F}}(\mathfrak{b}\mathfrak{a}) = \mathfrak{b}S_{\mathbf{E}\to\mathbf{F}}(\mathfrak{a})$.
6. The trace of an S ideal of $\mathbf{E}$ is an S ideal of $\mathbf{F}$.

Although $\mathbf{F}$ is not a field, vector space terminology may be carried over to the "extension" $\mathbf{E}/\mathbf{F}$.

5-6-3. Proposition. If $\{\omega_1, \ldots, \omega_n\}$ is a basis for E/F, then it is also a basis for $\mathbf{E}/\mathbf{F}$, and the topology of $\mathbf{E}$ is the product space topology of n copies of $\mathbf{F}$.

Proof. Let $\{\omega_1^*, \ldots, \omega_n^*\}$ be the complementary basis (see Section 3-7). Thus, $S_{E\to F}(\omega_i\omega_j^*) = \delta_{ij}$, and any $\alpha \in E$ has unique expansions

$$\alpha = \sum_1^n S_{E\to F}(\alpha\omega_i^*)\omega_i = \sum_1^n S_{E\to F}(\alpha\omega_i)\omega_i^*$$

For each $P \in \mathfrak{M}(F)$, consider the direct sum $E_P = \sum_{Q\supset P} \oplus E_Q$. Then both E and F_P may be imbedded in E_P along the diagonal (via the composition maps), and E_P is a vector space of dimension n over F_P. The generic element $x \in E_P$ may be written in the form

$$x = \sum_{Q\supset P} x_Q \qquad x_Q \in E_Q$$

We may define a trace function (with the standard properties) from E_P to F_P by putting

$$S_{E_P\to F_P}(x) = \sum_{Q\supset P} S_{E_Q\to F_P}(x_Q)$$

Note that according to (2-4-6), if $x \in E$, then $S_{E_P\to F_P}(x) = S_{E\to F}(x)$. Therefore, we have, in E_P/F_P, $S_{E_P\to F_P}(\omega_i\omega_j^*) = S_{E\to F}(\omega_i\omega_j^*) = \delta_{ij}$, and it follows that $\{\omega_1, \ldots, \omega_n\}$ and $\{\omega_1^*, \ldots, \omega_n^*\}$ are complementary bases for E_P/F_P with respect to the pairing $(x, y) \to S_{E_P\to F_P}(xy)$. This means, in particular, that

$$E_P = F_P\omega_1 \oplus \cdots \oplus F_P\omega_n$$

[A special case of this fact was proved in (5-4-3).]

Since $S_{\mathbf{E}\to\mathbf{F}}$ is essentially a listing of the $S_{E_P\to F_P}$ for all $P \in \mathfrak{M}(F)$, it follows that

$$S_{\mathbf{E}\to\mathbf{F}}(\omega_i\omega_j^*) = \delta_{ij}$$

It also follows that any $\alpha \in \mathbf{E}$ is of form

$$\alpha = \sum_1^n S_{\mathbf{E}\to\mathbf{F}}(\alpha\omega_i^*)\omega_i = \sum_1^n S_{\mathbf{E}\to\mathbf{F}}(\alpha\omega_i)\omega_i^*$$

and that if $0 = \Sigma a_i\omega_i$ with $a_i \in \mathbf{F}$, then $a_1 = \cdots = a_n = 0$. In other words,

$$\mathbf{E} = \mathbf{F}\omega_1 \oplus \cdots \oplus \mathbf{F}\omega_n$$

The remaining assertion is now a consequence of the continuity of the trace.

5-6-4. Exercise. If $\{\omega_1, \ldots, \omega_n\}$ is an integral basis for E_S/F_S, then it is an integral basis for $\mathbf{E}_S/\mathbf{F}_S$—that is,

$$E_S = F_S\omega_1 \oplus \cdots \oplus F_S\omega_n \Rightarrow \mathbf{E}_S = \mathbf{F}_S\omega_1 \oplus \cdots \oplus \mathbf{F}_S\omega_n$$

In order to discuss the different in the present framework, let us keep in mind the local differents (as in Section 3-7) $\mathfrak{D}_{E_Q/F_P} = (O_Q^*)^{-1}$, where $O_Q^* = \{\alpha \in E_Q | S_{E_Q \to F_P}(\alpha O_Q) \subset O_P\}$ and also, with a change of notation (to avoid confusion), the global differents (as in Section 4-8) $\mathfrak{D}_{E/F,S} = (E_S^\#)^{-1}$, where $E_S^\# = \{\alpha \in E | S_{E\to F}(\alpha E_S) \subset F_S\}$.

5-6-5. Proposition. If we put

$$\mathbf{E}_S^\# = \{\alpha \in \mathbf{E} | S_{\mathbf{E}\to\mathbf{F}}(\alpha \mathbf{E}_S) \subset \mathbf{F}_S\}$$

then $\mathbf{E}_S^\#$ is an S ideal of $\mathbf{E}$.

Proof. It is clear that $\mathbf{E}_S^\#$ is an $\mathbf{E}_S$ module which contains $\mathbf{E}_S$. Choose a basis $\{\omega_1, \ldots, \omega_n\}$ for E/F all of whose elements belong to E_S. If $\alpha \in \mathbf{E}_S^\#$, then $S_{\mathbf{E}\to\mathbf{F}}(\alpha\mathbf{E}_S) \subset \mathbf{F}_S$ and $S_{\mathbf{E}\to\mathbf{F}}(\alpha\omega_i) \in \mathbf{F}_S$ for $i = 1, \ldots, n$; therefore,

$$\mathbf{E}_S^\# \subset \mathbf{F}_S\omega_1^* + \cdots + \mathbf{F}_S\omega_n^*$$

Now choose $b \neq 0 \in F_{\mathfrak{M}_\infty(F)} \subset F_S$ such that $b\omega_i^* \in E_S$ for $i = 1, \ldots, n$. We have then

$$b\mathbf{E}_S \subset \mathbf{E}_S \subset \mathbf{E}_S^\# \subset \frac{1}{b}\mathbf{E}_S$$

and, according to (5-2-3), $\mathbf{E}_S^\#$ is an S ideal of $\mathbf{E}$.

Naturally, $\mathfrak{D}_{\mathbf{E}/\mathbf{F},S} = (\mathbf{E}_S^\#)^{-1}$ is called the S ***different*** for $\mathbf{E}/\mathbf{F}$. It is an integral S ideal in the sense that it is contained in $\mathbf{E}_S$. We may leave it to the reader to verify that this different has the standard properties [especially those in (3-7-16), (3-7-17), and (3-7-18)].

5-6-6. Proposition. Under the isomorphisms of $\mathfrak{J}_S(E)$ and $\mathbf{I}_S(E)$ the ideals $\mathfrak{D}_{E/F,S}$ and $\mathfrak{D}_{\mathbf{E}/\mathbf{F},S}$ correspond to each other. In other words,

$$\mathbf{E}_S^\# \cap E = E_S^\# \qquad \text{and} \qquad E_S^\#\mathbf{E}_S = \mathbf{E}_S^\#$$

Proof. By using (5-2-5) we observe that

$$E_S^\# = E_S^\#\mathbf{E}_S \cap E \subset \mathbf{E}_S^\# \cap E \subset E_S^\#$$

Thus, $E_S^\#\mathbf{E}_S \cap E = \mathbf{E}_S^\# \cap E$, and it follows that $E_S^\#\mathbf{E}_S = \mathbf{E}_S^\#$ and $\mathbf{E}_S^\# \cap E = E_S^\#$.

Now, $\mathfrak{D}_{E/F,S}$ is under control because it is easy to determine $\mathfrak{D}_{\mathbf{E}/\mathbf{F},S}$ in terms of local things; in fact, it is immediate that the following proposition holds.

5-6-7. Proposition.

$$\mathfrak{D}_{\mathbf{E}/\mathbf{F},S} = \prod_{Q \in S} E_Q \times \prod_{P \notin S} \mathfrak{D}_{E_Q/F_P}$$

5-6-8. Exercise. i. If F is an algebraic number field, then

$$\mathbf{F}/F \text{ is compact}$$

(*Hint:* Work first with $F = \mathbf{Q}$. Take $S = \mathfrak{M}_\infty(\mathbf{Q})$ and put $W = \{a \in \mathbf{F}_S | \phi_\infty(a) \leq \frac{1}{2}\}$. Then W is compact, $W \cap F = (0)$, and $\mathbf{F} = F + W$.)

ii. With Λ as in part vi of (5-5-4), we have $\Lambda(a) = 0$ for all $a \in F$. Under the identification of $\hat{\mathbf{F}}$ and $\mathbf{F}$, we know that $F^\perp = \{\chi \in \hat{\mathbf{F}} | \chi(F) = 1\} = \{a \in \mathbf{F} | \Lambda(aF) = 0\}$. Show that

$$F^\perp = F$$

EXERCISES

5-1. Suppose that F is an algebraic function field in one variable over a finite constant field and that E/F is a finite extension. Let P be any prime divisor of F, and let Q be any extension of P to E. In connection with (5-1-2), we wish to show that $g(Q/P) = 1$.

a. Because the g's are multiplicative, it suffices to consider the case where $F = k(x)$ (x is transcendental over the finite field k).

b. For some integer $s \geq 0$ the separable part E^S of the extension E/F is of form $E^{p^s} = \{\alpha^{p^s} | \alpha \in E\}$ and $[E:E^{p^s}] = p^s$.

c. E is separably generated over k; more precisely, with s as above, E is a finite separable extension of $k(x^{p^{-s}})$. This is the theorem of F. K. Schmidt.

d. All possible g's are 1 for the separable extension $E/k(x^{p^{-s}})$ and for the extension $k(x^{p^{-s}})/k(x)$.

5-2. In the situation of Exercise 5-1 suppose that the constant field k is arbitrary and that P is any discrete prime divisor of F which is trivial on k. If k has characteristic 0, then all the g's are 1. If k is a perfect field of characteristic $p \neq 0$, then all the g's are 1 (the arguments used in Exercise 5-1 go through). Furthermore, all g's are 1 for arbitrary k.

5-3. *a.* With respect to the topology $\mathfrak{J}_0$ induced from $\mathbf{F}$, the map $a \to a^{-1}$ in $\mathbf{F}^*$ is not continuous. (*b*) $\mathbf{F}^*$ is not complete with respect to $\mathfrak{J}_0$. (*c*) The restricted direct product topology of $\mathbf{F}^*$ is stronger than $\mathfrak{J}_0$.

5-4. The topological ring $\mathbf{F}$ has no nontrivial open or compact ring ideals.

5-5. In the situation of (5-6-3), if $\{\omega_1, \ldots, \omega_n\}$ is an integral basis for E_S/F_S, then $\mathbf{E}_S^\# = \mathbf{F}_S\omega_1^* + \cdots + \mathbf{F}_S\omega_n^*$.

5-6. Show that $\mathbf{E} = \mathbf{F}\omega_1 + \cdots + \mathbf{F}\omega_n$ [in the situation of (5-6-3)] even when E/F is not separable.

5-7. Suppose that F is an algebraic number field of degree n.

a. There exists a constant $\bar{A}$ (depending on F) such that given $\lambda \in F^*$ there exists a unit ε of F with

$$|\lambda^{(i)}\varepsilon^{(i)}| \leq \bar{A} \cdot |N\lambda|^{1/n} \qquad i = 1, \ldots, n$$

b. There exist units $\varepsilon_1, \ldots, \varepsilon_{r_1+r_2}$ of F with

$$|\varepsilon_j^{(i)}| < 1 \qquad j \neq i,\ i, j = 1, \ldots, r_1 + r_2$$

5-8. Find the class number of $\mathbf{Q}(\alpha)$, where $f(\alpha, \mathbf{Q}) = x^3 + 7x - 3$.

5-9. There exist only a finite number of algebraic number fields of degree n with given discriminant.

5-10. Suppose that E/F is a Galois extension. Define an action of $\mathcal{G}$ on $\mathbf{E}$ and $\mathbf{E}^*$ which extends the action of $\mathcal{G}$ on E. Discuss its properties; for example, which elements are fixed under $\mathcal{G}$? How is it related to the norm?

5-11. *a.* Suppose that F is a global field and that $P \in \mathfrak{M}_0(F)$. Then F_P^* is a group of automorphisms of F_P^+ and, as such, may be endowed with the compact-open topology. By this we mean that a fundamental system of neighborhoods of 1 in F_P^* is given by the set of all

$$N(C,U) = \{a \in F_P^* | (1 - a)C \subset U, \quad (1 - a^{-1})C \subset U\}$$

where C is a compact set and U an open neighborhood of 0 in F_P^+. This topology on F_P^* coincides with the usual one.

b. Show that the restricted direct product topology we have defined for $\mathbf{F}^*$ coincides with the compact-open topology of $\mathbf{F}^*$ viewed as a group of automorphisms of the additive group of $\mathbf{F}$.

5-12. If F is a global field, then the intersection of all closed maximal ideals of the topological ring $\mathbf{F}$ is (0).

5-13. Suppose that F is an algebraic number field of degree n with integral basis $\{\omega_1, \ldots, \omega_n\}$. So these elements are a lattice basis in $F_\infty = \sum_{P \supset \infty} \oplus F_P$ for the lattice $\mathfrak{L}$ determined in $\mathbf{R}^n$ by $O_F = F_{\mathfrak{M}_\infty(F)}$ [see Sec. 5-4 and (5-6-3)]. Let D denote the set of all $a \in \mathbf{F}_{\mathfrak{M}_\infty(F)}$ such that the infinite part of a (that is, $a_\infty = \sum_{P \supset \infty} a_P \in \mathbf{R}^n$) belongs to the fundamental domain (with respect to $\omega_1, \ldots, \omega_n$) of $\mathfrak{L}$. Then D is an additive fundamental domain for $\mathbf{F}$ in the sense that

$$\mathbf{F} = F + D \qquad F \cap D = (0)$$

It follows, in particular, that F is discrete and $\mathbf{F}/F$ is compact.

6

Quadratic Fields

In this chapter we provide applications and illustrations of the general theory of the preceding chapters by studying the simplest class of algebraic number fields—this is the class of all quadratic fields.

6-1. Integral Basis and Discriminant

Consider a quadratic field E. We know that there exists an element $\alpha \in O_E$ such that $E = \mathbf{Q}(\alpha)$. Since the minimum polynomial of α over $\mathbf{Q}$ is of form $x^2 + ax + b$ with $a, b \in \mathbf{Z}$, it follows from the high-school formula for the roots of a quadratic equation that we may write $E = \mathbf{Q}(\sqrt{m})$, where $m \neq 1$ is a square-free integer. If we view $\mathbf{Q}$ concretely as a subset of the complex field $\mathbf{C}$, then, by convention, $\sqrt{m}$ is the positive square root when $m > 0$ and $m = i\sqrt{-m}$ when $m < 0$.

It is clear that if m_1, m_2 are square-free integers then $\mathbf{Q}(\sqrt{m_1}) = \mathbf{Q}(\sqrt{m_2})$ if and only if $m_1 = m_2$. In general, it is easy to see that the quadratic extensions of $\mathbf{Q}$ are in 1-1 correspondence with the elements (excluding the identity) of the factor group $\mathbf{Q}^*/(\mathbf{Q}^*)^2$; in fact, the square-free integers $m \neq 1$ serve as a full set of representatives for these cosets.

The quadratic field $E = \mathbf{Q}(\sqrt{m})$ is a Galois extension of $\mathbf{Q}$; we denote its Galois group by $\mathcal{G} = \{1, \sigma\}$, where

$$\sigma(a + b\sqrt{m}) = a - b\sqrt{m}$$

for all $a, b \in \mathbf{Q}$.

In $E = \mathbf{Q}(\sqrt{m})$ let us consider the $\mathbf{Z}$ module $\mathbf{Z} \cdot 1 \oplus \mathbf{Z}\sqrt{m}$. It is, of course, a subring of O_E. The discriminant of this $\mathbf{Z}$ module is

$$\Delta(1, \sqrt{m}) = \begin{vmatrix} 1 & \sqrt{m} \\ 1 & -\sqrt{m} \end{vmatrix}^2 = 4m$$

Therefore, because m is square-free, the discriminant $\Delta = \Delta_E$ of the extension $E/\mathbf{Q}$ is either $4m$ or m. Moreover, according to the theorem of Stickelberger [see (4-8-19)], $\Delta \equiv 0$ or $1 \pmod 4$. In the case where $m \equiv 2$ or $3 \pmod 4$ it follows that $\Delta = 4m$ and that $\{1, \sqrt{m}\}$ is an integral basis. On the other hand, if $m \equiv 1 \pmod 4$ then $(1 - m)/4 \in \mathbf{Z}$ and the polynomial $x^2 - x + (1 - m)/4 \in \mathbf{Z}[x]$ has the roots $(1 \pm \sqrt{m})/2 \in O_E$. It follows that $\mathbf{Z} \cdot 1 \oplus \mathbf{Z}\left(\frac{1 + \sqrt{m}}{2}\right)$ is a subring of O_E, and also that it strictly contains $\mathbf{Z} \cdot 1 \oplus \mathbf{Z}\sqrt{m}$. This implies [see (3-7-15)] that $\Delta = m$ and that $\{1, (1 + \sqrt{m})/2\}$ is an integral basis. We have proved the following result.

6-1-1. Theorem. Consider the quadratic field $\mathbf{Q}(\sqrt{m})$, where $m \neq 1$ is square-free. In the case where $m \equiv 1 \pmod 4$, we have $\Delta = m$ and $\{1, (1 + \sqrt{m})/2\}$ is an integral basis. On the other hand, if $m \equiv 2$ or $3 \pmod 4$, then $\Delta = 4m$ and $\{1, \sqrt{m}\}$ is an integral basis.

It may be noted that in either case $\mathbf{Q}(\sqrt{\Delta}) = \mathbf{Q}(\sqrt{m})$ and $\{1, (\Delta + \sqrt{\Delta})/2\}$ is an integral basis.

6-2. Prime Ideals

In order to find the prime ideals of $E = \mathbf{Q}(\sqrt{m})$, we must describe how the primes p of $\mathbf{Q}$ decompose in $\mathbf{Q}(\sqrt{m})$—that is, we must factor $(p) = pO_E$.

To do this, observe first that because $\Delta(1, \sqrt{m}) = 4m$ it follows from (4-8-17) that $\{1, \sqrt{m}\}$ is an integral basis at all p *except* for the "exceptional situation" where both $m \equiv 1 \pmod 4$ and $p = 2$. According to the theorem of Kummer [see (4-9-1) and (4-9-2)] the decomposition of p into a product of prime ideals (for all but the exceptional situation) is completely determined by the factorization of $f(x) = x^2 - m$ modulo p. Now, over the field with p elements we have the factorizations

$$x^2 - m = \begin{cases} x^2 & \text{if } p \mid m \\ x^2 - m & \text{if } m \text{ is not a square} \pmod p \\ (x - a)(x + a) & \text{if } m \equiv a^2 \pmod p \text{ with } a \in \mathbf{Z} \end{cases}$$

Moreover, if p is an odd prime, then by Kummer

$$p|m \Rightarrow (p) = (p, \sqrt{m})^2$$

$$\left(\frac{m}{p}\right) = -1 \Rightarrow (p) \text{ is a prime ideal of } \mathbf{Q}(\sqrt{m})$$

$$\left(\frac{m}{p}\right) = +1 \Rightarrow (p) = (p, \sqrt{m} - a)(p, \sqrt{m} + a)$$
$$\text{where } a \in \mathbf{Z} \text{ and } a^2 \equiv m \pmod{p}$$

Note that the prime ideals $(p, \sqrt{m} - a)$ and $(p, \sqrt{m} + a)$ are distinct because $x - a \not\equiv x + a \pmod{p}$.

In the same way one treats the case where $p = 2$ and $m \equiv 2$ or $3 \pmod 4$. The prime 2 is then ramified; in fact,

$$m \equiv 2 \pmod 4 \Rightarrow (2) = (2, \sqrt{m})^2$$

$$m \equiv 3 \pmod 4 \Rightarrow (2) = (2, \sqrt{m} - 1)^2$$
$$\text{since } x^2 - m \equiv (x - 1)^2 \pmod 2$$

Thus, to find all the prime ideals of $\mathbf{Q}(\sqrt{m})$ explicitly, it remains to consider the exceptional situation. Of course, the prime 2 is now unramified. Since $\{1, (1 + \sqrt{m})/2\}$ is an integral basis, we must examine the factorization of $f(x) = x^2 - x + (1 - m)/4$ modulo 2. If $m \equiv 1 \pmod 8$, then $f(x) \equiv x(x + 1) \pmod 2$, and if $m \equiv 5 \pmod 8$, then $f(x) \equiv x^2 + x + 1 \pmod 2$. Since these are the only possibilities, we conclude that

$$m \equiv 1 \pmod 8 \Rightarrow (2) = (2, (1 + \sqrt{m})/2)(2, (1 - \sqrt{m})/2)$$
$$\text{and these are distinct prime ideals}$$

$$m \equiv 5 \pmod 8 \Rightarrow (2) \text{ is a prime ideal}$$

Let us summarize some of our results.

6-2-1. Theorem. The decomposition of the primes p of $\mathbf{Q}$ in $\mathbf{Q}(\sqrt{m})$ is as follows:

i. A prime p which divides Δ is the square of a prime ideal in $\mathbf{Q}(\sqrt{m})$.

ii. An odd prime p which does not divide Δ decomposes as the product of two distinct conjugate prime ideals of degree 1 or becomes a prime ideal of degree 2 according as Δ is a quadratic residue or a quadratic nonresidue (mod p).

iii. As for the prime 2, when $m \equiv 1 \pmod 4$, it is the product of two distinct conjugate prime ideals or is itself a prime according as $m \equiv 1$ or $5 \pmod 8$.

6-2-2. Exercise. An integer $n > 0$ can be written as the sum of two squares $\Leftrightarrow$ the factorization of n has no prime $p \equiv 3 \pmod 4$, which occurs with odd multiplicity. The number of solutions (even when $x^2 + y^2 = n$ is not solvable for $x, y \in \mathbf{Z}$) is $4(a - b)$, where a is the number of divisors d of n with $d \equiv 1 \pmod 4$ and b is the number of divisors d of n with $d \equiv 3 \pmod 4$. [*Hint:* Use the fact, to be proved in (6-4-1), that $\mathbf{Q}(\sqrt{-1})$ has class number 1. A quick way to do the counting is to use (6-2-6).]

The statement of (6-2-1) may be simplified with the aid of some notation. We recall that the Legendre symbol (a/p) is defined when p is an odd prime and a is an integer prime to p. This definition may be extended to give the ***Kronecker symbol*** (Δ/d), which is defined for all Δ which are discriminants of quadratic fields [so, in particular, $\Delta \equiv 0$ or $1 \pmod 4$] and for d any positive integer, as follows:

i. When the Legendre symbol is defined, the Kronecker symbol coincides with it. Of course, we shall write $(a/p) = \left(\frac{a}{p}\right)$ and $(\Delta/d) = \left(\frac{\Delta}{d}\right)$.

ii. If p is any prime with $p|\Delta$, then $(\Delta/p) = 0$.

iii. If Δ is odd, then

$$\left(\frac{\Delta}{2}\right) = \begin{cases} +1 & \text{if } \Delta \equiv 1 \pmod 8 \\ -1 & \text{if } \Delta \equiv 5 \pmod 8 \end{cases}$$

iv. For any positive integers a and b

$$\left(\frac{\Delta}{ab}\right) = \left(\frac{\Delta}{a}\right)\left(\frac{\Delta}{b}\right)$$

Note that $(\Delta/1) = 1$. We shall not be concerned with any further properties of the Kronecker symbol (except in the exercises); its definition is all that is needed for a concise restatement of (6-2-1).

6-2-3. Corollary. If p is any prime in $\mathbf{Q}$, then in $\mathbf{Q}(\sqrt{m})$ we have:

i. p is decomposed [meaning that (p) is the product of two distinct prime ideals] $\Leftrightarrow (\Delta/p) = 1$.

ii. p is inertial [meaning that (p) is a prime ideal] $\Leftrightarrow (\Delta/p) = -1$.

iii. p ramifies [meaning that (p) is the square of a prime ideal] $\Leftrightarrow (\Delta/p) = 0$.

6-2-4. Example. For specific values of m, the properties of the Legendre symbol make it possible to be quite precise. Thus, for example:

1. In the gaussian field $\mathbf{Q}(\sqrt{-1})$: p ramifies $\Leftrightarrow p = 2$, p remains prime $\Leftrightarrow p \equiv -1 \pmod 4$, p splits into two conjugate primes $\Leftrightarrow p \equiv 1 \pmod 4$.

2. In $\mathbf{Q}(\sqrt{2})$: p ramifies $\Leftrightarrow p = 2$, p remains prime $\Leftrightarrow p \equiv \pm 3 \pmod 8$, p splits $\Leftrightarrow p \equiv \pm 1 \pmod 8$.

3. In $\mathbf{Q}(\sqrt{-3})$: p ramifies $\Leftrightarrow p = 3$, p remains prime $\Leftrightarrow p \equiv -1 \pmod 3$, p splits $\Leftrightarrow p \equiv 1 \pmod 3$.

6-2-5. Exercise. Find the decomposition laws for the primes of $\mathbf{Q}$ in $\mathbf{Q}(\sqrt{-5})$ and $\mathbf{Q}(\sqrt{6})$. More generally, by making use of the quadratic reciprocity law [which we prove in (7-3-3)] show that for any quadratic field, if p and q are primes which do not divide $|\Delta|$ and $p \equiv q \pmod{|\Delta|}$, then $(\Delta/p) = (\Delta/q)$ so that the behavior of the unramified primes depends only on their residue class $\pmod{|\Delta|}$.

In the course of the proof it should fall out that, when Δ is odd, $(\Delta/p) = (p/|\Delta|)$. One may find an analogous formula when Δ is even.

The basic result of this section may also be phrased so as to distinguish three types of prime ideal in $\mathbf{Q}(\sqrt{m})$:

1. The primes of degree 1 which are distinct from their conjugates; $\mathbf{Q}(\sqrt{m})$ is the decomposition field for these primes, and the $p \in \mathbf{Z}$ lying below them are decomposed.

2. The primes of degree 2 of form (p) for some prime $p \in \mathbf{Z}$; $\mathbf{Q}(\sqrt{m})$ is the inertia field for these primes, and the primes below them are inertial.

3. The primes of degree 1 whose square is a rational prime dividing the discriminant; $\mathbf{Q}(\sqrt{m})$ is the ramification field for these primes, and the primes below them ramify.

Of course, the infinite prime of $\mathbf{Q}$ has one or two extensions to $\mathbf{Q}(\sqrt{m})$ according as $m < 0$ or $m > 0$.

6-2-6. Exercise. For any positive integer n let $F(n)$ denote the number of integral ideals of norm n in the quadratic field with discriminant Δ; then

$$F(n) = \sum_{d|n} \left(\frac{\Delta}{d}\right)$$

6-3. Units

Consider $E = \mathbf{Q}(\sqrt{m})$, and for it, as usual, let r_1 be the number of real archimedean primes and r_2 be the number of complex archimedean primes; so $r_1 + 2r_2 = 2$, and $r = r_1 + r_2 - 1$ is the number of independent units of E. It is clear that $m > 0 \Rightarrow r_1 = 2,\ r_2 = 0 \Rightarrow r = 1 \Rightarrow E$ has an infinite number of units, and that $m < 0 \Rightarrow r_1 = 0,\ r_2 = 1 \Rightarrow r = 0 \Rightarrow E$ has a finite number of units, which are, in fact, all the roots of unity belonging to E. In order to get more precise information about the units, it is convenient to make use of the fact that, for $\alpha \in O_E$,

$$\alpha \text{ is a unit} \Leftrightarrow N(\alpha) = \pm 1$$

Suppose that $m < 0$; then it is quite trivial to find all the units of E. To see this, suppose first that $m \equiv 2$ or $3 \pmod 4$, so that $\{1, \sqrt{m}\}$ is an integral basis. Therefore, $\alpha = x + y\sqrt{m}$ with $x, y \in \mathbf{Z}$ is a unit $\Leftrightarrow x^2 + |m|y^2 = \pm 1$, and we must find the integral points (x, y) on the ellipse

$$x^2 + |m|y^2 = 1$$

If $m < -1$, there are exactly two solutions, namely: $x = \pm 1,\ y = 0$. If $m = -1$, there are the four solutions: $x = \pm 1,\ y = 0$ and $x = 0,\ y = \pm 1$; so the units are the fourth roots of unity.

Suppose next that $m \equiv 1 \pmod 4$, so that $\{1, (1 + \sqrt{m})/2\}$ is an integral basis. Then $\alpha = x + y((1 + \sqrt{m})/2)$ with $x, y \in \mathbf{Z}$ is a unit $\Leftrightarrow x^2 + xy + y^2((1 - m)/4) = \pm 1$, and we must find the integral points on the ellipse

$$(2x + y)^2 + |m|y^2 = 4$$

If $m < -4$, we have the two solutions: $x = \pm 1,\ y = 0$. The only remaining possibility is $m = -3$, and here we get the six solutions: $x = \pm 1, y = 0; x = 0, y = \pm 1; x = 1, y = -1; x = -1, y = 1$—so the units are the sixth roots of unity. We have proved the following result.

6-3-1. Proposition. If $m > 0$, the real quadratic field $\mathbf{Q}(\sqrt{m})$ has an infinite number of units. If $m < 0$, the imaginary quadratic field $\mathbf{Q}(\sqrt{m})$ has only the units ± 1, except that $\mathbf{Q}(\sqrt{-1})$ has the four units ± 1, $\pm i$ and $\mathbf{Q}(\sqrt{-3})$ has the six units ± 1, $\pm \zeta$, $\pm \zeta^2$, where $\zeta \neq 1$ is a cube root of unity—say, $\zeta = (-1 + \sqrt{-3})/2$.

Now, let us turn to the case $m > 0$. Since $\mathbf{Q}(\sqrt{m})$ is real, the only roots of unity it contains are ± 1, and there exists a unit ε such that the group of units is

$$\{\pm \varepsilon^s | s = 0, \pm 1, \pm 2, \ldots\}$$

Since exactly one of $\pm \varepsilon$, $\pm 1/\varepsilon$ is greater than 1, we may assume that $\varepsilon > 1$. This ε is the unique smallest unit greater than 1 and is called the ***fundamental unit.*** A procedure for actually finding ε may be based on the next result.

6-3-2. Proposition. i. In the real quadratic field $\mathbf{Q}(\sqrt{m})$ consider the element $\eta = (x + y\sqrt{m})/2$ with $x, y \in \mathbf{Z}$; then

$$\eta \text{ is a unit} \Leftrightarrow x^2 - my^2 = \pm 4$$

Moreover, if η is a unit, then

$$\eta > 1 \Leftrightarrow x > 0 \text{ and } y > 0$$

ii. If $\eta_i = (x_i + y_i\sqrt{m})/2$, x_i, $y_i \in \mathbf{Z}$ $(i = 1, 2)$ are units greater than 1, then

$$\eta_1 < \eta_2 \Leftrightarrow x_1 < x_2 \text{ and } y_1 \leq y_2$$

In fact, if the pair $(1 + \sqrt{5})/2 < (3 + \sqrt{5})/2$ is excluded, the equality in $y_1 \leq y_2$ may be omitted.

Proof. The form chosen for η enables us to make the entire proof independent of the congruence class of m (mod 4). Of course, the ring of integers of $\mathbf{Q}(\sqrt{m})$ is contained in $(\mathbf{Z} \cdot 1 \oplus \mathbf{Z}\sqrt{m})/2$.

i. If η is a unit, then $N(\eta) = \pm 1$ and $x^2 - my^2 = \pm 4$. On the other hand, if $x, y \in \mathbf{Z}$ are such that $x^2 - my^2 = \pm 4$ and we denote the conjugate $(x - y\sqrt{m})/2$ of η by $\bar{\eta}$, then $N(\eta) = \eta\bar{\eta} = \pm 1 \in \mathbf{Z}$ and $S(\eta) = \eta + \bar{\eta} = x \in \mathbf{Z}$. It follows that η is integral and, therefore, a unit.

It is trivial that $x > 0$, $y > 0 \Rightarrow (x + y\sqrt{m})/2 > 1$. Conversely, suppose the unit $\eta = (x + y\sqrt{m})/2$ is >1. We have then $0 < \eta^{-1} < 1$ and

$$(*) \qquad \begin{aligned} \bar{\eta} > 0 &\Leftrightarrow N(\eta) = +1 \Leftrightarrow \bar{\eta} = \eta^{-1} \\ \bar{\eta} < 0 &\Leftrightarrow N(\eta) = -1 \Leftrightarrow \bar{\eta} = -\eta^{-1} \end{aligned}$$

Therefore, $\bar{\eta} > 0$ implies that $x = \eta + \bar{\eta} > 0$ and $y\sqrt{m} = \eta - \bar{\eta} = \eta - \eta^{-1} > 0$, while $\bar{\eta} < 0$ implies that $x = \eta - \eta^{-1} > 0$ and $y\sqrt{m} = \eta + \eta^{-1} > 0$.

ii. Clearly, $x_1 < x_2$ and $y_1 \leq y_2$ imply $\eta_1 < \eta_2$. To prove the converse, suppose first that $\bar{\eta}_1 > 0$. Since $x_i = \eta_i + \bar{\eta}_i$, $\sqrt{m}\, y_i = \eta_i - \bar{\eta}_i$ and

$$\eta_2 > \eta_1 > 1 \Leftrightarrow 0 < \eta_2^{-1} < \eta_1^{-1} < 1$$

it follows from $(*)$ that

$$\sqrt{m}\,(y_2 - y_1) = (\eta_2 - \eta_1) + (\bar{\eta}_1 - \bar{\eta}_2) > 0$$

for both $\bar{\eta}_2 > 0$ and $\bar{\eta}_2 < 0$. Therefore, $y_2 > y_1$. Now, by hypothesis, we have $N(\eta_1) = 1$ and $x_1^2 = my_1^2 + 4$. Consider the equation $x_2^2 = my_2^2 \pm 4$. If the plus sign holds, then surely $x_2 > x_1$. If the minus sign holds, we have

$$x_2^2 - x_1^2 = m(y_2^2 - y_1^2) - 8$$

Since $m \geq 2$, the right side can be negative only when $y_2^2 - y_1^2 \leq 4$. The only possibility is then $y_1 = 1$, $y_2 = 2$, $m = 2$—but this gives $x_2^2 = x_1^2 - 2$, which has no solutions in positive integers. We have proved that

$$\eta_2 > \eta_1, \bar{\eta}_1 > 0 \Rightarrow x_2 > x_1, y_2 > y_1$$

It remains to show that

$$\eta_2 > \eta_1, \bar{\eta}_1 < 0 \Rightarrow x_2 > x_1, y_2 \geq y_1$$

To do this, one observes that for both $\bar{\eta}_2 > 0$ and $\bar{\eta}_2 < 0$ we have $x_2 - x_1 = (\eta_2 - \eta_1) + (\bar{\eta}_2 - \bar{\eta}_1) > 0$. Then from $x_2 > x_1$ and $N(\eta_1) = -1$ it follows as above that $y_2 \geq y_1$, with equality holding only for $(1 + \sqrt{5})/2 < (3 + \sqrt{5})/2$. The details are left to the reader.

According to the above, one may search for the fundamental unit $\varepsilon = (\bar{x} + \bar{y}\sqrt{m})/2$ by testing $y = 1, 2, 3, \ldots$ consecutively in the Diophantine equation $x^2 - my^2 = \pm 4$. The first y that leads to an integral x determines ε. A slight simplification occurs when $m \equiv 2$ or $3 \pmod 4$ because then both $\bar{x}$ and $\bar{y}$ are even. In this case it suffices, therefore, to apply the same procedure to the equation $x^2 - my^2 = \pm 1$.

By using this technique it is easy to see, for example, that for $m = 2, 3, 5, 6, 7, 10, 11, 13, 14, 15, 17, 21, 34, 65, 69$ the fundamental units are, respectively, $1 + \sqrt{2}$, $2 + \sqrt{3}$, $(1 + \sqrt{5})/2$, $5 + 2\sqrt{6}$, $8 + 3\sqrt{7}$, $3 + \sqrt{10}$, $10 + 3\sqrt{11}$, $(3 + \sqrt{13})/2$, $15 + 4\sqrt{14}$, $4 + \sqrt{15}$, $4 + \sqrt{17}$, $(5 + \sqrt{21})/2$, $35 + 6\sqrt{34}$, $8 + \sqrt{65}$, $(25 + 3\sqrt{69})/2$. Of course, our simple-minded method is not practical for showing that the fundamental unit of $\mathbf{Q}(\sqrt{22})$ is $197 + 42\sqrt{22}$ and that of $\mathbf{Q}(\sqrt{73})$ is $1068 + 125\sqrt{73}$. These fundamental units and other complicated ones are accessible because there exists a straightforward method (which involves finding the convergents of the continued fraction expansion of $\sqrt{m}$) for finding the positive solutions of Pell's equation—which is, after all, the equation that we are trying to solve here.

6-3-3. Exercise. Find all rational integral solutions of $x^2 - 2y^2 = \pm 119$.

6-3-4. Exercise. The real quadratic fields may be divided into two classes according as the fundamental unit ε has norm $+1$ (in which case every unit has norm 1) or -1. Show that if $N(\varepsilon) = -1$

then m contains no prime factor $p \equiv 3 \pmod 4$. The converse is false since $N(\varepsilon) = 1$ in $\mathbf{Q}(\sqrt{34})$. Show further that if m is a prime $\equiv 1 \pmod 4$ then $N(\varepsilon) = -1$. *Hint:* To prove the last assertion, use the fact that if $N(\varepsilon) = 1$ there exists an integer α of $\mathbf{Q}(\sqrt{m})$ such that $\varepsilon = \alpha/\sigma\alpha$, where σ denotes conjugation in $\mathbf{Q}(\sqrt{m})$. This is a special case of "Hilbert theorem 90," which says that if E/F is an arbitrary cyclic extension and σ is a generator of $\mathcal{G}(E/F)$ then any element of E whose norm is 1 can be expressed in the form $\alpha/\sigma\alpha$ for some $\alpha \in E$. More generally, the reader may prove a result of E. Noether which says that, if E/F is a Galois extension and $f:\mathcal{G} \to E^*$ is a function such that

$$f(\sigma\tau) = [f(\sigma)][\sigma f(\tau)]$$

for all $\sigma, \tau \in \mathcal{G}$, then there exists an $\alpha \in E^*$ such that

$$f(\sigma) = \frac{\alpha}{\sigma\alpha}$$

for every $\sigma \in \mathcal{G}$. We are not in a position here to discuss the cohomological significance of these results.

6-4. Class Number

Perhaps the most obvious approach to the question of unique factorization in the quadratic field $E = \mathbf{Q}(\sqrt{m})$ is to try to show (in analogy with the case of the rationals) that its ring of integers is a euclidean domain. Let us recall that O_E is euclidean if to every $\alpha \neq 0 \in O_E$ there is assigned an integer $\lambda(\alpha) \geq 0$ such that:

i. If $\alpha|\beta$, then $\lambda(\alpha) \leq \lambda(\beta)$.

ii. If $\alpha\beta \neq 0$, then there exist elements μ, ε in O_E such that $\alpha = \mu\beta + \varepsilon$, where either $\varepsilon = 0$ or $\lambda(\varepsilon) < \lambda(\beta)$.

Of course, a euclidean domain is a principal ideal domain and, hence, a unique factorization domain. An obvious candidate for the function λ is

$$\lambda(\alpha) = |N(\alpha)|$$

and if O_E is euclidean for this λ, we say that E is ***euclidean.*** Since both norm and absolute value are multiplicative, condition i is always satisfied. Furthermore, λ is defined on all of E and $\lambda(0) = 0$; so the requirement of condition ii may be rewritten as $|N((\alpha/\beta) - \mu)| < 1$. Finally, since α/β runs over E, (ii) may be restated as follows: Given

$\gamma \in E$, there exists $\mu \in O_E$ such that

$$|N(\gamma - \mu)| < 1$$

Consider first the case $m \equiv 2, 3 \pmod 4$. Here we may write $\gamma = a + b\sqrt{m}$, $\mu = x + y\sqrt{m}$, with $a, b \in \mathbf{Q}$, $x, y \in \mathbf{Z}$ and put $r = a - x$, $s = b - y$. Then the condition becomes

$$(*) \qquad |N(\gamma - \mu)| = |N(r + s\sqrt{m})| = |r^2 - ms^2| < 1$$

Clearly x and y may be chosen so that $|r| \leq \frac{1}{2}$ and $|s| \leq \frac{1}{2}$. Thus (*) is satisfied for $m = -2, -1, 2, 3$, and the corresponding fields are euclidean. Moreover, it is impossible to satisfy (*) for any other negative m in this class.

Now let us turn to the case $m \equiv 1 \pmod 4$. In this case we write $\gamma = a + b((1 + \sqrt{m})/2)$, $\mu = x + y((1 + \sqrt{m})/2)$ with $a, b \in \mathbf{Q}$, $x, y \in \mathbf{Z}$, and put $r = a - x$, $s = b - y$. The condition becomes

$$(**) \quad |N(\gamma - \mu)| = \left|N\left(r + s\left(\frac{1 + \sqrt{m}}{2}\right)\right)\right| = \left|\left(r + \frac{s}{2}\right)^2 - m\left(\frac{s}{2}\right)^2\right| < 1$$

One may choose y first so that $|s| \leq \frac{1}{2}$ and then x so that

$$\left|r + \frac{s}{2}\right| = \left|a + \frac{b}{2} - \frac{y}{2} - x\right| \leq \frac{1}{2}$$

Thus, it follows that $\mathbf{Q}(\sqrt{m})$ is euclidean for $m = -11, -7, -3, 5, 13$. It is also clear that for no other negative m in this class can (**) be satisfied. We have proved the following result.

6-4-1. Proposition. $\mathbf{Q}(\sqrt{m})$ is euclidean (and, in particular, its class number is 1) for $m = -11, -7, -3, -2, -1, 2, 3, 5, 13$. Moreover, there are no other complex quadratic fields which are euclidean.

The problem of determining all euclidean quadratic fields has been solved completely; the final step was accomplished by Barnes and Swinnerton-Dyer in *Acta Mathematika* (vol. 87, pp. 259–323, 1952). The answer is that $\mathbf{Q}(\sqrt{m})$ is euclidean $\Leftrightarrow m = -1, \pm 2, \pm 3, 5, 6, \pm 7, \pm 11, 13, 17, 19, 21, 29, 33, 37, 41, 57, 73$.

A more powerful approach to the class number may be based on the fact [see (5-4-1)] that for each field there exists a constant B (which may be taken as the Minkowski bound) such that every ideal class contains an integral ideal $\mathfrak{A}$ with $\mathfrak{N}\mathfrak{A} \leq B$. In view of (5-4-7) we may take here

$$B = \frac{2}{\pi}\sqrt{|\Delta|} \qquad \text{when } \mathbf{Q}(\sqrt{m}) \text{ is imaginary}$$

$$B = \frac{1}{2}\sqrt{|\Delta|} \qquad \text{when } \mathbf{Q}(\sqrt{m}) \text{ is real}$$

Because B depends on Δ, it is convenient to index the fields $\mathbf{Q}(\sqrt{m}) = \mathbf{Q}(\sqrt{\Delta})$ in terms of their discriminants.

Consider the case $\Delta < 0$. If $\Delta \geq -9$, then $B < 2$, which implies that there is at most one ideal class. In other words, $h = 1$ for $\Delta = -3, -4, -7, -8$—that is, for $m = -1, -2, -3, -7$. Furthermore, if $\Delta = -11, -19$, then $B < 3$ and every ideal class contains an integral ideal of norm 1 or 2. Since $-11 \equiv -19 \equiv 5 \pmod 8$, we know from (6-2-1) that 2 remains prime in both $\mathbf{Q}(\sqrt{-11})$ and $\mathbf{Q}(\sqrt{-19})$. Thus, in either case, the prime ideal $(2) = 2O_E$ has norm 4, and we conclude that there is no integral ideal of norm 2. Therefore, $h = 1$ for $\Delta = -11, -19$.

6-4-2. Proposition. There are exactly nine imaginary quadratic fields of class number 1 with $\Delta > -520$—namely, for

$$\Delta = -3, -4, -7, -8, -11, -19, -43, -67, -163$$

Proof. The first six values of Δ are already under control. The rest of the proof will develop in several steps.

Suppose first that $\alpha \in O_E$, $\alpha \notin \mathbf{Z}$; then write $\alpha = (x + y\sqrt{\Delta})/2$ with $x, y \in \mathbf{Z}$ and note that

$$|N(\alpha)| = \left|\frac{x^2 - \Delta y^2}{4}\right| \geq \frac{|\Delta|}{4}$$

Suppose next that $h = 1$ and that p is a rational prime with $p < |\Delta|/4$. If $\mathfrak{P}$ is a prime ideal of E lying over p, then $\mathfrak{P} = \pi O_E$ for some $\pi \in O_E$ and $p^{f(\mathfrak{P}/p)} = \mathfrak{N}\mathfrak{P} = |N(\pi)| \geq |\Delta|/4$. Therefore, $f(\mathfrak{P}/p) = 2$, p remains prime in E, and $(\Delta/p) = -1$. This simple observation will enable us to win through.

If $h = 1$ and $\Delta < -9$, then $|\Delta|/4 > 2$; hence $(\Delta/2) = -1$, which says that

$$\Delta \equiv 5 \pmod 8$$

This implies, in particular, that for $-19 < \Delta < -11$ the class number cannot be 1. Thus far we have determined for which $\Delta \geq -19$ the class number is 1.

If $h = 1$ and $\Delta < -19$, then $|\Delta|/4 > 5$. The case $\Delta = -20$ does not occur because we know (from the remarks at the beginning of

Chapter 4) that $\mathbf{Q}(\sqrt{-5})$ has class number >1. [In fact, since $B = (2/\pi)\sqrt{20} = \sqrt{80}/\pi < 3$ and the prime 2 ramifies, it is immediate that $h = 2$.] From $(\Delta/3) = (\Delta/5) = -1$ it follows that

$$\Delta \equiv 2 \pmod{3}$$

$$\Delta \equiv 2 \text{ or } 3 \pmod{5}$$

Putting all the conditions together, we see that

$$\Delta \equiv -43 \text{ or } -67 \pmod{120}$$

The only possibilities for Δ are, therefore, -43, -67, -163, -187, -283, -307, -403, -427. Moreover, $|\Delta|$ must be a prime; for otherwise $|\Delta|$ has a prime factor $p < \sqrt{|\Delta|} < |\Delta|/4$, so that $(\Delta/p) = -1$ in contradiction to the fact that p ramifies. Because $403 = (13)(31)$, $427 = (7)(61)$, $-187 \equiv 3^2 \pmod{7}$, $-283 \equiv 2^2 \pmod{7}$, and $-307 \equiv 1^2 \pmod{7}$, only the possibilities $\Delta = -43, -67, -163$ remain.

Consider the case $\Delta = -163$. Here $B < 9$, and one checks trivially that $(-163/p) = -1$ for $p = 2, 3, 5, 7$. Thus 2, 3, 5, and 7 remain prime in $\mathbf{Q}(\sqrt{-163})$, and it follows that $h = 1$. The cases $\Delta = -43$, -67 go the same way. This completes the proof.

It is known (Heilbronn and Linfoot, *Quarterly Journal of Mathematics*, vol. 5, pp. 293–301, 1934) that there exists at most one more $\Delta < 0$ for which $h = 1$, and also that this Δ cannot be $> (-5)(10^9)$ (Lehmer, *Bulletin of the AMS*, vol. 39, p. 360, 1933).

6-4-3. Exercise. Suppose that $\mathbf{Q}(\sqrt{\Delta})$ has class number 1, where $\Delta = 1 - 4n$, $n > 0$ (for example, $n = 41$). If $x, y \in \mathbf{Z}$, $y \neq 0$, $(x, y) = 1$, and $x^2 + xy + ny^2 < n^2$, then $x^2 + xy + ny^2$ is a prime. In particular, $x^2 + x + 41$ is a prime for $x = 0, 1, 2, \ldots, 40$.

6-4-4. Example. By extending the techniques used thus far we may actually find the class number (when it is >1) and the structure of the ideal class group. Rather than describe the method in general terms, let us illustrate it by studying $E = \mathbf{Q}(\sqrt{-47})$.

Here $\Delta = -47$, and $B = (2/\pi)\sqrt{47} < 5$, so that only the primes 2 and 3 concern us. Since $(-47/2) = (-47/3) = 1$, we have in E the factorizations $2O_E = \mathfrak{P}_2\mathfrak{P}_2'$, $\mathfrak{P}_2 \neq \mathfrak{P}_2'$ and $3O_E = \mathfrak{P}_3\mathfrak{P}_3'$, $\mathfrak{P}_3 \neq \mathfrak{P}_3'$. The nonprincipal integral ideals of $\mathbf{Q}(\sqrt{-47})$ with norm ≤ 4 must come from the set $\{\mathfrak{P}_2, \mathfrak{P}_2', \mathfrak{P}_3, \mathfrak{P}_3', \mathfrak{P}_2^2, \mathfrak{P}_2'^2\}$; hence, $h \leq 7$.

Suppose that $\mathfrak{A}$ is an integral ideal. Since any element of O_E can be put in the form $(x + y\sqrt{-47})/2$ with $x, y \in \mathbf{Z}$, we see that if $\mathfrak{A}$ is to be principal then the equation

$$(*) \qquad x^2 + 47y^2 = 4(\mathfrak{N}\mathfrak{A})$$

must have an integral solution (x, y).

Because $\mathfrak{N}\mathfrak{P}_2 = 2$, it is immediate from $(*)$ that $\mathfrak{P}_2$ is not principal—that is, $\mathfrak{P}_2 \not\sim 1$. Furthermore, $\mathfrak{P}_2^2$ is not principal; for if it were, $(*)$ would imply that $\mathfrak{P}_2^2 = 2O_E = \mathfrak{P}_2\mathfrak{P}_2'$, a contradiction. In the same way one checks that $\mathfrak{P}_2^3 \not\sim 1$ and $\mathfrak{P}_2^4 \not\sim 1$. Thus we have representatives for five distinct ideal classes.

According to $(*)$, the only principal ideals of norm 32 are

$$\left(\frac{9 + \sqrt{-47}}{2}\right) \quad \text{and} \quad \left(\frac{9 - \sqrt{-47}}{2}\right)$$

Now, there are exactly six possible integral ideals of norm 32, namely,

$$\mathfrak{P}_2^5,\ \mathfrak{P}_2^4\mathfrak{P}_2',\ \mathfrak{P}_2^3\mathfrak{P}_2'^2,\ \mathfrak{P}_2^2\mathfrak{P}_2'^3,\ \mathfrak{P}_2\mathfrak{P}_2'^4,\ \mathfrak{P}_2'^5$$

Of these only $\mathfrak{P}_2^5$ and $\mathfrak{P}_2'^5$ can be principal, because, for example, if $\mathfrak{P}_2^4\mathfrak{P}_2'$ is principal, then so is $\mathfrak{P}_2^3$, a contradiction. It follows that $\mathfrak{P}_2^5$ is $\left(\frac{9 + \sqrt{-47}}{2}\right)$ or $\left(\frac{9 - \sqrt{-47}}{2}\right)$. In particular $\mathfrak{P}_2^5 \sim 1$, and we conclude that $h = 5$.

6-4-5. Exercise. Show that $h = 2$ for $\Delta = -15, -20, -24, -35, -40$, $h = 3$ for $\Delta = -23, -31$, and $h = 4$ with cyclic ideal class group for $\Delta = -39$. This takes care of all negative $\Delta > -50$.

6-4-6. Exercise. Consider the real quadratic field $\mathbf{Q}(\sqrt{m}) = \mathbf{Q}(\sqrt{\Delta})$; show that its class number is 1 for all $\Delta < 100$ except possibly $\Delta = 40$, 60, 65, 85. (*Hint:* One must show that the prime ideals lying over 2 and 3 are principal.)

It can be shown that for each of the four possible exceptions the class number is 2. The tools needed for this are described, for example, in J. Sommer's "Vorlesungen über Zahlentheorie" (Leipzig-Berlin, 1907). This useful book contains tables of fundamental units and class numbers for quadratic fields, and many concrete examples.

There exist straightforward procedures, going back to Gauss (with transcendental derivations), for the determination of the class number of any quadratic field (see [13], pp. 193–201); however, it is still not known, for example, whether or not there exist an infinite number of real quadratic fields with $h = 1$.

6-5. The Local Situation

In this section we are concerned with the quadratic extensions of the p-adic fields $\mathbf{Q}_p$. The results, which are of interest in themselves, will be of use in Section 6-6.

Fix $p \in \mathfrak{M}(\mathbf{Q})$. From the simple arguments employed in Section 6-1, it follows that any quadratic extension of $\mathbf{Q}_p$ can be expressed in the form $\mathbf{Q}_p(\sqrt{a})$ for some $a \in \mathbf{Q}_p^*$ and that $\mathbf{Q}_p(\sqrt{a}) = \mathbf{Q}_p(\sqrt{b})$ for $a, b \in \mathbf{Q}_p^*$ if and only if $a/b \in \mathbf{Q}_p^{*2}$. Thus, the quadratic extensions of $\mathbf{Q}_p$ are in 1-1 correspondence with the elements (excluding the identity) of the group $\mathbf{Q}_p^*/\mathbf{Q}_p^{*2}$. It is not hard to do the counting; in fact, we have the following result.

6-5-1. Proposition. We have

$$(\mathbf{Q}_p^*:\mathbf{Q}_p^{*2}) = \begin{cases} 2 & \text{for } p = \infty \\ 4 & \text{for } p \text{ odd} \\ 8 & \text{for } p = 2 \end{cases}$$

Proof. The case $p = \infty$ is trivial. The remaining assertions may be proved easily on the basis of (3-1-4), but we prefer to prove a more general result.

6-5-2. Proposition. Let F be a field of characteristic p which is complete with respect to the discrete prime divisor P and with finite-residue class field $\bar{F}$; let m be a natural number with $(m, p) = 1$ (no restriction on m when $p = 0$); let m_0 be the number of mth roots of unity in F; let $\varphi_P \in P$ be the normalized valuation given by $\varphi_P(a) = (\mathfrak{N}P)^{-\nu_P(a)}$; then

$$(F^*:F^{*m}) = \frac{mm_0}{\varphi_P(m)}$$

Moreover, this index relation holds when F is complete under an archimedean prime P and φ_P is the normalized valuation as given in (5-1-2).

Proof. The archimedean cases are easily checked. For the discrete case, we shall make use of an elementary lemma which says that, if A, B, C are abelian groups with $B \subset A$ and $\Theta:A \to C$ is a homomorphism, then

$$(A:B) = (\Theta A:\Theta B)(\ker(\Theta|A):\ker(\Theta|B))$$

From the homomorphism $\nu_P:F^* \twoheadrightarrow \mathbf{Z}$ with kernel $U = U_P$, we see that

$$(F^*:F^{*m}) = (\mathbf{Z}:m\mathbf{Z})(U:U^m) = m(U:U^m)$$

If we put $s = \nu_P(m)$ and fix any $i \geq s + 1$, then (3-1-6) applies (for this i) to the map $a \to a^m$ of $U \to U^m$. Since the kernel on U_i is trivial, it follows that

$$(U:U_i) = (U^m:U_i^m)m_0$$

Because $(U:U_i^m)$ is equal to both $(U:U^m)(U^m:U_i^m)$ and $(U:U_i)(U_i:U_i^m)$, we conclude that

$$\begin{aligned}(F^*:F^{*m}) = m(U:U^m) &= m\,\frac{(U:U_i)(U_i:U_i^m)}{(U^m:U_i^m)}\\ &= mm_0\,\frac{(U^m:U_i^m)(U_i:U_{i+s})}{(U^m:U_i^m)}\\ &= mm_0(\mathfrak{N}P)^s\\ &= \frac{mm_0}{\varphi_P(m)}\end{aligned}$$

6-5-3. Exercise. Let the hypotheses be as in (6-5-2), and suppose that $m = 2 = m_0$. Let E/F be a quadratic extenson; then:

(i) The complete inverse image of F^{*2} with respect to the norm from E into F is $E^{*2}F^*$—that is, $N^{-1}(F^{*2}) = E^{*2}F^*$.

(ii) $(E^{*2} \cap F^*:F^{*2}) = 2$

(iii) $(E^*:E^{*2}) = (F^*:F^{*2})^2/2(F^*:NE^*)$

Hint: Start from $NE^*/F^{*2} \approx E^*/E^{*2}F^* \approx (E^*/E^{*2})/(E^{*2}F^*/E^{*2})$.

(iv) $(F^*:NE^*) = 2$

The quadratic extensions of $\mathbf{Q}_p$ having been counted it remains to exhibit them—and this may be done without reference to (6-5-1).

Suppose that $p \neq 2$. Any $a \in \mathbf{Q}_p^*$ can be written uniquely in the form $a = p^{\nu_p(a)}\eta$ with $\eta \in U_p$. Let r be an ordinary integer such that $\bar{r}$ is a quadratic nonresidue (mod p); in particular, $r \in U_p$. Thus exactly one of $\bar{\eta}$, $\overline{\eta r^{-1}}$ is a quadratic residue (mod p), and, according to (3-1-4), exactly one of η, ηr^{-1} is a square in U_p—write it as ε^2. Therefore, any $a \in \mathbf{Q}_p^*$ may be put in the form

$$a = p^{\nu_p(a)}r^\delta\varepsilon^2$$

where ε is a unit and $\delta = 0$ or 1. Since elements of $\mathbf{Q}_p^*$ that differ multiplicatively by a square lead to the same quadratic extension, it follows that there are exactly three quadratic extensions of $\mathbf{Q}_p$, namely,

$$\mathbf{Q}_p(\sqrt{p}) \qquad \mathbf{Q}_p(\sqrt{rp}) \qquad \mathbf{Q}_p(\sqrt{r})$$

The first two of these are generated by roots of Eisenstein polynomials, so they are fully ramified. From (3-2-6) and the choice of r it is clear that $\mathbf{Q}_p(\sqrt{r})/\mathbf{Q}_p$ is unramified. We have proved the following result.

6-5-4. Proposition. For any $p \neq 2$ there are exactly three quadratic extensions of $\mathbf{Q}_p$; they are of form

$$\mathbf{Q}_p(\sqrt{p}) \qquad \mathbf{Q}_p(\sqrt{rp}) \qquad \mathbf{Q}_p(\sqrt{r})$$

where r is an integer which is a quadratic nonresidue (mod p). The first two extensions are totally ramified, while the third is unramified.

Now suppose that $p = 2$. As before, any $a \in \mathbf{Q}_2^*$ is of form $a = 2^{\nu_2(a)}\eta$ with $\eta \in U_2$. Consider the four units

$$\begin{aligned} \varepsilon_1 &= +1 \\ \varepsilon_2 &= -1 = 1 + 2 + 2^2 + 2^3 + \cdots \\ \varepsilon_3 &= +3 = 1 + 2 + 0 + 0 + \cdots \\ \varepsilon_4 &= -3 = 1 + 0 + 2^2 + 2^3 + \cdots \end{aligned}$$

The unit η is congruent to exactly one of these (mod 8)—say, $\eta \equiv \varepsilon_i$ (mod 8). Then

$$\nu_2\left(\frac{\eta}{\varepsilon_i} - 1\right) = \nu_2(\eta - \varepsilon_i) \geq 3$$

and, according to (3-1-4), η/ε_i is a square in U_2—call it ε^2. Thus, any $a \in \mathbf{Q}_2^*$ can be put in the form

$$a = 2^{\nu_2(a)}\varepsilon_i\varepsilon^2$$

where ε is a unit and $\varepsilon_i = \pm 1, \pm 3$. It follows that there are exactly seven quadratic extensions of $\mathbf{Q}_2$, namely, $\mathbf{Q}_2(\sqrt{-1})$, $\mathbf{Q}_2(\sqrt{3})$, $\mathbf{Q}_2(\sqrt{-3})$, $\mathbf{Q}_2(\sqrt{-2})$, $\mathbf{Q}_2(\sqrt{6})$, $\mathbf{Q}_2(\sqrt{-6})$, $\mathbf{Q}_2(\sqrt{2})$. By Eisenstein, the last four of these are totally ramified. Furthermore, the change of variable $x \to x + 1$ shows that the first two are also totally ramified. It remains to consider $\mathbf{Q}_2(\sqrt{-3})$. Because $\mathbf{Q}_2(\sqrt{-3}) = \mathbf{Q}_2((-1 + \sqrt{-3})/2)$ and the minimum polynomial of $(-1 + \sqrt{-3})/2$ over $\mathbf{Q}_2$ is $x^2 + x + 1$, it follows that $(-1 + \sqrt{-3})/2$ is a unit and [from (3-2-6)] that the extension is unramified. We have proved the following proposition.

6-5-5. Proposition. There are exactly seven quadratic extensions of $\mathbf{Q}_2$; they are of form

$$\mathbf{Q}_2(\sqrt{-2}) \qquad \mathbf{Q}_2(\sqrt{2}) \qquad \mathbf{Q}_2(\sqrt{-6}) \qquad \mathbf{Q}_2(\sqrt{6})$$
$$\mathbf{Q}_2(\sqrt{-1}) \qquad \mathbf{Q}_2(\sqrt{3}) \qquad \mathbf{Q}_2(\sqrt{-3})$$

All of these are totally ramified, except the last, which is unramified.

6-6. Norm Residue Symbol

For every prime divisor p of $\mathbf{Q}$, including the infinite prime, we define the ***norm residue symbol*** (a notion introduced by Hilbert) by putting, for $a, b \in \mathbf{Q}_p^*$,

$$(a, b) = \begin{cases} +1 & \text{if } a \text{ is the norm of an element of } \mathbf{Q}_p(\sqrt{b}) \\ -1 & \text{otherwise} \end{cases}$$

When it is necessary to keep p in mind, we write $(a, b)_p$, and if $(a, b)_p = 1$, we say that a is a ***norm residue*** of $\mathbf{Q}_p(\sqrt{b})$.

Note that when $p = \infty$ the value of the norm residue symbol is given by

$$(a, b)_\infty = -1 \Leftrightarrow a < 0 \text{ and } b < 0$$

The problem of computing $(a, b)_p$ when $p \neq \infty$ will be handled in several steps.

6-6-1. Proposition. Let $F = \mathbf{Q}_p$ (any p), and consider $a, b, \in F^*$; then

(i) $(a^2, b) = (a, b^2) = 1$

(ii) $(a, b) = 1 \Leftrightarrow$ there exist $x, y \in F$ such that $ax^2 + by^2 = 1$. In particular, $(a, b) = (b, a)$.

(iii) $(a, -a) = 1$

(iv) $(a, 1 - a) = 1$ for $a \neq 0, 1$

Proof. (i) Trivial.

(ii) $\Rightarrow$: If $b \in F^{*2}$, then $1/\sqrt{b} \in F^*$ and $a \cdot 0 + b(1/\sqrt{b})^2 = 1$. If $b \notin F^{*2}$, then there exists, by hypothesis, an element $\alpha = c + d\sqrt{b}$ $c, d \in F$ such that $a = N\alpha = c^2 - d^2b$. In the case $c \neq 0$ we have $a(1/c)^2 + b(d/c)^2 = 1$; and if $c = 0$, then $d \neq 0$ and

$$a\left(\frac{1 - b}{2bd}\right)^2 + b\left(\frac{1 + b}{2b}\right)^2 = 1$$

$\Leftarrow$: In view of (i) we may assume that $b \notin F^{*2}$; then $x \neq 0$, and $ax^2 + by^2 = 1$ implies that $a = (1/x)^2 - (y/x)^2b = N(1/x + (y/x)\sqrt{b})$.

(iii) Immediate from $a = N(\sqrt{-a})$.

(iv) Immediate from $a \cdot 1^2 + (1 - a) \cdot 1^2 = 1$.

6-6-2. Proposition. Let $F = \mathbf{Q}_p$ $(p \neq \infty)$; then for $a_1, a_2, b \in F^*$ we have

$$(a_1a_2, b) = (a_1, b)(a_2, b)$$

Proof. Because of (6-6-1(i)) we need consider only the case $b \notin F^{*2}$. Putting $E = F(\sqrt{b})$, we recall [see (6-5-3)] that $(F^*:NE^*) = 2$, so that for any $a \in F^*$ both a and a^{-1} belong to the same coset of NE^* in F^*. Therefore, $(a_1a_2, b) = 1 \Leftrightarrow a_1a_2 \in NE^* \Leftrightarrow a_1^{-1}$ and a_2 are in the same coset of NE^* in $F^* \Leftrightarrow a_1$ and a_2 are in the same coset of NE^* in $F^* \Leftrightarrow (a_1, b) = (a_2, b) \Leftrightarrow (a_1, b)(a_2, b) = 1$.

It follows from the preceding results that for $p \neq \infty$ the norm residue symbol determines a pairing of $\mathbf{Q}_p^*/\mathbf{Q}_p^{*2} \times \mathbf{Q}_p^*/\mathbf{Q}_p^{*2} \to \{\pm 1\}$ which is both symmetric and multiplicative. In particular, if $b \notin \mathbf{Q}_p^{*2}$, then because $(\mathbf{Q}_p^*:N[(\mathbf{Q}_p(\sqrt{b}))^*]) = 2$, the map $a \to (a, b)$ is a homomorphism of $\mathbf{Q}_p^*$ (or of $\mathbf{Q}_p^*/\mathbf{Q}_p^{*2}$) onto $\{\pm 1\}$.

When $p = 2$, the information at hand suffices for the computation of $(a, b)_2$. We recall [see (6-5-5)] that $(\mathbf{Q}_2^*:\mathbf{Q}_2^{*2}) = 8$ and that the elements ± 1, ± 3, ± 2, ± 6 may be taken as the coset representatives. The complete table of values of $(a, b)_2$ is then as follows:

$a \backslash b$	1	−1	3	−3	2	−2	6	−6
1	1	1	1	1	1	1	1	1
−1	1	−1	−1	1	1	−1	−1	1
3	1	−1	−1	1	−1	1	1	−1
−3	1	1	1	1	−1	−1	−1	−1
2	1	1	−1	−1	1	1	−1	−1
−2	1	−1	1	−1	1	−1	1	−1
6	1	−1	1	−1	−1	1	−1	1
−6	1	1	−1	−1	−1	−1	1	1

The reader should have no difficulty in finding all the entries in the table; one uses (6-6-1), (6-6-2), and the fact that any row or column except the first has exactly four $+1$s (with corresponding cosets forming a group) and four -1s.

Let us note an important application of the table. Consider any odd prime p. Then, as noted in the proof of (6-5-5), p and its residue (mod 8) are in the same coset of $\mathbf{Q}_2^{*2}$ in $\mathbf{Q}_2^*$. By checking the entries of the table (note that only the 5×4 rectangle in the upper left-hand corner is needed) we observe that the following proposition applies.

6-6-3. Proposition. Let p and q be odd primes, $p \neq q$; then

(i) $$(-1, p)_2 = \begin{cases} 1 & \text{if } p \equiv 1 \pmod 4 \\ -1 & \text{if } p \equiv 3 \pmod 4 \end{cases}$$

(ii) $$(2, p)_2 = \begin{cases} 1 & \text{if } p \equiv \pm 1 \pmod 8 \\ -1 & \text{if } p \equiv \pm 3 \pmod 8 \end{cases}$$

(iii) $$(q, p)_2 = \begin{cases} -1 & \text{if } p \equiv q \equiv 3 \pmod 4 \\ 1 & \text{otherwise} \end{cases}$$

In order to compute $(a, b)_p$ when p is odd, it is convenient to introduce some notation. Suppose that u is a unit of $\mathbf{Q}_p$. Then its image $\bar{u}$ in the residue class field $\bar{\mathbf{Q}}_p = \{0, 1, 2, \ldots, p-1\}$ is not zero, and one may speak of the value of the Legendre symbol $(\bar{u}/p)$. We put

$$\left(\frac{u}{p}\right) = \left(\frac{\bar{u}}{p}\right)$$

and observe [see (3-1-4)] that

$$\left(\frac{u}{p}\right) = 1 \Leftrightarrow u \in \mathbf{Q}_p^{*2}$$

6-6-4. Proposition. Suppose that u and v are units of $\mathbf{Q}_p$, p odd; then

(i) $$(p, p) = \left(\frac{-1}{p}\right)$$

(ii) $$(u, p) = \left(\frac{u}{p}\right)$$

(iii) $$(u, v) = 1$$

Proof. (i) If $(-1/p) = 1$, then $-1 \in \mathbf{Q}_p^{*2}$ and $(-1, p) = 1$; therefore, $(p, p) = (-1, p)(-p, p) = 1$. Furthermore, we know that $\mathbf{Q}_p(\sqrt{p})/\mathbf{Q}_p$ is totally ramified and that $\{1, \sqrt{p}\}$ is an integral basis [see (6-5-4)]. It follows that if $(p, p) = 1$ then there exist $c, d \in O_p$ such that $p = N(c + d\sqrt{p}) = c^2 - d^2p$. It is clear that $\nu_p(c) \geq 1$, so that $c_1 = c/p \in O_p$. Thus, $-1 = -cc_1 + d^2$, and $-1 = \bar{d}^2$ in $\bar{\mathbf{Q}}_p$—that is, $(-1/p) = 1$.

(ii) The proof goes like that of (i).

(iii) We need to consider only the case where both u and v are not in $\mathbf{Q}_p^{*2}$ and show that there exist $x, y \in \mathbf{Q}_p$ such that

$$u = N(x + y\sqrt{v}) = x^2 - y^2v$$

If $(-v/p) = -1$ [that is, if $(-1/p) = 1$], then, because our generalized Legendre symbol is multiplicative, $(-v^{-1}u/p) = 1$ and there exists

$y \in \mathbf{Q}_p$ with $y^2 = u/-v$. Take this y, and $x = 0$. On the other hand, suppose that $(-v/p) = 1$. Let $a \in \{1, 2, \ldots, p-1\}$ be the smallest nonresidue (mod p); so $a - 1 \not\equiv 0$ and $((a-1)/p) = 1$. It follows that, for some $b \in \mathbf{Q}_p^*$, we have $(a-1)/-v = b^2$ and $((1 - vb^2)/p) = 1$. Consequently, $u/(1 - vb^2) \in \mathbf{Q}_p^{*2}$, and if we call this element x^2 and take $y = bx$, then the proof is complete.

Now that we can compute $(a, b)_p$ for all primes p and all $a, b \in \mathbf{Q}_p^*$, it is easy to connect the norm residue symbol with the quadratic reciprocity law, which says, of course, that if p and q are odd primes then

$$\left(\frac{-1}{p}\right) = (-1)^{(p-1)/2} = \begin{cases} 1 & \text{if } p \equiv 1 \pmod 4 \\ -1 & \text{if } p \equiv 3 \pmod 4 \end{cases}$$

$$\left(\frac{2}{p}\right) = (-1)^{(p^2-1)/8} = \begin{cases} 1 & \text{if } p \equiv \pm 1 \pmod 8 \\ -1 & \text{if } p \equiv \pm 3 \pmod 8 \end{cases}$$

$$\left(\frac{p}{q}\right)\left(\frac{q}{p}\right) = (-1)^{\left(\frac{p-1}{2}\right)\left(\frac{q-1}{2}\right)} = \begin{cases} -1 & \text{if } p \equiv q \equiv 3 \pmod 4 \\ 1 & \text{otherwise} \end{cases}$$

6-6-5. Theorem (Hilbert). The quadratic reciprocity law holds if and only if $\prod_{\text{all } p} (a, b)_p = 1$ for all $a, b \in \mathbf{Q}^*$.

Proof. It is understood that in the infinite product almost all terms have to be 1, and it is easy to check that the "product formula" holds for all $a, b \in \mathbf{Q}^*$ if and only if it holds for the pairs $(-1, -1)$, $(-1, 2)$, $(2, 2)$, $(-1, p)$, $(2, p)$, and (q, p), where p and q are odd primes.

Observe first that for any p we have $(-1, -1)_p = 1$, $(-1, 2)_p = (-1, 1 - (-1))_p = 1$ and $(2, 2)_p = (2, -2)_p(2, -1)_p = 1$. For the remainder, let p and q be odd. Since $(-1, p)_\infty = 1$, $(-1, p)_p = (-1/p)$ and $(-1, p)_{p'} = 1$ for $p' \neq p$ with p' odd, it follows from (6-6-3) that

$$\prod_{\text{all } p'} (-1, p)_{p'} = 1 \Leftrightarrow (-1, p)_2 = \left(\frac{-1}{p}\right) \Leftrightarrow \left(\frac{-1}{p}\right) = (-1)^{(p-1)/2}$$

The two remaining cases go exactly the same way.

Because of this result, the product formula $\Pi(a, b)_p = 1$ may be viewed as a general reciprocity law which is related to the decomposition law for ideals. Since the quadratic reciprocity law is well known and easily accessible, it is natural to view this product formula as a consequence of it. However, the product formula can be proved directly without difficulty. This is the approach that is taken in the generalizations.

6-6-6. Remark. Consider the field $\mathbf{Q}(\sqrt{m})$, and let $p_1, \ldots, p_t$ be the distinct rational primes which divide its discriminant. For each $i = 1, \ldots, t$ we have a homomorphism $\chi_{p_i}:\mathbf{Q}^* \to \{\pm 1\}$ given by

$$\chi_{p_i}(a) = (a, m)_{p_i}$$

The t-tuple

$$\chi_{p_1}(a), \ldots, \chi_{p_t}(a)$$

is called the ***character system of*** a.

Suppose further that $\mathfrak{A}$ is an ideal of $\mathbf{Q}(\sqrt{m})$. If $\mathbf{Q}(\sqrt{m})$ is imaginary, we put $\overline{\mathfrak{N}\mathfrak{A}} = \mathfrak{N}\mathfrak{A}$, $s = t$, and $\chi_{p_i}(\mathfrak{A}) = (\overline{\mathfrak{N}\mathfrak{A}}, m)_{p_i}$ and call the s-tuple

$$(*) \qquad \chi_{p_1}(\mathfrak{A}), \ldots, \chi_{p_s}(\mathfrak{A})$$

the ***character system of*** $\mathfrak{A}$. In the case where $\mathbf{Q}(\sqrt{m})$ is real, we begin by considering the character system of -1. If -1 does not appear, we put $\overline{\mathfrak{N}\mathfrak{A}} = \mathfrak{N}\mathfrak{A}$, $s = t$, and $\chi_{p_i}(\mathfrak{A}) = (\overline{\mathfrak{N}\mathfrak{A}}, m)_{p_i}$ and call (*) the ***character system of*** $\mathfrak{A}$. On the other hand, if -1 does appear, we may reindex so that $\chi_{p_t}(-1) = -1$. We then put $s = t - 1$ and $\chi_{p_i}(\mathfrak{A}) = (\overline{\mathfrak{N}\mathfrak{A}}, m)_{p_i}$ (for $i = 1, \ldots, t$), where $\overline{\mathfrak{N}\mathfrak{A}}$ is the member of $\{\mathfrak{N}\mathfrak{A}, -\mathfrak{N}\mathfrak{A}\}$ for which $\chi_{p_t}(\overline{\mathfrak{N}\mathfrak{A}}) = 1$, and call (*) the ***character system of*** $\mathfrak{A}$. These definitions are formulated in such a way that ideals belonging to the same ideal class have the same character system.

The mapping

$$\chi:\mathfrak{A} \to (\chi_{p_1}(\mathfrak{A}), \ldots, \chi_{p_s}(\mathfrak{A}))$$

of $\mathfrak{J}$ into the direct product of s copies of the two-element group $\{\pm 1\}$ is a homomorphism (we can also view χ as a mapping defined on the ideal class group $\mathfrak{J}/\mathfrak{J}^*$). Its kernel $\mathfrak{G}$ contains $\mathfrak{J}^*$ and is called the ***principal genus*** (*Hauptgeschlecht*); each coset of $\mathfrak{G}$ in $\mathfrak{J}$ is called a ***genus*** (*Geschlecht*). Each genus is the union of ideal classes, and the number of ideal classes in each genus is the same. The square of any ideal (or ideal class) is contained in $\mathfrak{G}$.

Obviously, knowing the number of genera provides some information about the class number. For any $\mathfrak{A} \in \mathfrak{J}$, it is easy to see that

$$\prod_{i=1}^{s} \chi_{p_i}(\mathfrak{A}) = 1$$

so that $(\mathfrak{J}:\mathfrak{G}) \leq 2^{s-1}$. Actually, it is known (see [8]) that

$$(\mathfrak{J}:\mathfrak{G}) = 2^{s-1}$$

but we do not give a proof. The standard proof involves the notion of ambient ideal and the following facts, among others:

1. If $a, b \in \mathbf{Q}^*$ and $b \notin \mathbf{Q}^{*2}$, then a is norm from $\mathbf{Q}(\sqrt{b}) \Leftrightarrow (a, b)_p = 1$ for all $p \Leftrightarrow a$ is norm from $\mathbf{Q}_p(\sqrt{b})$ for all p; in other words, the equation $ax^2 + by^2 = 1$ has a solution in $\mathbf{Q} \Leftrightarrow$ it has a solution in every $\mathbf{Q}_p$.

2. If an ideal class is contained in $\mathfrak{G}$, then it is the square of some ideal class.

3. Any set of s plus or minus ones lies in the image of χ.

7

Cyclotomic Fields

In this chapter we present further illustrations of the general theory by studying the class of cyclotomic fields. This is a vast subject and we shall barely scratch the surface.

7-1. Elementary Facts

Consider an algebraic closure Ω of the rational field $\mathbf{Q}$, and view Ω as a subset of some concrete copy $\mathbf{C}$ of the complex field. For any integer $m > 2$ (unless explicit mention is made to the contrary, we do not consider the case $m = 2$, because, as a rule, everything is then trivial) the polynomial $x^m - 1$ has m distinct roots $\xi_1, \ldots, \xi_m$ in Ω; they are all the mth roots of unity in Ω. These elements form a multiplicative group, which must, therefore, by cyclic. Any generator of the group of mth roots of unity is called a ***primitive mth root of unity*** and is denoted by ζ—or by ζ_m if the context so requires. For any integer a, ζ^a is a primitive mth root of unity $\Leftrightarrow$ $(a, m) = 1$; so there are $\phi(m)$ primitive mth roots of unity.

The field $E = \mathbf{Q}(\xi_1, \ldots, \xi_m) = \mathbf{Q}(\zeta)$ is known as the ***field of mth roots of unity.*** Any such E (which may be denoted by E_m when necessary) is called a ***cyclotomic field.***

7-1-1. Proposition. $E/\mathbf{Q}$ is an abelian extension of degree $\phi(m)$; in fact, $\mathcal{G}(E/\mathbf{Q})$ is isomorphic to the multiplicative group of integers (mod m) which are relatively prime to m [that is, to the multiplicative group of units of the ring $\mathbf{Z}/(m)$]. The conjugates of $\zeta = \zeta_m$ are precisely the primitive mth roots of unity, and

$$f(\zeta, \mathbf{Q}) = \prod_{(a,m)=1} (x - \zeta^a)$$

(meaning that the product is taken over the residue classes prime to m). We have also

$$(-1)^{(m-1)}m^m = \prod_{j \neq i} (\xi_i - \xi_j)$$

Proof. $E/\mathbf{Q}$ is the splitting field of the separable polynomial $x^m - 1$; so it is a Galois extension. Consider the minimum polynomial $f(x) = f(\zeta, \mathbf{Q}) \in \mathbf{Q}[x]$. It is easy to see that every root of $f(x)$ is a primitive mth root of unity and that for $\sigma \in \mathcal{G} = \mathcal{G}(E/\mathbf{Q})$ we have $\sigma\zeta = \zeta^a$, where $1 \leq a < m$ and $(a, m) = 1$. It is clear that the map $\sigma \rightarrow a$ is an isomorphism of $\mathcal{G}$ into the group of units of $\mathbf{Z}/(m)$. Of course, $[E:\mathbf{Q}] = \deg f(x) \leq \phi(m)$.

The crucial step is to show that

$$(a, m) = 1 \Rightarrow f(\zeta^a) = 0$$

To accomplish this, it suffices to prove that, if $\xi \in O_E$ satisfies $f(\xi) = 0$ and p is a prime not dividing m, then $f(\xi^p) = 0$ (as a may then be built up from such primes one step at a time). Now, $f(x) \in \mathbf{Z}[x]$, and by Fermat's theorem $f(x^p) \equiv [f(x)]^p \pmod{p}$ (meaning that the polynomial $f(x^p) - f(x)^p \in \mathbf{Z}[x]$ has all coefficients divisible by p). Therefore, $f(\xi^p) \in pO_E$, and p divides $f(\xi^p)$ in O_E. Suppose that $f(\xi^p) \neq 0$; we shall soon see that this assumption leads to $f(\xi^p)|m^m$—a contradiction, since $p \nmid m^m$.

From

$$x^m - 1 = \prod_{j=1}^{m} (x - \xi_j)$$

we have

$$\prod_{j=1}^{m} \xi_j = (-1)^{m-1}$$

and by differentiating and setting $x = \xi_i$ we get

$$m\xi_i^{m-1} = \prod_{j \neq i} (\xi_i - \xi_j)$$

Therefore,

$$(-1)^{(m-1)}m^m = \prod_{j \neq i} (\xi_i - \xi_j)$$

Since $f(x)|(x^m - 1)$ and ξ^p is one of the ξ_i, it follows that $f(\xi^p)$ can be written as a product of certain of the $(\xi_i - \xi_j)$. Hence, $f(\xi^p)|m^m$, and the entire proof is complete.

The polynomial

$$\Phi_m(x) = \prod_{(a,m)=1} (x - \zeta_m^a)$$

is known as the ***cyclotomic polynomial*** of index m. We have proved that it has coefficients in $\mathbf{Z}$ and is irreducible over $\mathbf{Q}$—that is, $\Phi_m(x) = f(\zeta_m, \mathbf{Q})$. The definition applies just as well to $m = 1, 2$ so that $\Phi_1(x) = x - 1$ and $\Phi_2(x) = x + 1$. It is clear that

$$x^m - 1 = \prod_{d|m} \Phi_d(x)$$

and then the Möbius inversion formula facilitates the computation of $\Phi_m(x)$, namely,

$$\Phi_m(x) = \prod_{d|m} (x^{m/d} - 1)^{\mu(d)}$$

where μ is the Möbius function.

From the well-known structure of the group of units of $\mathbf{Z}/(m)$, we conclude that if $m = \Pi p_i^{s_i}$ is the factorization of m then $\mathcal{G}(E_m/\mathbf{Q})$ is isomorphic to the direct product of the $\mathcal{G}(E_{p_i^{s_i}}/\mathbf{Q})$. Note that there is no difficulty if some $p_i^{s_i} = 2$. Furthermore, we have the following corollary.

7-1-2. Corollary. If m is a power of an odd prime, then $\mathcal{G}(E/\mathbf{Q})$ is cyclic. If $m = 2^2$, then $\mathcal{G}(E/\mathbf{Q})$ is cyclic. If $m = 2^s$ $(s \geq 3)$, then $\mathcal{G}(E/\mathbf{Q})$ is the direct product of two cyclic groups; the first is of order 2 with the element $\sigma \in \mathcal{G}(E/\mathbf{Q})$ for which $\sigma\zeta_m = \zeta_m^{-1}$ as generator, and the second is of order 2^{s-2} with the element $\tau \in \mathcal{G}(E/\mathbf{Q})$ for which $\tau\zeta_m = \zeta_m^5$ as generator.

If it is not clear already, it will surely become clear in Section 7-6 that when $m = 2^s$ $(s \geq 3)$ E_m is the composite of $\mathbf{Q}(i)$ and $\mathbf{Q}(\zeta_m + \zeta_m^{-1})$.

The appropriate part of the proof of (7-1-1) obviously carries over to give the following proposition.

7-1-3. Proposition. Let F be an arbitrary algebraic number field, and put $E = F(\zeta_m) = F(\xi_1, \ldots, \xi_m)$. Then E/F is an abelian extension of degree $\leq \phi(m)$; in fact, $\mathcal{G}(E/F)$ is isomorphic to a subgroup of the group of units of $\mathbf{Z}/(m)$.

7-1-4. Exercise (Kronecker). In the situation of (7-1-3) we have $[E:F] = \phi(m)$—that is, $\Phi_m(x)$ is irreducible over F—if the discriminant of F is prime to m. [*Hint:* Look ahead and use (7-2-1).]

A famous theorem of Kronecker (which is a consequence of class field theory and is, therefore, beyond the scope of this book) says that an algebraic number field is an abelian extension of $\mathbf{Q}$ if and only if it is contained in a cyclotomic field.

7-2. Unramified Primes

One way to decide which primes $p \in \mathbf{Z}$ are unramified in E is to find Δ_E. This will be done eventually. For the moment, it is sufficient for our purposes to locate a finite set of primes from which the divisors of Δ_E must come. More precisely, we have the following result.

7-2-1. Proposition. Δ_E divides m^m.

Proof. By (4-8-17), Δ_E divides $\Delta(1, \zeta, \ldots, \zeta^{\phi(m)-1})$, and using (3-7-9), we see that $\Delta(1, \zeta, \ldots, \zeta^{\phi(m)-1})$ divides

$$\prod_{i \neq j} (\xi_i - \xi_j) = \pm m^m$$

Thus, any prime p with $p \nmid m$ is unramified (we shall see later that with one possible exception any p which divides m is ramified), and there is associated with it the Frobenius automorphism $[E/p]$ of the abelian extension $E/\mathbf{Q}$. $[E/p]$ is characterized, as usual, by the property

$$\left[\frac{E}{p}\right] \alpha \equiv \alpha^p \pmod{pO_E} \qquad \forall \alpha \in O_E$$

However, one can go further and exhibit $[E/p]$ explicitly.

7-2-2. Proposition. If $p \nmid m$, then $[E/p]$ is the element of $\mathcal{G}(E/\mathbf{Q})$ which maps $\zeta \to \zeta^p$.

Proof. As noted in the proof of (7-1-1), $[E/p]\zeta = \zeta^a$ for some integer a with $(a, m) = 1$. Therefore, $\zeta^a \equiv \zeta^p \pmod{pO_E}$, and the proof is complete as soon as we verify the following lemma.

7-2-3. Lemma. Suppose that $p \nmid m$ and that $a, b \in \mathbf{Z}$ are such that $\zeta^a \equiv \zeta^b \pmod{pO_E}$; then $\zeta^a = \zeta^b$.

Proof. Suppose $\zeta^a \neq \zeta^b$. Since ζ is a unit, we have $\zeta^{a-b} \equiv 1 \pmod{pO_E}$ and there exists j with $0 < j < m$ such that $p|(1 - \zeta^j)$. On the other hand, the factorization

$$x^{m-1} + x^{m-2} + \cdots + x + 1 = \prod_{i=1}^{m-1} (x - \zeta^i)$$

implies that

$$m = \prod_{i=1}^{m-1} (1 - \zeta^i)$$

contradicting $p \nmid m$.

The decomposition law for the unramified primes is now immediate.

7-2-4. Theorem. If $p \nmid m$, then p factors in $E = E_m$ into the product of r distinct prime ideals of degree f, where $rf = \phi(m)$ and f is the smallest positive integer such that $p^f \equiv 1 \pmod{m}$.

Proof. Let $\mathfrak{P}_1, \ldots, \mathfrak{P}_r$ be the prime ideals of O_E lying over p; for each of them we have $e(P_i/p) = 1$ and $\mathfrak{N}\mathfrak{P}_i = p^f$, where $f = f(P_i/p)$. In particular,

$$pO_E = \mathfrak{P}_1 \cdots \mathfrak{P}_r$$

and $rf = [E:\mathbf{Q}] = \phi(m)$. Since f is the order of $[E/p]$ and

$$\left[\frac{E}{p}\right]^\nu = 1 \in \mathcal{G}\left(\frac{E}{\mathbf{Q}}\right) \Leftrightarrow \left[\frac{E}{p}\right]^\nu \zeta = \zeta \Leftrightarrow \zeta^{p^\nu} = \zeta \Leftrightarrow p^\nu \equiv 1 \pmod{m}$$

it follows that f is indeed the smallest positive integer such that $p^f \equiv 1 \pmod{m}$.

7-2-5. Corollary. If $p \nmid m$, then p splits completely in E if and only if $p \equiv 1 \pmod{m}$.

From (7-2-2) we see that if p_1 and p_2 are primes which do not divide m, then they have the same Frobenius substitution $\Leftrightarrow$ they are in the same arithmetic progression with difference m. In fact,

$$\left[\frac{E}{p_1}\right] = \left[\frac{E}{p_2}\right] \Leftrightarrow \zeta^{p_1} = \zeta^{p_2} \Leftrightarrow \zeta^{p_1 - p_2} = 1 \Leftrightarrow m|(p_1 - p_2)$$

Thus, two primes in the same arithmetic progression with difference m have the same decomposition law. Of course, the converse is false; for example, in E_8, the field of eighth roots of unity, the primes congruent to 3, 5, or 7 (mod 8) have the same decomposition law. We may also note in passing that, as soon as (7-4-1) has been proved, it will be clear that no rational prime remains prime in E_8.

7-2-6. Exercise. There are an infinite number of primes p with $p \equiv 1 \pmod{m}$. *Hint:* Consider $f(x) = f(\zeta_m, \mathbf{Q}) \in \mathbf{Z}[x]$, and let $M = \{f(n) | n \in \mathbf{Z}\}$. It is easy to see that an infinite number of distinct primes p appear as factors of some element of M. By using (4-9-2) it is not hard to see that if $p | f(n)$ and $(p, m) = 1$ then $p \equiv 1 \pmod{m}$. In particular, a weak form of the theorem of Dirichlet on the infinitude of primes in an arithmetic progression may be proved by strictly algebraic techniques.

7-3. Quadratic Reciprocity Law

We digress from our study of the properties of cyclotomic fields to give an elegant and conceptual proof of the quadratic reciprocity law.

7-3-1. Proposition. Let q be an odd prime, and let ζ be a primitive qth root of unity; then $E = \mathbf{Q}(\zeta)$ contains exactly one quadratic field E', namely,

$$E' = \mathbf{Q}(\sqrt{(-1)^{(q-1)/2}q}\,)$$

Proof. We know that $\mathcal{G} = \mathcal{G}(E/\mathbf{Q})$ is cyclic of order $q - 1$; so $\mathcal{G}$ contains exactly one subgroup $\mathcal{G}'$ of index 2. If E' denotes the fixed field of $\mathcal{G}'$, then E' is the unique quadratic subfield of E.

Now, there exists a square-free integer n such that $E' = \mathbf{Q}(\sqrt{n})$, and then $\Delta_{E'}$ is either n or $4n$. From the chain rule for discriminants and from (7-2-1), we see that $\Delta_{E'}|\Delta_E|q^q$. It follows that $\Delta_{E'} = n$, so that $n \equiv 1 \pmod 4$ and also $n = \pm q$. If $q \equiv 1 \pmod 4$, then $n = q$, and if $q \equiv -1 \pmod 4$, then $n = -q$; consequently,

$$n = (-1)^{(q-1)/2}q$$

The heart of the matter is to connect the decomposition of a prime p in E' with its decomposition in E.

7-3-2. Lemma. Let p be a prime, $p \neq q$; then:

i. p splits in $E' \Leftrightarrow p$ decomposes in E into an even number of factors.

ii. p splits in $E' \Leftrightarrow (p/q) = 1$.

Proof. i. $\Rightarrow$: By hypothesis, there exist distinct prime ideals $\mathfrak{P}$ and $\mathfrak{P}'$ of $O_{E'}$ such that $pO_{E'} = \mathfrak{P}\mathfrak{P}'$. Since $\mathfrak{P}$ and $\mathfrak{P}'$ are conjugate, there exists $\sigma' \in \mathcal{G}(E'/\mathbf{Q})$ such that $\sigma'\mathfrak{P} = \mathfrak{P}'$. Let $\sigma \in \mathcal{G}(E/\mathbf{Q})$ be an extension of σ'. Suppose $\mathfrak{P}O_E = \mathfrak{Q}_1 \cdots \mathfrak{Q}_s$, where the $\mathfrak{Q}_i$ are prime ideals of O_E; then $\mathfrak{P}'O_E = \sigma(\mathfrak{P}O_E) = (\sigma\mathfrak{Q}_1) \cdots (\sigma\mathfrak{Q}_s)$. Therefore, $pO_E = \mathfrak{P}\mathfrak{P}'O_E = (\mathfrak{Q}_1) \cdots (\mathfrak{Q}_s)(\sigma\mathfrak{Q}_1) \cdots (\sigma\mathfrak{Q}_s)$, and these are distinct because p is unramified in E.

$\Leftarrow$: Suppose that $pO_E = \mathfrak{Q}_1 \cdots \mathfrak{Q}_r$ with r even. Let $\mathcal{Z}$ be the decomposition group of $\mathfrak{Q}_1$ over $\mathbf{Q}$, so that $(\mathcal{G}:\mathcal{Z}) = r$. Since r is even and $\mathcal{G}$ is cyclic, it follows that $\mathcal{Z} \subset \mathcal{G}'$. Therefore, $(\mathcal{G}':\mathcal{Z}) = r/2$, and $\mathcal{Z}$ is the decomposition group of $\mathfrak{Q}_1$ over E'. The prime ideal $\mathfrak{P} = \mathfrak{Q}_1 \cap O_{E'}$ of $O_{E'}$ has, consequently, $r/2$ prime factors in O_E. Thus, $pO_{E'} = \mathfrak{P}$ is impossible, and we conclude that p splits in E'.

ii. For this we need to know that, when $(a, q) = 1$,

$$\left(\frac{a}{q}\right) \equiv a^{(q-1)/2} \pmod{q}$$

A proof of this elementary fact starts from the assertion of Fermat's theorem, that is, from $a^{q-1} \equiv 1 \pmod{q}$. More precisely, we have then either (*) $a^{(q-1)/2} - 1 \equiv 0 \pmod{q}$ or $a^{(q-1)/2} + 1 \equiv 0 \pmod{q}$. Now, there are $(q - 1)/2$ distinct quadratic residues (mod q), and each one clearly satisfies (*). Since (*) can have no more than $(q - 1)/2$ solutions, we have them all—that is,

$$a^{(q-1)/2} \equiv 1 \pmod{q} \Leftrightarrow \left(\frac{a}{q}\right) = 1$$

Consider the factorization $pO_E = \mathfrak{Q}_1 \cdots \mathfrak{Q}_r$ with $rf = q - 1$; then p splits in $E' \Leftrightarrow r$ is even $\Leftrightarrow f \mid \frac{q-1}{2} \Leftrightarrow p^{(q-1)/2} \equiv 1 \pmod{q}$ [by (7-2-4)] $\Leftrightarrow \left(\frac{p}{q}\right) \equiv 1 \pmod{q} \Leftrightarrow \left(\frac{p}{q}\right) = 1$.

7-3-3. Quadratic reciprocity law. Let p and q be distinct odd primes; then

(i) $$\left(\frac{-1}{q}\right) = (-1)^{(q-1)/2}$$

(ii) $$\left(\frac{2}{q}\right) = (-1)^{(q^2-1)/8}$$

(iii) $$\left(\frac{p}{q}\right)\left(\frac{q}{p}\right) = (-1)^{[(p-1)/2][(q-1)/2]}$$

Proof

(i) $\left(\frac{-1}{q}\right) \equiv (-1)^{(q-1)/2} \pmod{q} \Leftrightarrow \left(\frac{-1}{q}\right) = (-1)^{(q-1)/2}$

(ii) We have $(2/q) = 1 \Leftrightarrow 2$ splits in $E' = \mathbf{Q}(\sqrt{(-1)^{(q-1)/2}q}) \Leftrightarrow (-1)^{(q-1)/2}q \equiv 1 \pmod{8}$ [see (6-2-1)] $\Leftrightarrow q \equiv (-1)^{(q-1)/2} \pmod{8} \Leftrightarrow q \equiv \pm 1 \pmod{8} \Leftrightarrow q^2 \equiv 1 \pmod{16} \Leftrightarrow \frac{q^2-1}{8}$ is even $\Leftrightarrow (-1)^{(q^2-1)/8} = 1$.

(iii) $\left(\frac{p}{q}\right) = 1 \Leftrightarrow p$ splits in the quadratic field $\mathbf{Q}(\sqrt{(-1)^{(q-1)/2}q}) \Leftrightarrow \left(\frac{(-1)^{(q-1)/2}q}{p}\right) = 1$ [see (6-2-1)]. Therefore,

$$\left(\frac{p}{q}\right) = \left(\frac{(-1)^{(q-1)/2}q}{p}\right) = \left(\frac{-1}{p}\right)^{(q-1)/2}\left(\frac{q}{p}\right) = (-1)^{[(p-1)/2][(q-1)/2]}\left(\frac{q}{p}\right)$$

There exist alternative proofs of the quadratic reciprocity law, based on the properties of quadratic and cyclotomic fields, but we do not discuss them.

7-4. Ramified Primes

Let us return to the notation of Section 7-2 and examine the primes p for which $p|m$.

7-4-1. Proposition. Suppose that m is a power of a prime, say, $m = p^s$. Then p has a unique extension P to E—its degree is 1 (that is, $\mathfrak{N}P = p$) and its ramification index is $\phi(p^s)$ [that is, $e(P/p) = \phi(p^s)$]. In fact, the principal ideal $(1 - \zeta)$ in O_E is prime, and

$$pO_E = (1 - \zeta)^{\phi(p^s)}$$

Proof. Every primitive p^sth root of unity is a root of $x^{p^s} - 1$ but not of $x^{p^{s-1}} - 1$. We have, therefore,

$$\prod_{(i,p^s)=1} (x - \zeta^i) = \frac{x^{p^s} - 1}{x^{p^{s-1}} - 1} = x^{p^{s-1}(p-1)} + x^{p^{s-1}(p-2)} + \cdots + x^{p^{s-1}} + 1$$

and, in particular,

$$p = \prod_{(i,p^s)=1} (1 - \zeta^i)$$

For each such i, $(1 - \zeta^i)/(1 - \zeta) = 1 + \zeta + \cdots + \zeta^{i-1}$ belongs to O_E. In the same way, if we choose j such that $(\zeta^i)^j = \zeta$, then $(j, p^s) = 1$ and $(1 - \zeta)/(1 - \zeta^i) = [1 - (\zeta^i)^j]/[1 - \zeta^i] \in O_E$. Therefore, $1 - \zeta^i$ and $1 - \zeta$ are associates in O_E, and we may write

$$p = \varepsilon(1 - \zeta)^{\phi(p^s)}$$

where ε is some unit in O_E.

If P is any extension of p to E, then $e\left(\frac{P}{p}\right) = e\left(\frac{P}{p}\right) \nu_p(p) = \nu_P(p) = \phi(p^s)\nu_P(1 - \zeta)$. Since $1 \le e(P/p) \le [E:\mathbf{Q}] = \phi(p^s)$, it follows that $\nu_P(1 - \zeta) = 1$ and $e(P/p) = \phi(p^s)$. This completes the proof.

The above proof also yields the following result.

7-4-2. Proposition. Suppose that F is an algebraic number field and that P is a prime divisor of F with $\nu_P(p) = 1$. Let ζ be a primitive p^sth root of unity, and put $E = F(\zeta)$. Then P has a unique extension Q to E, and $f(Q/P) = 1$, $e(Q/P) = \phi(p^s) = [E:F]$. In particular, the cyclotomic polynomial $\Phi_{p^s}(x)$ is irreducible over F.

7-4-3. Theorem. If $p|m$ and we write $m = p^s m'$ with $(p, m') = 1$, then p factors in $E = E_m$ in the form

$$pO_E = (\mathfrak{P}_1 \cdots \mathfrak{P}_r)^{\phi(p^s)}$$

where $\mathfrak{P}_1, \ldots, \mathfrak{P}_r$ are distinct prime ideals of O_E of degree f, $fr = \phi(m')$, and f is the smallest positive integer such that $p^f \equiv 1 \pmod{m'}$.

Proof. First, we observe that, in general, if $m = m_1 \cdots m_t$ with the m_i relatively prime in pairs, then $\zeta_{m_1} \cdots \zeta_{m_t}$ is a primitive mth root of unity and, conversely, any ζ_m can be written in the form $\zeta_{m_1} \cdots \zeta_{m_t}$, where each ζ_{m_i} is a power of ζ_m. In our case, we have, therefore, $\mathbf{Q}(\zeta_m) = \mathbf{Q}(\zeta_{m'}, \zeta_{p^s})$. By moving p up into $\mathbf{Q}(\zeta_{m'})$ and from there into $\mathbf{Q}(\zeta_m)$, we see that the desired result is an immediate consequence of (7-2-4) and (7-4-2)—at least when $m' > 2$ and $p^s > 2$. The case where $m' = 2$ or $p^s = 2$ is trivial.

In summary, our results show that, if p is an odd prime, then p is unramified in $E_m \Leftrightarrow p \nmid m$, and the prime $p = 2$ is unramified in $E_m \Leftrightarrow \nu_2(m) \leq 1$. Moreover, the various constants associated with the factorization of p in E_m are easily determined.

7-4-4. Exercise. The quadratic fields contained in certain cyclotomic fields were described in (7-3-1). A related question is the following: Given a quadratic field, in which cyclotomic fields is it contained?

Any quadratic field is of form $\mathbf{Q}(\sqrt{m})$, where m is a square-free integer. The factorization of m may therefore be put in the form

$$m = \eta 2^t \prod_j (-1)^{(p_j-1)/2} p_j$$

where the p_j are the odd primes appearing in m, $t = 0$ or 1, and $\eta = +1$ or -1, whichever is needed to make the right side equal m. Let us put

$$m' = \prod_j p_j$$

Because we may take $\zeta_4 = i = \sqrt{-1}$ and $\zeta_8 = (1 + i)/\sqrt{2}$ (so that $\zeta_8 + \zeta_8^{-1} = \sqrt{2}$), it is easy to see that

$$\mathbf{Q}(\sqrt{m}) \subset \mathbf{Q}(\zeta_{8m'})$$

Moreover, the smallest cyclotomic field containing $\mathbf{Q}(\sqrt{m})$ is

$$\begin{array}{ll} \mathbf{Q}(\zeta_{m'}) & \text{if } m \equiv 1 \pmod 4 \\ \mathbf{Q}(\zeta_{4m'}) & \text{if } m \equiv -1 \pmod 4 \\ \mathbf{Q}(\zeta_{8m'}) & \text{if } m \equiv 2 \pmod 4 \end{array}$$

If the discriminant of $\mathbf{Q}(\sqrt{m})$ is denoted by Δ_m, then for all m, the smallest cyclotomic field containing $\mathbf{Q}(\sqrt{m}) = \mathbf{Q}(\sqrt{\Delta_m})$ is $\mathbf{Q}(\zeta_{|\Delta_m|})$.

7-5. Integral Basis and Discriminant

The powers of ζ_m provide a basis for the cyclotomic field E_m over $\mathbf{Q}$; in order to show that they are, in fact, an integral basis, we make use of some simple lemmas.

7-5-1. Lemma. If $E = E_m$, $\zeta = \zeta_m$, and $p \nmid m$, then the set of elements $\{1, \zeta, \ldots, \zeta^{\phi(m)-1}\}$ is an integral basis at p for $E/\mathbf{Q}$—that is, $O_{p,E} = O_p[\zeta]$.

Proof. It is clear that $f(x) = f(\zeta, \mathbf{Q}) \in \mathbf{Z}[x] \subset O_p[x]$ and, in fact, that all irreducible factors of $x^m - 1$ belong to $O_p[x]$. We may, therefore, write $x^m - 1 = f(x)g(x)$ with $f(x), g(x) \in O_p[x]$. By differentiating and setting $x = \zeta$ we have

$$m\zeta^{m-1} = f'(\zeta)g(\zeta)$$

with $f'(\zeta), g(\zeta) \in O_p[\zeta] \subset O_{p,E}$. Of course, ζ^{m-1} is a unit of $O_p[\zeta]$. Since $p \nmid m$, m is a unit of O_p and of $O_p[\zeta]$. Thus $f'(\zeta)$ is a unit of $O_p[\zeta]$. Now, an application of (3-7-14) yields

$$\begin{aligned} O_p[\zeta] &= O_p \cdot 1 + \cdots + O_p\zeta^{\varphi(m)-1} \subset O_{p,E} \\ &\subset O^*_{p,E} \subset [O_p(\zeta)]^* = \frac{O_p[\zeta]}{f'(\zeta)} = O_p[\zeta] \end{aligned}$$

7-5-2. Lemma. Suppose that F is an algebraic number field, $\zeta = \zeta_m$, $E = F(\zeta)$, and P is a prime divisor of F such that $\nu_P(m) = 0$; then $O_{P,E} = O_p[\zeta]$.

Proof. Like the above.

7-5-3. Lemma. If $E = E_{p^s}$ and $\zeta = \zeta_{p^s}$, then $\{1, \zeta, \ldots, \zeta^{\phi(p^s)-1}\}$ is an integral basis at p for $E/\mathbf{Q}$.

Proof. In view of (7-4-1), if P is the unique extension of p to E, then the local extension $E_P/\mathbf{Q}_p$ is totally ramified and $1 - \zeta$ is a prime element. According to (3-3-2), $\{1, 1 - \zeta, \ldots, (1 - \zeta)^{\phi(p^s)-1}\}$ is an integral basis for this local extension. From (4-8-13ii) and (3-7-15) it follows that $\{1, 1 - \zeta, \ldots, (1 - \zeta)^{\phi(p^s)-1}\}$ is an integral basis at p for $E/\mathbf{Q}$. Of course, $O_p[1 - \zeta] = O_p[\zeta]$.

7-5-4. Theorem. Suppose that $E = E_m$ and $\zeta = \zeta_m$; then the set $\{1, \zeta, \ldots, \zeta^{\phi(m)-1}\}$ is an integral basis for $E/\mathbf{Q}$—that is, $O_E = \mathbf{Z}[\zeta]$.

Proof. It suffices to show that the powers of ζ are an integral basis at every prime p of $\mathbf{Q}$. If $p \nmid m$, we apply (7-5-1). If $p|m$, we write $m = m'p^s$ with $(m', p) = 1$ and let P be the unique extension of p to E_{p^s}. From (7-5-2) and (7-5-3) it follows that $O_P = O_{p,E_{p^s}} = O_p[\zeta_{p^s}]$ and $O_{p,E} = O_{P,E} = O_P[\zeta_{m'}] = O_p[\zeta_{p^s}, \zeta_{m'}] = O_p[\zeta_m]$.

7-5-5. Corollary. Suppose that $m = m'p^s$ with p prime, $p^s > 2$, and $(m', p) = 1$; then $\{1, \zeta_{p^s}, \ldots, \zeta_{p^s}^{\phi(p^s)-1}\}$ is an integral basis for E_m over $E_{m'}$—that is, $O_{E_m} = O_{E_{m'}}[\zeta_{p^s}]$.

Proof. The given powers of ζ_{p^s} are surely a basis for $E_m/E_{m'}$ since $[E_m:E_{m'}] = \phi(p^s)$. Furthermore, $O_{E_m} = \mathbf{Z}[\zeta_m] = \mathbf{Z}[\zeta_{m'}, \zeta_{p^s}] = O_{E_{m'}}[\zeta_{p^s}]$.

7-5-6. Remark. In view of (7-5-4) the theorem of Kummer (4-9-1) may be applied to the cyclotomic polynomial $\Phi_m(x) = f(\zeta, \mathbf{Q})$ to exhibit the prime ideals of E_m in concrete form. For example, by (7-2-4) we know that 2 factors in E_{23} into two prime ideals $\mathfrak{P}_1$ and $\mathfrak{P}_2$ of degree 11. Because $\Phi_{23}(x) = x^{22} + x^{21} + \cdots + x + 1$ has mod 2, the two irreducible factors

$$f_1(x) = x^{11} + x^{10} + x^6 + x^5 + x^4 + x^2 + 1$$
$$f_2(x) = x^{11} + x^9 + x^7 + x^6 + x^5 + x + 1$$

it follows that

$$\mathfrak{P}_1 = (p, f_1(\zeta)) \qquad \mathfrak{P}_2 = (p, f_2(\zeta))$$

Now that we have an integral basis, let us use it to compute the discriminant. The discriminant of a cyclotomic field is easily determined in the case where m is an odd prime q. In fact we then have

$$\prod_{1}^{q-1} (\zeta^i - 1) = \prod_{1}^{q-1} (1 - \zeta^i) = q$$

and

$$\prod_{\substack{i,j=0\\ i\neq j}}^{q-1} (\zeta^i - \zeta^j) = (-1)^{q-1}q^q = q^q$$

[see (7-1-1)], so that

$$\Delta(1, \zeta \ldots, \zeta^{q-2}) = \prod_{\substack{i,j=1\\ i>j}}^{q-1} (\zeta^i - \zeta^j)^2 \qquad \text{[see (3-7-9)]}$$

$$= \frac{(-1)^{[(q-1)(q-2)]/2} \prod_{\substack{i,j=0\\ i\neq j}}^{q-1} (\zeta^i - \zeta^j)}{\prod_{1}^{q-1} (\zeta^i - 1) \prod_{1}^{q-1} (1 - \zeta^j)}$$

$$= (-1)^{(q-1)/2} q^{q-2}$$

In order to treat the case of general m, this result must first be extended to the power of a prime.

7-5-7. Proposition. If $E = E_{p^s}$ (p any prime, $p^s > 2$), then

$$\Delta_E = (-1)^{\phi(p^s)/2} \frac{p^{s\phi(p^s)}}{p^{\phi(p^s)/p-1}}$$

Proof. We recall from (7-4-1) that, for $\zeta = \zeta_{p^s}$, we have $f(\zeta, \mathbf{Q}) = f(x) = x^{p^{s-1}(p-1)} + x^{p^{s-1}(p-2)} + \cdots + x^{p^{s-1}} + 1 = \prod_{(i,p^s)=1} (x - \zeta^i)$ and $(x^{p^{s-1}} - 1)f(x) = x^{p^s} - 1$. It follows immediately that

$$(\zeta^{p^{s-1}} - 1)f'(\zeta) = p^s\zeta^{p^s-1}$$

so that

$$f'(\zeta) = \frac{-p^s\zeta^{p^s-1}}{1 - \zeta^{p^{s-1}}}$$

By (3-7-11) we know that

$$\Delta_E = \Delta(1, \zeta, \ldots, \zeta^{\phi(p^s)-1}) = (-1)^{\frac{[\phi(p^s)][\phi(p^s)-1]}{2}} N_{E\to\mathbf{Q}}[f'(\zeta)]$$

Now, it is trivial that $N_{E\to\mathbf{Q}}(-p^s) = (-p^s)^{\phi(p^s)}$ and that $N_{E\to\mathbf{Q}}(\zeta) = \prod_{(i,p^s)=1} \zeta^i = (-1)^{\phi(p^s)}$. Furthermore, we note that $\xi = \zeta^{p^{s-1}}$ is a primitive pth root of unity, so that

$$\begin{aligned} N_{E\to\mathbf{Q}}(1 - \xi) &= [N_{E_p\to\mathbf{Q}}(1 - \xi)]^{[E:E_p]} \\ &= \left[\prod_{i=1}^{p-1} (1 - \xi^i)\right]^{\phi(p^s)/(p-1)} \\ &= p^{\phi(p^s)/(p-1)} \end{aligned}$$

The remaining details are easily checked.

7-5-8. Theorem. If $E = E_m$, then

$$\Delta_E = (-1)^{\phi(m)/2} \frac{m^{\phi(m)}}{\prod_{p|m} p^{\phi(m)/(p-1)}}$$

Proof. We may proceed by induction on m and suppose that m is not a power of a prime. Let us write $m = m'q^s$, q odd prime, $(q, m') = 1$—and we need to consider only the situation where $m' > 2$.

Because sign $\Delta_E = (-1)^{\phi(m)/2}$ [see (4-8-19)], our concern is with $|\Delta_E|$. From (4-8-12) we conclude that

$$\begin{aligned} |\Delta_{E/\mathbf{Q}}| &= |\Delta_{E_{m'}/\mathbf{Q}}|^{[E:E_{m'}]} |N_{E_{m'}\to\mathbf{Q}}(\Delta_{E/E_{m'}})| \\ &= |\Delta_{E_{m'}/\mathbf{Q}}|^{\phi(q^s)} |\Delta_{E_{q^s}/\mathbf{Q}}|^{\phi(m')} \end{aligned}$$

because from (7-5-5) it follows that $\Delta_{E/E_{m'}} = \Delta_{E_{q^s}/\mathbf{Q}}$. The remaining details are straightforward.

7-6. Units

7-6-1. Proposition. The number of independent units in E_m is $(\phi(m)/2) - 1$, while the number of roots of unity is m when m is even and $2m$ when m is odd.

Proof. The first part is immediate from (5-3-10) and the fact [already used in the proof of (7-5-8)] that the infinite prime of $\mathbf{Q}$ has $\phi(m)/2$ extensions to E_m, all of which are complex.

To prove the second part, let W denote the set of all roots of unity that belong to E_m. Thus, W is a cyclic group—of order n, say. Clearly, $E_m = E_n$ so that $\phi(m) = \phi(n)$; and since $\zeta_m \in W$, it follows that $m|n$. If m is even, it is easy to check that $m|n$ and $\phi(m) = \phi(n)$ imply that $m = n$. If m is odd, then $E_{2m} = E_m$ and the even case applies.

The roots of unity in E_m are now completely known. Furthermore, there are other units of E_m which may be exhibited explicitly.

7-6-2. Proposition. Consider $E = E_m$, $\zeta = \zeta_m$ and any positive integer a such that $(a, m) = 1$.

i. If m is a power of a single prime, then $(1 - \zeta^a)/(1 - \zeta)$ is a unit.

ii. If m has two or more distinct prime factors, then $1 - \zeta^a$ is a unit.

Proof. i. See the proof of (7-4-1).

ii. Since ζ^a is a primitive mth root of unity, it suffices to show that $1 - \zeta$ is a unit. Let us write $m = p^s m'$ with $(m', p) = 1$, $p^s > 2$ and also $\zeta = \zeta_{p^s}\zeta_{m'}$. From

$$x^{p^s} - 1 = \prod_{i=0}^{p^s-1} (x - \zeta_{p^s}^i)$$

it follows that

$$\prod_{i=0}^{p^s-1} (x - \zeta_{p^s}^i \zeta_{m'}) = x^{p^s} - \zeta_{m'}^{p^s}$$

Consequently,

$$\prod_{i=1}^{p^s-1} (1 - \zeta_{p^s}^i \zeta_{m'}) = \frac{1 - \zeta_{m'}^{p^s}}{1 - \zeta_{m'}}$$

If $m' > 2$ is a power of a single prime, the right side is a unit of $E_{m'}$ (and of E_m) by (i); therefore, every factor on the left, and in particular $1 - \zeta = 1 - \zeta_{p^s}\zeta_{m'}$, is a unit of E_m. The case $m' = 2$ is trivial.

The proof may now be completed by an induction on the number of distinct prime factors of m.

In order to get further information about units, let us denote conjugation in $\mathbf{C}$ by ρ—that is, $\rho(\alpha) = \bar{\alpha}$ for every $\alpha \in \mathbf{C}$. For any root of unity $\zeta = \zeta_m$ we have $\rho(\zeta) = \bar{\zeta} = \zeta^{-1}$, and ρ therefore determines an automorphism (also denoted by ρ) belonging to $\mathcal{G}(E_m/\mathbf{Q})$. Since ρ has order 2, the fixed field of ρ, which we denote by $E_m^{\#}$, is a real field and $[E_m:E_m^{\#}] = 2$, $[E_m^{\#}:\mathbf{Q}] = \phi(m)/2$. Now, $\mathbf{Q}(\zeta + \zeta^{-1})$ is a real field which is pointwise fixed under ρ, so that $\mathbf{Q}(\zeta + \zeta^{-1}) \subset E_m^{\#}$. Since ζ is a root of the polynomial $x^2 - (\zeta + \zeta^{-1})x + 1$, we conclude that

$$E_m^{\#} = \mathbf{Q}(\zeta + \zeta^{-1})$$

There are two roots of unity in $E_m^{\#}$, namely, ± 1. All the archimedean primes of $E_m^{\#}$ are real, and there are $\phi(m)/2$ of them. In particular, E_m and $E_m^{\#}$ have the same number of independent units. One may note that the group of units of $E_m^{\#}$ is of finite index in the group of units of E_m.

7-6-3. Proposition. If m is a power of an odd prime, $m = p^s$, then any fundamental system of units of $E_{p^s}^{\#}$ is a fundamental system of units of E_{p^s}.

Proof. We must show that if η is any unit of E_{p^s} then it may be expressed as the product of a root of unity from E_{p^s} and a unit from $E_{p^s}^{\#}$.

First, let us observe that for *any* prime P of E_{p^s} we have

$$\varphi_P[\rho(\eta)] = \varphi_P(\eta)$$

Thus, $\rho(\eta)/\eta$ is an absolute unit of E_{p^s}, and it is, therefore, a root of unity. Now, if ζ_{p^s} is a primitive p^sth root of unity, then so is $\zeta_{p^s}^2$, and $\zeta_{2p^s} = \zeta_2\zeta_{p^s}^2$ is a primitive $2p^s$th root of unity which generates the group of roots of unity of E_{p^s}. Consequently, there exists an integer a such that

$$\frac{\rho(\eta)}{\eta} = \zeta_{2p^s}^a = \zeta_2^a\zeta_{p^s}^{2a} = \pm\zeta_{p^s}^{2a}$$

It follows that (with $\zeta_{p^s} = \zeta$)

$$\rho(\eta\zeta^a) = \pm\eta\zeta^a$$

To complete the proof, it suffices to show that the minus sign cannot hold. Let P be the unique extension of p to E_{p^s}, and let $P^\#$ be the restriction of P to $E^\#_{p^s}$. We know that $e(P/p) = \phi(p^s)$, so that $e(P^\#/p) = \phi(p^s)/2$ and $e(P/P^\#) = 2$. Furthermore, if the minus sign holds, then the minimum polynomial of $\zeta^a\eta$ over $E^\#_{p^s}$ is $f(x) = x^2 - \eta^2\zeta^{2a}$. Thus, the different of $\eta\zeta^a$ is $f'(\eta\zeta^a) = 2\eta\zeta^a$ and $\nu_P(2\eta\zeta^a) = 0$. In view of (4-8-18) the different of $E_{p^s}/E^\#_{p^s}$ has $\nu_P = 0$, so that $e(P/P^\#) = 1$, a contradiction.

As a simple example of this result we may note that E_5 has the same fundamental unit as $E^\#_5 = \mathbf{Q}(\zeta_5 + \zeta_5^{-1}) = \mathbf{Q}(\sqrt{5})$, namely, $(1 + \sqrt{5})/2$.

7-6-4. Exercise. i. If m is even and η is a unit of E_m, then in E_{2m} we can express η as the product of a root of unity and a real unit—that is, $\eta = \zeta^a_{2m}\varepsilon$ with ε in $E^\#_{2m}$.

ii. If m is odd and η is a unit of E_m, then in E_{4m} we can express η as the product of a root of unity and a real unit—that is, $\eta = \zeta^a_{4m}\varepsilon$ with ε in $E^\#_{4m}$.

7-7. Class Number

The situation with regard to the class number h_m of the cyclotomic field E_m is not particularly satisfactory. Because we have no analytic tools at our disposal, all that can be done without additional effort is to apply the procedure that was used to get information about the class number of quadratic fields. Thus, we shall prove only the following simple result, even though there exists an analytically derived formula for h_m (this formula leaves something to be desired since it does not even enable us to answer the question of when $h_m = 1$).

7-7-1. Proposition. For $3 \leq m \leq 10$ we have $h_m = 1$.

Proof. Since $E_3 = \mathbf{Q}(\sqrt{-3})$ and $E_4 = \mathbf{Q}(\sqrt{-1})$, we know (from the theory of quadratic fields) that $h_3 = h_4 = 1$. If m is odd, then $E_m = E_{2m}$; so it suffices to prove our assertion for $m = 5, 7, 8, 9$. Because all these go the same way, we do only the case $m = 9$.

In view of (7-5-7), E_9 has discriminant $\Delta = -3^9$. Therefore [see (5-4-7)]

$$B = \left(\frac{4}{\pi}\right)^{\phi(9)/2} \frac{[\phi(9)]!}{[\phi(9)]^{\phi(9)}} \sqrt{|\Delta|} < 5$$

and every ideal class contains an integral ideal of norm <5. If $\mathfrak{P}$ is a prime ideal of E_9 lying over 2, then $\mathfrak{N}\mathfrak{P} = 2^f$ where f is the smallest positive integer such that $2^f \equiv 1 \pmod 9$; so $\mathfrak{N}\mathfrak{P} = 64 > 5$. On the other hand, the prime 3 is completely ramified in $E = E_9$; in fact, $3O_E = [(1 - \zeta_9)O_E]^6$. Thus, it follows that $h_9 = 1$.

Cyclotomic fields and their class numbers have been studied rather extensively (see, for example, Algebraic Numbers, Part II, Report of the Committee on Algebraic Numbers, *Bulletin of the National Research Council No.* 62, February, 1928) beginning with the work of Kummer; this interest is, of course, largely due to their connection with Fermat's last theorem.

Symbols and Notation

$\exists$	there exists
$\ni:$	such that
$\Rightarrow$	implication
$\Leftrightarrow$	if and only if
$\Rightarrow:$	we prove the implication $\Rightarrow$
$\subset$	inclusion
$\prec$	proper inclusion
$\sim$	equivalence
$\notin$	is not a member of
$\twoheadrightarrow$	onto map
$\rightarrowtail$	1-1 map
$\rightarrowtail\!\!\!\twoheadrightarrow$	1-1 onto map
$\forall$	for all
$\emptyset$	empty set
$\approx$	isomorphism
#	number of elements
Q	field of rational numbers
Z	ring of rational integers
R	field of real numbers
C	field of complex numbers
$[E:F]$	degree of the field E over the subfield F
$[E:F]_{sep}$	separable factor of the degree $[E:F]$
$[E:F]_{ins}$	inseparable factor of the degree $[E:F]$
$(G:H)$	group index of H in G
$f'(x)$	derivative of the polynomial $f(x)$
$N_{E\to F}$	norm from E to F
$S_{E\to F}$	trace from E to F

$f(\alpha,F)$	minimum polynomial of α over F
$f(\alpha,E/F)$	field polynomial of α in the extension E/F
$\mathcal{G}(E/F)$	Galois group of E/F
$\Gamma(F, E \to \Omega)$	set of F isomorphisms of E into Ω
$R[\alpha]$	ring generated by α over R
$F(x)$	field generated by x over F
$F\langle x\rangle$	field of formal power series in x over F
ϕ	Euler ϕ function
(a,b)	greatest common divisor of a and b
i	inclusion map
φ	valuation
P, Q	prime divisors
ν	exponential valuation
ν_P	normalized exponential valuation belonging to P
F_P	completion of F at P
$\bar{F}$	residue class field
ψ	residue class map
O	ring of integers
U	groups of units
$e(Q/P)$, $e(E/F)$	ramification index
$f(Q/P)$, $f(E/F)$	residue class degree
$\mathfrak{N}$	counting norm
$\mathcal{P}$, $\mathcal{Q}$	local prime ideals
$\mathfrak{P}$, $\mathfrak{Q}$	global prime ideals
$\mathbf{Q}_p$	p-adic completion of $\mathbf{Q}$
$\mathfrak{I}$	inertia group
$\mathfrak{V}$	ramification group
T	inertia field
$\mathfrak{Z}$	decomposition group
$\mathfrak{M}(F)$	set of nontrivial prime divisors of the global field F
Δ	discriminant
$\boldsymbol{\Delta}$	discriminant ideal
$\mathfrak{D}$	different
$\mathbf{F}$	adèle ring
$\mathbf{F}^*$	idèle group
$\mathbf{D}$	divisor group
$\mathbf{D}^*$	group of principal divisors
$\mathfrak{J}$	ideal group of an OAF
$\mathbf{I}_S$	group of S ideals of a global field
$\left(\frac{a}{p}\right) = (a/p)$	Legendre or Kronecker symbol
h	class number

Index

A CATALOG OF SELECTED

DOVER BOOKS

IN SCIENCE AND MATHEMATICS

THE ELECTROMAGNETIC FIELD, Albert Shadowitz. Comprehensive undergraduate text covers basics of electric and magnetic fields, builds up to electromagnetic theory. Also related topics, including relativity. Over 900 problems. 768pp. 5⅝ x 8¼. 65660-8 Pa. $18.95

FOURIER SERIES, Georgi P. Tolstov. Translated by Richard A. Silverman. A valuable addition to the literature on the subject, moving clearly from subject to subject and theorem to theorem. 107 problems, answers. 336pp. 5⅜ x 8½. 63317-9 Pa. $9.95

THEORY OF ELECTROMAGNETIC WAVE PROPAGATION, Charles Herach Papas. Graduate-level study discusses the Maxwell field equations, radiation from wire antennas, the Doppler effect and more. xiii + 244pp. 5⅜ x 8½. 65678-0 Pa. $6.95

DISTRIBUTION THEORY AND TRANSFORM ANALYSIS: An Introduction to Generalized Functions, with Applications, A.H. Zemanian. Provides basics of distribution theory, describes generalized Fourier and Laplace transformations. Numerous problems. 384pp. 5⅜ x 8½. 65479-6 Pa. $11.95

THE PHYSICS OF WAVES, William C. Elmore and Mark A. Heald. Unique overview of classical wave theory. Acoustics, optics, electromagnetic radiation, more. Ideal as classroom text or for self-study. Problems. 477pp. 5⅜ x 8½. 64926-1 Pa. $13.95

CALCULUS OF VARIATIONS WITH APPLICATIONS, George M. Ewing. Applications-oriented introduction to variational theory develops insight and promotes understanding of specialized books, research papers. Suitable for advanced undergraduate/graduate students as primary, supplementary text. 352pp. 5⅜ x 8½. 64856-7 Pa. $9.95

A TREATISE ON ELECTRICITY AND MAGNETISM, James Clerk Maxwell. Important foundation work of modern physics. Brings to final form Maxwell's theory of electromagnetism and rigorously derives his general equations of field theory. 1,084pp. 5⅜ x 8½. 60636-8, 60637-6 Pa., Two-vol. set $25.90

AN INTRODUCTION TO THE CALCULUS OF VARIATIONS, Charles Fox. Graduate-level text covers variations of an integral, isoperimetrical problems, least action, special relativity, approximations, more. References. 279pp. 5⅜ x 8½. 65499-0 Pa. $8.95

HYDRODYNAMIC AND HYDROMAGNETIC STABILITY, S. Chandrasekhar. Lucid examination of the Rayleigh-Benard problem; clear coverage of the theory of instabilities causing convection. 704pp. 5⅝ x 8¼. 64071-X Pa. $14.95

CALCULUS OF VARIATIONS, Robert Weinstock. Basic introduction covering isoperimetric problems, theory of elasticity, quantum mechanics, electrostatics, etc. Exercises throughout. 326pp. 5⅜ x 8½. 63069-2 Pa. $9.95

DYNAMICS OF FLUIDS IN POROUS MEDIA, Jacob Bear. For advanced students of ground water hydrology, soil mechanics and physics, drainage and irrigation engineering and more. 335 illustrations. Exercises, with answers. 784pp. 6⅛ x 9¼. 65675-6 Pa. $19.95

CATALYSIS IN CHEMISTRY AND ENZYMOLOGY, William P. Jencks. Exceptionally clear coverage of mechanisms for catalysis, forces in aqueous solution, carbonyl- and acyl-group reactions, practical kinetics, more. 864pp. 5⅜ x 8½.
65460-5 Pa. $19.95

PROBABILITY: An Introduction, Samuel Goldberg. Excellent basic text covers set theory, probability theory for finite sample spaces, binomial theorem, much more. 360 problems. Bibliographies. 322pp. 5⅜ x 8½. 65252-1 Pa. $10.95

LIGHTNING, Martin A. Uman. Revised, updated edition of classic work on the physics of lightning. Phenomena, terminology, measurement, photography, spectroscopy, thunder, more. Reviews recent research. Bibliography. Indices. 320pp. 5⅜ x 8¼. 64575-4 Pa. $8.95

PROBABILITY THEORY: A Concise Course, Y.A. Rozanov. Highly readable, self-contained introduction covers combination of events, dependent events, Bernoulli trials, etc. Translation by Richard Silverman. 148pp. 5⅜ x 8¼. 63544-9 Pa. $7.95

AN INTRODUCTION TO HAMILTONIAN OPTICS, H. A. Buchdahl. Detailed account of the Hamiltonian treatment of aberration theory in geometrical optics. Many classes of optical systems defined in terms of the symmetries they possess. Problems with detailed solutions. 1970 edition. xv + 360pp. 5⅜ x 8½.
67597-1 Pa. $10.95

STATISTICS MANUAL, Edwin L. Crow, et al. Comprehensive, practical collection of classical and modern methods prepared by U.S. Naval Ordnance Test Station. Stress on use. Basics of statistics assumed. 288pp. 5⅜ x 8½. 60599-X Pa. $7.95

DICTIONARY/OUTLINE OF BASIC STATISTICS, John E. Freund and Frank J. Williams. A clear concise dictionary of over 1,000 statistical terms and an outline of statistical formulas covering probability, nonparametric tests, much more. 208pp. 5⅜ x 8½. 66796-0 Pa. $7.95

STATISTICAL METHOD FROM THE VIEWPOINT OF QUALITY CONTROL, Walter A. Shewhart. Important text explains regulation of variables, uses of statistical control to achieve quality control in industry, agriculture, other areas. 192pp. 5⅜ x 8½. 65232-7 Pa. $7.95

METHODS OF THERMODYNAMICS, Howard Reiss. Outstanding text focuses on physical technique of thermodynamics, typical problem areas of understanding, and significance and use of thermodynamic potential. 1965 edition. 238pp. 5⅜ x 8½.
69445-3 Pa. $8.95

STATISTICAL ADJUSTMENT OF DATA, W. Edwards Deming. Introduction to basic concepts of statistics, curve fitting, least squares solution, conditions without parameter, conditions containing parameters. 26 exercises worked out. 271pp. 5⅜ x 8½.
64685-8 Pa. $9.95

TENSOR CALCULUS, J.L. Synge and A. Schild. Widely used introductory text covers spaces and tensors, basic operations in Riemannian space, non-Riemannian spaces, etc. 324pp. 5⅜ x 8¼. 63612-7 Pa. $9.95

CATALOG OF DOVER BOOKS